SINGULAR PERTURBATION PROBLEMS IN CHEMICAL PHYSICS

ADVANCES IN CHEMICAL PHYSICS

VOLUME XCVII

EDITORIAL BOARD

SINGULAR PERTURBATION PROBLEMS IN CHEMICAL PHYSICS Analytic and Computational Methods

Edited by

JOHN J. H. MILLER
Department of Mathematics
Trinity College
Dublin, Ireland

ADVANCES IN CHEMICAL PHYSICS
VOLUME XCVII

Series Editors

I. PRIGOGINE
Center for Studies in Statistical Mechanics
and Complex Systems
The University of Texas
Austin, Texas
and
International Solvay Institutes
Université Libre de Bruxelles
Brussels, Belgium

STUART A. RICE
Department of Chemistry
and
The James Franck Institute
The University of Chicago
Chicago, Illinois

AN INTERSCIENCE® PUBLICATION
JOHN WILEY & SONS
NEW YORK • CHICHESTER • WEINHEIM • BRISBANE • SINGAPORE • TORONTO

This text is printed on acid-free paper.

An Interscience® Publication

Library of Congress Catalog Number: 58-9935

ISBN 0-471-11531-2

Printed in the United States of America

10 9 8 7 6 5 4 3 2 1

CONTRIBUTORS TO VOLUME XCVII

V. F. Butuzov, Department of Physics, Moscow State University, Moscow, Russia

A. M. Il'in, Institute of Mathematics and Mechanics, Ural Branch of the Russian Academy of Sciences, Ekaterinburg, Russia

L. A. Kalyakin, Institute of Mathematics, Ufa, Russia

V. L. Kolmogorov, Institute of Engineering Science, Ural Branch of the Russian Academy of Sciences, Ekaterinburg, Russia

S. I. Maslennikov, Institute of Organic Chemistry, Ufa, Russia

G. I. Shishkin, Institute of Mathematics and Mechanics, Ural Branch of the Russian Academy of Sciences, Ekaterinburg, Russia

A. B. Vasilieva, Department of Physics, Moscow State University, Moscow, Russia

PREFACE

Since boundary layers were first introduced by Prandtl at the start of the twentieth century, rapid strides have been made in the analytic and numerical investigation of such phenomena. It has also been realized that boundary and interior layer phenomena are ubiquitous in the problems of chemical physics. Nowhere have developments in this area been more notable than in the Russian school of singular perturbation theory and its application. The three chapters in this book are representative of the best analytic and computational work in this field in the second half of the century.

This volume is concerned with singular perturbation problems that occur in many areas of chemical physics. When singular perturbations are present, various kinds of boundary and interior layers appear. In these layers the physical variables change extremely rapidly over small domains in space or short intervals of time. Such phenomena give rise to significant numerical difficulties that can be overcome only by using specially designed numerical methods. It is important to appreciate the fact that some of these computational problems cannot in principle be overcome by the brute force solution of throwing more computing power at the problem (for example, by using ever-finer uniform meshes). For some layer phenomena it can be proved rigorously that the error in solving a family of singular perturbation problems cannot be reduced below a certain fixed limit unless specially designed nonuniform meshes are used. The design of such meshes depends on *a priori* knowledge of the location and nature of the boundary layers under investigation. For these reasons the study of these phenomena is vital, if robust and accurate solutions of such problems are required.

The three chapters in this volume deal with various aspects of singular perturbations and their numerical solution. The first chapter is concerned with the analysis of some singular perturbation problems that arise in chemical kinetics. In it the matching method is applied to find asymptotic solutions of some dynamical systems of ordinary differential equations whose solutions have multiscale time dependence. The second chapter contains a comprehensive overview of the theory and application of asymptotic approximations for many different kinds of problems in chemical physics, with boundary and interior layers governed by either ordinary or partial differential equations. In the final chapter the numerical difficulties arising in the solution of the problems described in the previous chapters are discussed. In addition, rigorous criteria are proposed for determining whether or not a numerical method is satisfactory for such problems. Some methods satisfying these criteria are constructed using specially designed meshes on which the numerical solution is defined. These methods are then applied to obtain numerical solutions for a range of sample problems.

JOHN J. H. MILLER

Dublin

INTRODUCTION

Few of us can any longer keep up with the flood of scientific literature, even in specialized subfields. Any attempt to do more and be broadly educated with respect to a large domain of science has the appearance of tilting at windmills. Yet the synthesis of ideas drawn from different subjects into new, powerful, general concepts is as valuable as ever, and the desire to remain educated persists in all scientists. This series, *Advances in Chemical Physics*, is devoted to helping the reader obtain general information about a wide variety of topics in chemical physics, a field that we interpret very broadly. Our intent is to have experts present comprehensive analyses of subjects of interest and to encourage the expression of individual points of view. We hope that this approach to the presentation of an overview of a subject will both stimulate new research and serve as a personalized learning text for beginners in a field.

I. Prigogine
Stuart A. Rice

CONTENTS

SINGULAR PERTURBATION PROBLEMS IN CHEMICAL PHYSICS

ADVANCES IN CHEMICAL PHYSICS

VOLUME XCVII

THE MATCHING METHOD FOR ASYMPTOTIC SOLUTIONS IN CHEMICAL PHYSICS PROBLEMS

A. M. IL'IN

Institute of Mathematics and Mechanics, Ural Branch of the Russian Academy of Sciences, 620219 Ekaterinburg, Russia

L. A. KALYAKIN

Institute of Mathematics, 450000 Ufa, Russia

S. I. MASLENNIKOV

Institute of Organic Chemistry, 450054 Ufa, Russia

CONTENTS

Singular Perturbation Problems in Chemical Physics: Analytic and Computational Methods, Edited by John J. H. Miller, Advances in Chemical Physics Series, Vol. XCVII.
ISBN 0-471-11531-2

I. INTRODUCTION

One of the main problems of chemical kinetics is the study of the time evolution of chemical reactions. The quantitative approach to chemical kinetics leads to mathematical models in the form of dynamical systems similar to classical mechanics [1, 2]. The specific chemical nature of the system appears here often as multiscale time dependence of the solutions of the mathematical problems. This feature occurs whenever fairly complex chemical processes are considered, examples of which are the subject matter of this chapter.

From a formal point of view, time multiscaling arises due to our attempts to simulate complex chemical processes by means of simple reactions. This gives rise to a strong difference in the activity of different components in the reaction mixture. Products appearing in the output either have a short lifetime or are rapidly stabilized. For example, the concentrations of active particles such as radicals, ions, and so on change noticeably during 10^{-6} s, while a number of hours is required for changes of a stable substance. These labile products play a significant role in the total process in spite of either their short lifetime or fast stabilization. In brief, the results of the fast reactions have an effect on the slow ones [2].

Thus, complex chemical processes are represented as a number of simple reactions that are very inhomogeneous on a time scale. Generally, it is impossible to separate the fast processes and the slow ones from each other, so that a continuous time monitoring of the total kinetic process is needed to understand the essence of the phenomenon. Mathematical models provide an adequate tool for the scanning of the kinetic curves. Fig. 1(a) shows a typical example of curves where two time scales are present. These time scales differ up to an order of 10^{-1} from each other. If one considers the process on the logarithmic scale, then just three different time scales may be identified, see Fig. 1(b). The presence of both fast and slow variables is explained by the occurrence of either large or small factors in the dynamical equations. For example, this is the case for so-called stiff systems of differential equations.

The small factors, responsible for the multiscale effects, play a dual role in the analysis of the mathematical problems. The first one is negative. Indeed, because of the large dimension of the system of

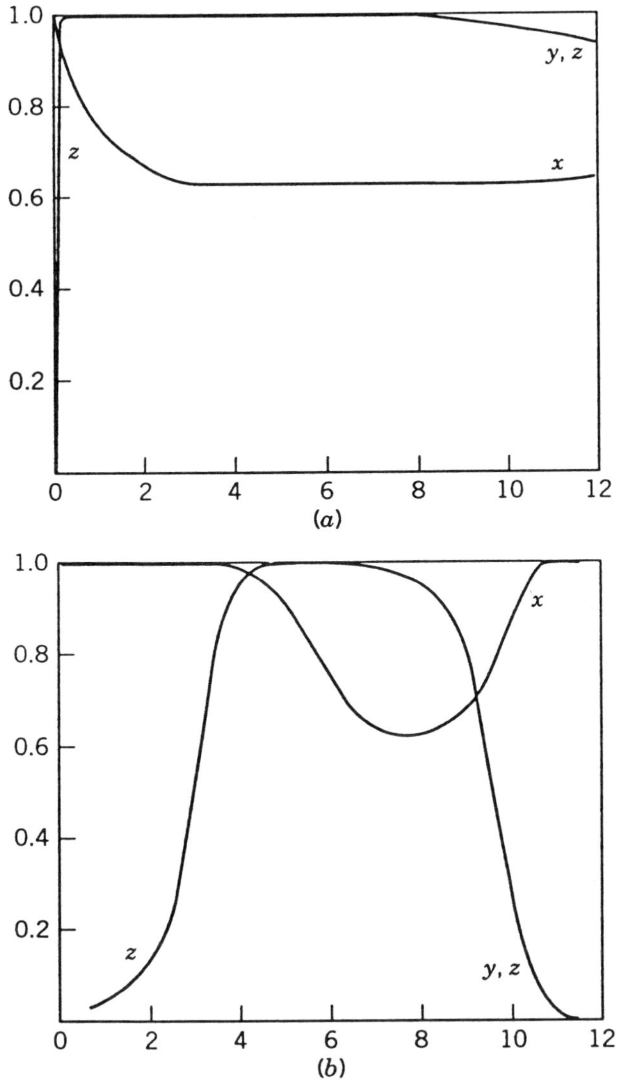

Figure 1. (*a*) Typical curves with two time scales. (*b*) Typical curves with three time scales.

nonlinear differential equations it is hard to write out an exact solution. Therefore, heroic efforts were undertaken to develop effective methods of numerical simulation on a computer [3, 4]. The problems that arise here due to multiscaling are well known. A small time mesh width in the difference scheme is needed to capture the fast processes. But calculations for stiff systems with a small mesh width are unstable over a long time.

There are different approaches to overcome the difficulties. The most efficient one is the use of a nonuniform difference scheme with a varying mesh width [5]. However, the structure of the time scales for the solution is needed to construct a good difference scheme. Thus a preliminary analysis of the equations is required in this approach, which is precisely the subject matter of this chapter.

Note that a single computer run gives highly incomplete information. Therefore, a comprehensive study of this process demands a great number of runs, and this feature is one of the main disadvantages of numerical simulations.

There is another approach, coming from the theory of dynamical systems, that deals with the phase portraits of all of the solutions, which gives the complete picture for the two-dimensional case. Unfortunately, the multidimensional systems, which generally arise in chemical kinetics, make this approach too complicated [6, 7].

Thus, both the small parameters and the multidimensional nature complicate the investigation of mathematical models of chemical kinetics. On the other hand, the small parameters often lead to a simplification of similar problems of classical mechanics by means of asymptotic approximations. From this point of view, the role of the small parameters is positive.

Among the known asymptotic tools the matching method seems to be the most powerful, because it is applicable to practically any problem where the separation of either time or spatial scales takes place. Used first for hydrodynamic problems, this method was later extended to different fields of mechanics, physics, and mathematics [8]. In particular, the matching method is well suited to the study of problems of chemical kinetics, where separation of the fast and slow processes occurs [9]. The main result of this approach is a simplification of the original problem up to the level where either explicit formulas or standard numerical simulations give valuable results.

In this chapter, the matching method is applied to solve two problems that deal with processes involving time multiscaling. The power of the method emerges for the second case, where the fast process arises as a background to the slow one. No other method seems to cope with this type of problem.

II. ELEMENTARY EXAMPLES

In principle, the basic concept of every asymptotic method is very simple [10]. The original problem, which seems to be too complicated, is replaced by a simpler approximate one, by eliminating small (inessential) terms. For instance, if the factor ε is small ($0<\varepsilon\ll 1$), the differential equation

$$d_t x = -\varepsilon x \qquad x|_{t=0} = 1$$

can be reduced to the simpler one,

$$d_t x = 0 \qquad x|_{t=0} = 1$$

In this approach, the function $x^0(t)\equiv 1$, found from the last two equations, can be considered to be a first approximation of the exact solution $x(t;\varepsilon)=\exp(-\varepsilon t)$. The error of the approximation is small if the parameter ε is small.

$$|x(t;\varepsilon) - x^0(t)| \le \varepsilon t$$

One can construct a more precise approximation. To this end, the first correction $\varepsilon x^1(t)$ is calculated from the equations

$$d_t x^1 = -x^0 \qquad x^1|_{t=0} = 0$$

So the function

$$x^0(t) + \varepsilon x^1(t) \equiv 1 - \varepsilon t$$

gives a more precise approximation (up to the order of the small parameter ε) as we see

$$|x(t;\varepsilon) - x^0(t) - \varepsilon x^1(t) \le \varepsilon^2 t^2/2$$

It is clear that all of these approximations are Taylor expansions of the exact solution $x(t;\varepsilon)=\exp(-\varepsilon t)$ as $\varepsilon\to 0$.

In this example, the parameter ε can be removed completely by a change of the independent variable $\tau=\varepsilon t$. In fact, a change of the time scale is made here. The exact problem for the new dependent variable $x_s(\tau;\varepsilon)=x(t;\varepsilon)$ now reads

$$d_\tau x_s = -x_s \qquad x_s|_{\tau=0} = 1$$

In this form the differential equation has no terms with small factors, hence it cannot be simplified. The exact solution reads

$$x_s(\tau) = \exp(-\tau)$$

Approximate formulas of the kind

$$x_s(\tau) \cong 1 + \mathcal{O}(\tau)$$

are suitable only for small (slow) times $0 \le \tau \ll 1$, (i.e., $0 \le t \ll 1/\varepsilon$) because of the error of order $\mathcal{O}(\tau)$.

The above way of excluding the small parameter terms may not be possible for more complicated equations. A simple instance of this type is given by a system of two differential equations, which are not coupled with each other. This example is the case when various time scales occur.

$$d_t x = -\varepsilon x \qquad x|_{t=0} = 1$$

$$d_t y = 1 - y \qquad y|_{t=0} = 2$$

Formulas for the exact solution have the small parameter in both the fast

$$x(t;\varepsilon) = \exp(-\varepsilon t) \qquad y(t;\varepsilon) = 1 + \exp(-t)$$

and the slow time scale

$$x_s(\tau;\varepsilon) = \exp(-t) \qquad y_s(\tau;\varepsilon) = 1 + \exp(-\tau/\varepsilon) \qquad (\tau = \varepsilon t)$$

In fact, there are two processes differing in their rates. The fast process is

$$y(t;\varepsilon) = 1 + \exp(-t)$$

whereas the slow one is

$$x(t;\varepsilon) = \exp(-\varepsilon t)$$

The rates of the processes are either 1 or ε, respectively. The slowness of the second process with respect to the first one is measured by the quantity ε. So the question of different scales may be discussed if only the value of ε is small.

If the parameter ε is small, the explicit formulas for the solution can be simplified. These simplifications are different in the various scales. They read either

$$x(t;\varepsilon) = 1 + \mathcal{O}(\varepsilon t) \qquad y(t;\varepsilon) = 1 + \exp(-t) \tag{2.1}$$

or

$$x_s(\tau;\varepsilon) = \exp(-\tau) \qquad y_s(\tau;\varepsilon) = 1 + \mathcal{O}((\varepsilon/\tau)^N) \qquad \forall N \tag{2.2}$$

Both of the remainder terms $\mathcal{O}(\varepsilon t)$ and $\mathcal{O}((\varepsilon/\tau)^N)$ give the errors of the approximation and are small if ε is small. Further corrections and more precise expansions can be derived as above.

These approximate formulas explain the concept of the well-known method of steady-state concentration. In the first time scale, the first component is steady-state whereas the second component is exponentially decreasing from the value 2 to 1

$$x(t;\varepsilon) \cong 1$$

$$y(t;\varepsilon) = 1 + \exp(-t)$$

In the second (slow) time scale, the y component is steady-state, whereas the x component is exponentially decreasing to zero

$$y_s(\tau;\varepsilon) \cong 1$$

$$x_s(\tau;\varepsilon) = \exp(-\tau)$$

As we see, the results of the approximate (asymptotic) analysis are different in the various scales, and cannot be interchanged with each other. The structure of the first correction and remainders allows us to see this feature in detail. The accuracy of the approximation depends on both the small parameter ε and the time t. There are terms both in the first correction and in the remainders that grow like t, as t tends to infinity. Such terms, occurring in the asymptotic formulas, are sometimes called secular terms. Due to this result the first formula (2.1) is suitable for times that are not very long $t \ll 1/\varepsilon$. For long times, $t \cong 1/\varepsilon$, the order of the remainder $\mathcal{O}(\varepsilon t)$ is the same as that of the leading term. Hence, the approximation turns out to be false for $t \cong 1/\varepsilon$. The second formula (2.2) is valid for (slow) times that are not very small $\tau \gg \varepsilon$. In this case, the order of the remainder $\mathcal{O}((\varepsilon/\tau)^N)$ is the same as that of the leading term for small times $\tau \cong \varepsilon$. Hence, the approximation turns out to be false for very small (slow) times $\tau \cong \varepsilon$.

Thus, the time intervals of the different asymptotic approximations do not coincide. Nevertheless, they can be chosen in such a manner that the intersection is not empty, for example,

$$0 \le t \le M/\sqrt{\varepsilon} \qquad \text{and} \qquad m/\sqrt{\varepsilon} \le t \le \infty \qquad (m\sqrt{\varepsilon} \le \tau < \infty) \qquad m \le M$$

In this way, the formulas (2.1) and (2.2) represent the exact solution up to order $\mathcal{O}(\sqrt{\varepsilon})$ uniformly for all times. The representation is different on different intervals.

Of course, the above discussion does not seem to make much sense, to say the least. Indeed, the explicit form of the solution is in many respects simpler, more clearcut, and easier to grasp than the two different asymptotics. But this is not the case for more complicated problems. As a rule, no explicit solution of a system of nonlinear equations is available. By being unable to write out an exact solution, one can, naturally, try to find functions satisfying the equations approximately. Such asymptotic (approximate) solutions can be constructed in an explicit form for a large class of the problems. The asymptotic solutions are often described by different formulas on different time intervals.

Let us consider the system of two coupled equations.

$$
\begin{aligned}
d_t x &= -\varepsilon y \qquad x|_{t=0} = 1 \\
d_t y &= x^2 - y \qquad y|_{t=0} = 2
\end{aligned} \tag{2.3}
$$

We are unable to write out the explicit solution in this case. Instead of solving the original problem, we try to obtain a simpler one, by using small factors. In this way, the following equations are obtained for the leading terms:

$$
\begin{aligned}
d_t x &= 0 \qquad x|_{t=0} = 1 \\
d_t y &= x^2 - y \qquad y|_{t=0} = 2
\end{aligned}
$$

Hence, the leading terms read

$$
\begin{aligned}
x^0(t) &\equiv 1 \\
y^0(t) &= 1 + \exp(-t)
\end{aligned}
$$

One can define the next correction of order $\mathcal{O}(\varepsilon)$. To this end, an anzatz in the form of a series in powers of ε

$$
\begin{aligned}
x(t;\varepsilon) &= x^0(t) + \varepsilon x^1(t) + \cdots \\
y(t;\varepsilon) &= y^0(t) + \varepsilon y^1(t) + \cdots
\end{aligned}
$$

is substituted into Eq. (2.3), and coefficients of the same power of ε are equated. For the functions x^1, $y^1(t)$ the equations are obtained as follows:

$$
\begin{aligned}
d_t x &= -y^0 \qquad x|_{t=0} = 0 \\
d_t y &= 2x^0 x - y \qquad y|_{t=0} = 0
\end{aligned}
$$

The solution of this problem allows us to write out a more precise approximate solution in the explicit form

$$x \cong 1 - \varepsilon[t + 1 - \exp(-t)]$$
$$y \cong 1 + \exp(-t) - \varepsilon 2t[1 - \exp(-t)]$$

These formulas represent not only the formal solution, which provides the small residuals in the equations, but also the exact solution of the original problem up to remainder terms. The errors of the approximation can be evaluated (it is a purely mathematical problem). After that, one can write equalities that are called asymptotic expansions of the exact solution. The leading terms of these are

$$x(t; \varepsilon) = 1 + \mathcal{O}(\varepsilon t) \qquad y(t; \varepsilon) = 1 + \exp(-t) + \mathcal{O}(\varepsilon t) \tag{2.4}$$

Thus, by eliminating the small (inessential) terms from the original equations one is able to construct an approximate solution in explicit form. The question is what terms are small, which is not trivial, as was seen above. Indeed, the structure of the first corrections and remainders shows that the approximation (2.4) is suitable only for the times $t \ll 1/\varepsilon$. For long times, $t \cong 1/\varepsilon$, the order of the first correction $\varepsilon x^1(t)$, $\varepsilon y^1(t)$ is the same as the leading one. Hence, the approximation (2.4) fails for $t \cong 1/\varepsilon$.

For long times, $t \cong 1/\varepsilon$, another asymptotic solution must be constructed. To this end, we make the change of independent variable $\tau = \varepsilon t$ in Eqs. (2.3), so they are rewritten for the new dependent variables $x_s(\tau; \varepsilon) = x(t; \varepsilon)$, $y_s(\tau; \varepsilon) = y(t; \varepsilon)$ as follows:

$$d_\tau x_s = -y_s$$
$$\varepsilon d_\tau y_s = x_s^2 - y_s$$

An asymptotic solution of the problem is constructed in a similar way. For the leading terms, two equations are obtained.

$$d_\tau x_s = -y_s$$
$$0 = x_s^2 - y_s$$

Since the second equation is an algebraic one, the variable y is excluded and the problem is reduced to the single nonlinear differential equation

$$d_\tau x_s = -x_s^2 \qquad x_s|_{t=0} = 1$$

The last is easily solved

$$x_s^0(\tau) = 1/(C+\tau)$$

The constant C can be found from the initial data in the general case.

So, in this case the asymptotic solution can also be written in the explicit form

$$x_s(\tau;\varepsilon) \cong x_s^0(\tau) = 1/(C+\tau) \qquad y_s(\tau;\varepsilon) \cong y_s^0(\tau) = 1/(C+\tau)^2 \quad (2.5)$$

However, this approximation is not suitable for small times, because there is no $C = \text{const}$ satisfying the two original initial conditions

$$1/(C+\tau)|_{\tau=0} = 1/C = 1 \qquad 1/(C+\tau)^2|_{\tau=0} = 1/C^2 = 2$$

Formulas (2.5) can be considered as an asymptotic solution on the interval

$$m/\sqrt{\varepsilon} \le t \le \infty \qquad (\sqrt{\varepsilon} \le \tau < \infty)$$

which excludes zero, whereas the approximation (2.4) is used on the interval

$$0 \le t \le M/\sqrt{\varepsilon} \qquad (m, M = \text{const})$$

which includes zero. It is easy to see that in the common domain, where $t = \mathcal{O}(1/\sqrt{\varepsilon})$, the various solutions do not coincide with each other. Nevertheless, if we set the constant $C = 1$, the agreement between Eqs. (2.4) and (2.5) does hold up to order $\mathcal{O}(\sqrt{\varepsilon})$

$$\begin{aligned} x^0(t)|_{t=1/\sqrt{\varepsilon}} &= x_s^0(\tau)|_{\tau=\sqrt{\varepsilon}} + \mathcal{O}(\sqrt{\varepsilon}) \\ y^0(t)|_{t=1/\sqrt{\varepsilon}} &= y_s^0(\tau)|_{\tau=\sqrt{\varepsilon}} + \mathcal{O}(\sqrt{\varepsilon}) \end{aligned} \qquad (2.6)$$

If one takes into account the subsequent corrections of order $\mathcal{O}(\varepsilon^n)$, $n = 1, 2, \ldots$, then the asymptotic solutions can be matched up to order $\mathcal{O}(\varepsilon^{(n+1)/2})$ in the common domain. Incidentally, similar relations to Eq. (2.6) apply for the previous trivial example.

The equalities (2.6) are usually called the matching requirements. It is very nice that the two equations (2.6) are satisfied by the choice of a single constant C. In fact, this astonishing property is an essential feature of any problem that can be solved by the matching method.

Thus, the formulas (2.5), with $C = 1$, represent an appropriate continuation of the asymptotic solution (2.4) on the long time interval (on the slow time scale). Moreover, using the matching (2.6) one can both prove the existence theorem for the exact solution and estimate the accuracy of the approximation.

As to the initial values $x_s(0; \varepsilon)$, $y_s(0; \varepsilon)$ in the slow time scale, it is necessary to understand that the original initial data from (2.3) have no sense here. The true initial values of the functions $x_s(\tau; \varepsilon)$, $y_s(\tau; \varepsilon)$ are obtained from the matching requirements (2.6). To this end, the relations (2.6) are taken in the limit as $\varepsilon \to 0$. This relation gives the equalities

$$\begin{aligned} \lim_{t \to \infty} x^0(t) &= \lim_{\tau \to 0} x_s^0(\tau) \\ \lim_{t \to \infty} y^0(t) &= \lim_{\tau \to 0} y_s^0(\tau) \end{aligned} \tag{2.7}$$

which are usually called the matching conditions. These conditions give the initial data just on the slow scale

$$x_s^0(0) = 1 \qquad y_s^0(0) = 1$$

The meaning of the relations (2.7) has been widely discussed, and it is now quite clear: The asymptotics at infinity (on the fast scale) give the initial data on the subsequent slow scale [8, 10].

Note that only one of the two initial data obtained is needed to construct the asymptotic solution on the slow scale; for example, $x_s^0(0) = 1$. The additional relation $y_s^0(0) = 1$ is then satisfied automatically, and this is a crucial part of the matching.

The matching relations are used here, and almost everywhere, to write the initial conditions in the next time scale. Sometimes other (asymptotic) conditions, obtained from the matching requirements, are used. In any case, they determine the indefinite constants [8].

It is possible to vary the common domain of the different asymptotic solutions, up to the order of a small parameter, as follows:

$$t \cong 1/\varepsilon^{\gamma} \qquad \forall \gamma \in (0, 1)$$

One has to note that the continuation of any asymptotic solution into the domain of another increases the error. The best choice for the above examples is given by $\gamma = \frac{1}{2}$, so that the error has the order $\mathcal{O}(\sqrt{\varepsilon})$.

In the common domain, each asymptotic solution can be replaced by its asymptotics

$$x = 1 + \mathcal{O}(\sqrt{\varepsilon}) \qquad y = 1 + \mathcal{O}(\sqrt{\varepsilon}) \qquad \text{as } t = \mathcal{O}(1/\sqrt{\varepsilon})$$

These intermediate asymptotics are sometimes used to express the approximate solutions in the form of single expressions valid everywhere

$$x(t,\varepsilon) = x^0(t) + x_s^0(\varepsilon t) - 1 + \mathcal{O}(\sqrt{\varepsilon})$$
$$y(t,\varepsilon) = y^0(t) + y_s^0(\varepsilon t) - 1 + \mathcal{O}(\sqrt{\varepsilon}) \tag{2.8}$$

Composite asymptotics of this type are obtained immediately by the boundary layer method, if that method is valid [11–13].

It is clear that, for the examples given above, there is no sense in mixing both fast and slow processes in the form (2.8). Of course, there are other problems where the additive separation of time scales, as was done in (2.8), is completely impossible. This situation occurs in the case for fast oscillations with slow modulation, for example. Other approaches, such as the well known WKB method, have to be applied in such cases. We will not dwell on these problems here.

III. THE EQUATIONS OF INHIBITED LIQUID-PHASE OXIDATION

The mechanism of liquid-phase chain oxidation of organic substrata RH with molecular oxygen (O_2) in the presence of the inhibiting agent InH is given as follows:

$$\mathrm{I} \to \dot{\mathrm{r}} \cdots \to \mathrm{R\dot{O}_2}$$

$$\mathrm{R\dot{O}_2} + \mathrm{R\dot{O}_2} \xrightarrow{K_6} \mathrm{P_6}$$

$$\mathrm{R\dot{O}_2} + \mathrm{InH} \xrightarrow{K_7} \mathrm{ROOH} + \mathrm{\dot{I}n}$$

$$\mathrm{R\dot{O}_2} + \mathrm{\dot{I}n} \xrightarrow{K_8} \mathrm{P_8}$$

$$\mathrm{\dot{I}n} + \mathrm{\dot{I}n} \xrightarrow{K_9} \mathrm{P_9}$$

This scheme is a rather widely known case of a complex mechanism, mentioned in [14, 15], and it is realized under the following restrictions:

1. Radicals $\dot{\mathrm{r}}$ resulting from the decomposition of the initiator I, take no part in reactions with either InH or free radicals.

   If the oxidation occurs with long chains, this requirement is realized. In the case of short chains, one has to use initiators, generating active radicals such as $\mathrm{R\dot{O}}$, $\mathrm{H\dot{O}}$, $\mathrm{\dot{C}l}$, and so on, that react totally with RH even if the substrata concentrations are low.

2. The concentration of oxygen in the liquid phase is so high that it is possible to take into account only reactions with radicals RO_2.
3. The rate of oxidation is so small that neither the ROOH nor the products P_6, P_8, P_9 are involved in the reaction.

The quantitative description of the process results in a mathematical model in the form of a dynamical system, consisting of three differential equations.

$$d_T X = W_i - 2K_6 X^2 - K_7 XY - K_8 XZ$$

$$d_T Y = -K_7 XY$$

$$d_T Z = K_7 XY - K_8 XZ - 2K_9 Z^2$$

Here T is the independent variable (the time) and d_T denotes the derivative d/dT; the unknown functions $X(T), Y(T), Z(T)$ are the concentrations of the substances $X = [RO_2]$, $Y = [InH]$, $Z = [In]$ under the initial values X_0, Y_0, Z_0; the constant W_i is the initiator rate; K_j $(6 \leq j \leq 9)$ are the effective rate constants.

The given form of the equations is not suitable for both numerical simulation and asymptotic analysis. In order to detect small terms and to define this smallness, the dependent variables X, Y, Z have to be scaled to quantities of like order. To this end typical values X^*, Y^*, Z^* are extracted from X, Y, Z as factors. It is clear that near the initial moment the initial values X_0, Y_0, Z_0 can be taken as typical, if they are not zero. If any initial data are zero, then another method is needed to find a suitable typical value. In addition, the independent variable (time) can be scaled.

Thus, the process of scaling is a change of variables as follows:

$$T = T^* t \qquad X(T) = X^* x(t) \qquad Y(T) = Y^* y(t) \qquad Z(T) = Z^* z(t)$$

The equations for the new unknown functions $x, y, z(t)$ are similar to the original ones

$$d_t x = w_i - 2k_6 x^2 - k_7 xy - k_8 xz$$

$$d_t y = -k_3 xy$$

$$d_t z = k_4 xy - k_5 xz - 2k_9 z^2$$

The initial values $x(0), y(0), z(0)$ are now either ones or zeros. The coefficients are determined by the basic given constants

$$k_6 = T^*X^*K_6 \qquad k_7 = T^*Y^*K_7 \qquad k_8 = T^*Z^*K_8$$

$$k_3 = T^*X^*K_7 \qquad k_4 = T^*X^*Y^*K_7/Z^* \qquad k_5 = T^*X^*K_8$$

$$k_9 = T^*Z^*K_9 \qquad w_i = T^*W_i/X^*$$

We are now in a position to compare various terms of the equations by means of the magnitudes of the coefficients. Due to the time normalization factor T^* we can assume here that there are no large coefficients in the equations, so that any k_j is either order one or is small.

In the case when all coefficients are $\mathcal{O}(1)$ no asymptotic simplification of the problem is available. The efficiency of an asymptotic tool depends on the disposition of small factors in the equations. They determine both the asymptotic structure of the solution and the various scales. In principle, very different cases are possible for the form of the equations and it is impossible to obtain a general form of the asymptotic solution that is always suitable. An asymptotic method can be common only to a number of problems.

We will consider two situations that are significant for chemical kinetics. In both cases, the coefficient k_9 is so small that it does not affect the leading term of the approximation. Therefore, without loss of generality, we assume hereafter that $k_9 = 0$. In addition, one of the coefficients of order $\mathcal{O}(1)$ can be taken equal to unity by a proper choice of the time normalization factor T^*.

Thus, two problems are considered that differ from each other in the location of the small factors.

$$\begin{aligned} d_t x &= \varepsilon[W - Ax^2 - x(By + Cz)] \qquad x(0) = 1 \\ d_t y &= -\varepsilon^2 Dxy \qquad y(0) = 1 \\ d_t z &= x(Ey - z) \qquad z(0) = 0 \end{aligned} \tag{I}$$

$$\begin{aligned} d_t x &= \varepsilon^4(W - Ax^2) - \varepsilon^3 x(By + Cz) \qquad x(0) = 1 \\ d_t y &= -\varepsilon^4 Dxy \qquad y(0) = 1 \\ d_t z &= x(Ey - z) \qquad z(0) = 0 \end{aligned} \tag{II}$$

Here $0 < \varepsilon \ll 1$ is a small parameter. The coefficients A, B, C, D, E, W are positive and do not depend on ε and t.

The first system of equations describes the process of inhibited liquid-phase oxidation under the assumptions

$$W_i = 2 \times 10^{-8}\,\text{mol}\,\text{L}^{-1}\,\text{s}^{-1} \qquad K_6 = 10^6\,\text{L}\,\text{mol}^{-1}\,\text{s}^{-1}$$
$$K_7 = 2 \times 10^4\,\text{L}\,\text{mol}^{-1}\,\text{s}^{-1} \qquad K_8 = 5 \times 10^8\,\text{L}\,\text{mol}^{-1}\,\text{s}^{-1} \tag{3.1}$$
$$2K_9 = 10^3\,\text{L}\,\text{mol}^{-1}\,\text{s}^{-1} \qquad X_0 = 10^{-7}\,\text{mol}\,\text{L}^{-1} \qquad Y_0 = 5 \times 10^{-6}\,\text{mol}\,\text{L}^{-1}$$

The small parameter ε has a value of about 10^{-1}.

The second system describes the same process under the assumptions

$$W_i = 10^{-8}\,\text{mol}\,\text{L}^{-1}\,\text{s}^{-1} \qquad 2K_6 = K_7 = 10^4\,\text{L}\,\text{mol}^{-1}\,\text{s}^{-1}$$
$$K_8 = 10^8\,\text{L}\,\text{mol}^{-1}\,\text{s}^{-1} \qquad 2K_9 = 10^3\,\text{L}\,\text{mol}^{-1}\,\text{s}^{-1} \tag{3.2}$$
$$X_0 = 10^{-6}\,\text{mol}\,\text{L}^{-1} \qquad Y_0 = 10^{-5}\,\text{mol}\,\text{L}^{-1}$$

In this case, the small parameter ε has a value of about 10^{-1} as well.

IV. ASYMPTOTIC SOLUTION OF PROBLEM I

The task is to find the asymptotic approximation of the solution as $\varepsilon \to 0$ uniformly over a long time interval $0 < t \leq \mathcal{O}(\varepsilon^{-2})$. A leading order term of $\mathcal{O}(1)$ is a main goal. The higher order corrections of $\mathcal{O}(\varepsilon)$ will be defined in order to identify both the secular terms and the time intervals of asymptotic fitness.

By looking at the location of the small parameters in the equations, one can guess three time scales t, εt, and $\varepsilon^2 t$. The original equations (I) were written for the very fast scale t

$$\begin{aligned} d_t x &= \varepsilon[W - Ax^2 - x(By + Cz)] \qquad x(0) = 1 \\ d_t y &= -\varepsilon^2 Dxy \qquad y(0) = 1 \\ d_t z &= x(Ey - z) \qquad z(0) = 0 \end{aligned} \tag{4.0}$$

There are many coefficients in the equations and each of them affects the exact solution. However, the leading terms of the asymptotics depend only on two combinations of the input constants, namely,

$$\lambda = [(B + CE)^2 + 4AW]^{1/2} \qquad X^+ = (1/2A)[\lambda - (B + CE)]$$

In addition, two other combinations will be used to simplify the calculation

$$\beta = (B + CE)/2A \qquad \gamma = (W/A)^{1/2}$$

A. The Fast Time Scale

In this case, we seek the asymptotic solution in the form of power series with coefficients, depending on the fast time t, as follows:

$$\begin{aligned} x(t;\varepsilon) &= x_1^0(t) + \varepsilon x_1^1(t) + \varepsilon^2 x_1^2(t) + \cdots \\ y(t;\varepsilon) &= y_1^0(t) + \varepsilon y_1^1(t) + \varepsilon^2 y_1^2(t) + \cdots \\ z(t;\varepsilon) &= z_1^0(t) + \varepsilon z_1^1(t) + \varepsilon^2 z_1^2(t) + \cdots \end{aligned} \tag{4.1}$$

The coefficients of these series will be found from the system of recurrence equations that are derived in the standard way. To this end, the series (4.1) are substituted into Eqs. (4.0) and the expressions for like powers of ε are equated. Ordinary differential equations and corresponding initial conditions are obtained.

At the first step, the problem is very simple.

$$d_t x = 0 \qquad d_t y = 0 \qquad d_t z + xz = Exy$$
$$x(0) = 1 \qquad y(0) = 1 \qquad z(0) = 0$$

So the leading term reads

$$x_1^0(t) \equiv 1 \qquad y_1^0(t) \equiv 1 \qquad z_1^0(t) = E[1 - \exp(-t)] \tag{4.2}$$

At the next step, the functions $x_1^1(t)$, $y_1^1(t)$, $z_1^1(t)$ are determined by the equations

$$d_t x = W - A - B - Cz_1^0(t) \qquad x(0) = 0$$
$$d_t y = 0 \qquad y(0) = 0$$
$$d_t z + z = Ex \qquad z(0) = 0$$

Because the function $z_1^0(t)$ is now known, explicit formulas can be written out for the first corrections as well

$$y_1^1(t) \equiv 0$$
$$x_1^1(t) = [W - A - B - CE]t + CE[1 - \exp(-t)]$$
$$z_1^1(t) = E[W - A - B - CE](t - 1 + \exp(-t)) + CE^2[1 - t\exp(-t)]$$

Note that the functions $x_1^1(t)$, $z_1^1(t)$ grow like $\mathcal{O}(t)$ as $t \to \infty$. Hence, for long times of order $\mathcal{O}(1/\varepsilon)$, the first terms of the series (4.1) have the same order as the leading terms. At later steps, the growth becomes stronger, for example,

$$x_1^2(t) = \mathcal{O}(t^2) \qquad t \to \infty$$

Thus, the series (4.1) turn out to be asymptotic only for times that are not too large $0 \le t \ll \mathcal{O}(1/\varepsilon)$.

This analysis shows that even the rough approximation

$$x_1(t) \cong 1 \qquad y_1(t) \cong 1 \qquad z_1(t) \cong E[1 - \exp(-t)]$$

is suitable as long as $t \ll \mathcal{O}(1/\varepsilon)$, and it becomes poor for long times $t \cong 1/\varepsilon$.

This situation is standard for problems with a small parameter. The interval of fitness of the trivial asymptotic solution (4.1) cannot be too long. The small terms affect the solution for long times. On the long time scale $t \cong 1/\varepsilon$ one has to construct a new asymptotic solution depending on another typical time. A suitable new independent variable is $\tau = \varepsilon t$ as one can guess by looking at the secular terms.

B. The First Slow Scale

After the change of variable $\tau = \varepsilon t$ the equations for the new unknown functions

$$x_2(\tau; \varepsilon) = x(t; \varepsilon) \qquad y_2(\tau; \varepsilon) = y(t; \varepsilon) \qquad z_2(\tau; \varepsilon) = z(t; \varepsilon)$$

differ from (4.0) in the location of the small parameters

$$\begin{aligned} d_\tau x_2 &= W - Ax_2^2 - x_2(By_2 + Cz_2) \\ d_\tau y_2 &= -\varepsilon D x_2 y_2 \\ \varepsilon d_\tau z_2 &= x_2(Ey_2 - z_2) \end{aligned} \tag{4.3}$$

We seek the asymptotic solution similar to (4.1) with coefficients, depending on the new time τ.

$$\begin{aligned} x_2(\tau; \varepsilon) &= \overset{0}{x_2}(\tau) + \varepsilon \overset{1}{x_2}(\tau) + \varepsilon^2 \overset{2}{x_2}(\tau) + \cdots \\ y_2(\tau; \varepsilon) &= \overset{0}{y_2}(\tau) + \varepsilon \overset{1}{y_2}(\tau) + \varepsilon^2 \overset{2}{y_2}(\tau) + \cdots \\ z_2(\tau; \varepsilon) &= \overset{0}{z_2}(\tau) + \varepsilon \overset{1}{z_2}(\tau) + \varepsilon^2 \overset{2}{z_2}(\tau) + \cdots \end{aligned} \tag{4.4}$$

There is some distinction at this stage. Indeed, the initial data are taken from the matching of the asymptotic series. The process is as follows. The known asymptotics of the coefficients $\overset{n}{x_1}(t)$, $\overset{n}{y_1}(t)$, $\overset{n}{z_1}(t)$ at infinity ($t \to \infty$) are substituted into the series of the fast asymptotic solution (4.1). After that the change of variable $\tau = \varepsilon t$ is made in the resulting double series, and the expressions in like powers of ε are gathered together. The series in powers of τ, obtained in such a manner, are now interpreted as the asymptotic conditions at zero (as $\tau \to 0$) of the coefficients $x_2^n(\tau)$, $y_2^n(\tau)$, $z_2^n(\tau)$ of the new (slow) asymptotic solution (4.4). This interpretation is the correct matching requirement.

In the case under consideration, a simpler form of matching can be used. Indeed, the nondecreasing terms (consts) of the asymptotics at infinity on the fast scale ($t \to \infty$) give the nonzero terms (consts) of the asymptotics at zero on the slow scale ($\tau \to 0$). This requirement results in initial data that naturally differ from the original ones

$$\{x_2, y_2, z_2\}(\tau; \varepsilon)|_{\tau=0} = \{1, 1, E\} + \varepsilon\{CE, 0, E[A + B + 2CE - W]\} + \mathcal{O}(\varepsilon^2)$$

The equations for the coefficients of the asymptotic solutions (4.4) are modified too. In particular, one algebraic and two differential equations determine the leading order terms

$$d_\tau x = W - Ax^2 - x[By + Cz]$$

$$d_\tau y = 0$$

$$x[Ey - z] = 0$$

In this case, the initial data for x, y, taken from matching, coincide by chance with the original ones

$$x_2^0(0) = 1 \qquad y_2^0(0) = 1 \qquad z_3^0(0) = E$$

Because the solutions for y, z are trivial

$$y_2^0(\tau) \equiv 1 \qquad z_2^0(\tau) \equiv E$$

the problem reduces to one for the first component

$$d_\tau x = -A[x^2 + 2\beta x - \gamma^2] \qquad x(0) = 1$$

Here $\beta = (B + CE)/2A$, $\gamma^2 = W/A$.

The differential equation is easy to solve and the solution $x_2^0(\tau)$ is either

$$\ln|(x_2^0 - X^+)/(x_2^0 - X^-)| = -\lambda\tau + \ln|(1 + X^+)/(1 - X^-)|$$

or is of the form

$$(x_2^0 - X^+)/(x_2^0 - X^-) = [(1 + X^+)/(1 - X^-)]\exp(-\lambda\tau)$$
$$\text{where } \lambda = [(B + CE)^2 + 4AW]^{1/2}$$

Here the constants $X^{\pm} = -\beta \pm (\beta^2 + \gamma^2)^{1/2}$ are the roots of the algebraic equation

$$x^2 + 2\beta x - \gamma^2 = 0$$

One can see from the explicit formula that the function $x_2^0(\tau)$ tends to the constant X^+ at infinity. Thus, the leading approximation on this scale has the asymptotics

$$y_2^0(\tau) \equiv 1 \qquad z_2^0(\tau) \equiv E \qquad x_2^0(\tau) = X^+ + \mathcal{O}(\exp(-\lambda\tau)) \qquad \tau \to \infty \quad (4.5)$$

In the case $X^+ = 1$ (i.e., as $\gamma^2 = 2\beta + 1$) the first component is stationary: $x_2^0(\tau) \equiv 1$.

At subsequent steps, the coefficients of the asymptotic solution are found from inhomogeneous equations where the right sides are determined as before. Growing terms emerge already in the first correction on this scale. Indeed, the problem for the first correction x_2^1, y_2^1, $z_2^1(\tau)$ is as follows:

$$d_\tau x = -2Ax_2^0 x - x(B + CE) - x_2^0(By + Cz) \qquad x(0) = CE$$
$$d_\tau y = -Dx_2^0 \qquad y(0) = 0$$
$$Ey - z = 0$$

The solution can be written out in explicit form, hence one can analyze the asymptotics at infinity. For example, the second component is given by the integral

$$y_2^1(\tau) = -D\int_0^\tau x_2^0(\zeta)\, d\zeta$$

It is easy to see that this integral grows like $\mathcal{O}(\tau)$ as $\tau \to \infty$ because the integrand tends to a nonzero constant X^+ at infinity.

$$y_2^1(\tau) = -DX^+\tau + \mathcal{O}(1) \qquad \tau \to \infty$$

The third component, determined by $z_2^1(\tau) = Ey_2^1(\tau)$, has a similar behavior.

$$z_2^1(\tau) = -DEX^+\tau + \mathcal{O}(1) \qquad \tau \to \infty$$

The first component $x_2^1(\tau)$ is found from the linear inhomogeneous equation.

$$d_\tau x + 2Ax_2^0(\tau)x + (B + CE)x = -x_2^0(\tau)[By_2^1(\tau) + Cz_2^1(\tau)]$$
$$x(0) = CE$$

The solution of this problem can be written in integral form as well. If we take into account the fact that the derivative $d_\tau x_2^0(\tau)$ is a solution of a homogeneous equation, we obtain

$$x_2^1(\tau) = d_\tau x_2^0(\tau)\left\{CE - (B + CE)\int_0^\tau [x_2^0(\zeta)y_2^1(\zeta)/d_\tau x_2^0(\zeta)]\, d\zeta\right\}$$

One can see, either from this formula or straight from the differential equation, that the function $x_2^1(\tau)$ grows at infinity

$$x_2^1(\tau) = \tau\beta DX^+/(X^+ + \beta) + \mathcal{O}(1) \qquad \tau \to \infty$$

Thus, there are secular terms in the asymptotic solution on the slow scale, which indicate that the approximation (4.4) is false for the long times $\tau \cong 1/\varepsilon$.

Notice that, for very small times $\tau \cong \varepsilon$, the approximation (4.4) is false as well, because then another asymptotic solution (4.1) is valid. For long times $\tau \cong 1/\varepsilon$, correct asymptotics have to be constructed.

C. The Second Slow Scale

After the change of the independent variable $\theta = \varepsilon\tau = \varepsilon^2 t$, the equations for the new unknown functions

$$x_3(\theta; \varepsilon) = x(t; \varepsilon) \qquad y_3(\theta; \varepsilon) = y(t; \varepsilon) \qquad z_3(\theta; \varepsilon) = z(t; \varepsilon)$$

take the form

$$\begin{aligned} \varepsilon d_\theta x_3 &= W - Ax_3^2 - x_3(By_3 + Cz_3) \\ d_\theta y_3 &= -Dx_3 y_3 \\ \varepsilon^2 d_\theta z_3 &= x_3(Ey_3 - z_3) \end{aligned} \tag{4.6}$$

The initial data are taken from the matching condition and have the asymptotic form

$$\{x_3, y_3, z_3\}(\theta;\varepsilon)|_{\theta\to 0} = \{X^+, 1, E\} + \mathcal{O}(\varepsilon)$$

We seek asymptotic solutions similar to (4.4) with coefficients depending on the new slow time

$$\begin{aligned} x_3(\theta;\varepsilon) &= \overset{0}{x}_3(\theta) + \varepsilon \overset{1}{x}_3(\theta) + \varepsilon^2 \overset{2}{x}_3(\theta) + \cdots \\ y_3(\theta;\varepsilon) &= \overset{0}{y}_3(\theta) + \varepsilon \overset{1}{y}_3(\theta) + \varepsilon^2 \overset{2}{y}_3(\theta) + \cdots \\ z_3(\theta;\varepsilon) &= \overset{0}{z}_3(\theta) + \varepsilon \overset{1}{z}_3(\theta) + \varepsilon^2 \overset{2}{z}_3(\theta) + \cdots \end{aligned} \tag{4.7}$$

The problems for the coefficients of the asymptotic solutions $\overset{0}{x}_3$, $\overset{0}{y}_3$, $\overset{0}{z}_3$ are again modified. In particular, one differential and two algebraic equations determine the leading order terms

$$\begin{aligned} 0 &= W - Ax^2 - x(By + Cz) \\ d_\theta y &= -Dxy \\ 0 &= x(Ey - z) \end{aligned}$$

The initial data for the leading terms, taken from matching, are

$$x_3(0) = X^+ \qquad y_3(0) = 1 \qquad z_3(0) = E$$

The algebraic equations allow us to eliminate either the second and third components $y_3, z_3(\theta)$ or the first one $x_3(\theta)$

$$\begin{gathered} z = Ey \qquad 2\beta y = \gamma^2/x - x \qquad x = [\beta^2 y^2 + \gamma^2]^{1/2} - \beta y \\ \text{where } \beta = (B + CE)/2A \qquad \gamma^2 = W/A \end{gathered} \tag{4.8}$$

The differential equations can be written out for either the component $\overset{0}{x}_3(\theta)$ or $\overset{0}{y}_3(\theta)$. Each of them can be solved. For example, solving the equation

$$(\gamma^2/x^2 + 1)d_\theta x = D(\gamma^2 - x^2) \qquad x(0) = X^+$$

leads to the implicit function $\overset{0}{x}_3(\theta)$ in the form

$$\ln|(\gamma + \overset{0}{x}_3)/(\gamma - \overset{0}{x}_3)| - \gamma/\overset{0}{x}_3 = D\gamma\theta + \overset{0}{C}_3 \tag{4.9}$$

where

$$C_3^0 = \ln|(\gamma + X^+)/(\gamma - X^+)| - \gamma/X^+ = \text{const}$$

Notice that

$$X^+ - \gamma = (\beta^2 + \gamma^2)^{1/2} - \beta - \gamma = \text{const} \neq 0$$

A similar expression for the function $y_3^0(\theta)$ can be obtained both from the formula (4.8) and from the similar differential equation for $y_3^0(\theta)$.

$$d_\theta y = -Dy[(\beta^2 y^2 + \gamma^2)^{1/2} - \beta y] \qquad y(0) = 1$$

To obtain the solution in implicit form it is expedient here to take into account the differential identity

$$d[(\beta^2 y^2 + \gamma^2)^{1/2} - \beta y]^{-1} = \beta(\beta^2 y^2 + \gamma^2)^{-1/2}[(\beta^2 y^2 + \gamma^2)^{1/2} - \beta y]^{-1}\, dy$$

Then, the implicit form of the solution reads

$$\ln|[\gamma + (\beta^2 y_3^0 + \gamma^2)^{1/2}]/\beta y_3^0| - \gamma/[(\beta^2 y_3^0 + \gamma^2)^{1/2} - \beta y_3^0] = \gamma D\theta + D_3^0 \tag{4.10}$$

Here

$$D_3^0 = \ln|[\gamma + (\beta^2 + \gamma^2)^{1/2}/\beta]| - \gamma/[(\beta^2 + \gamma^2)^{1/2} - \beta] = \text{const}$$

Formulas (4.9), (4.10) provide the approximate solution on the slow time scale. From these asymptotics one can see that the second and third components tend to zero, whereas the first one tends to a nonzero constant $\gamma = (W/A)^{1/2}$ at infinity. If we use Taylor expansions of the left sides of Eqs. (4.9) and (4.10), the error estimate of the asymptotic behavior at infinity can easily be derived as follows:

$$x_3^0(\theta) = \gamma + \mathcal{O}(\exp(-D\gamma\theta)) \qquad \theta \to \infty$$

$$y_3^0(\theta) = \exp(-D\gamma\theta + \delta) + \mathcal{O}(\exp(-2D\gamma\theta)) \qquad \theta \to \infty \ (\delta = \text{const}) \tag{4.11}$$

$$z_3^0(\theta) = \mathcal{O}(\exp(-D\gamma\theta)) \quad \theta \to \infty$$

These asymptotics are differentiable, hence the asymptotic estimates of the derivatives read

$$d_\theta x_3^0(\theta) = \mathcal{O}(\exp(-D\gamma\theta))$$

$$d_\theta y_3^0(\theta) = -D\gamma \exp(-D\gamma\theta + \delta) + \mathcal{O}(\exp(-2D\gamma\theta)) \qquad \theta \to \infty$$

The last results can be used to analyze the asymptotic behavior of the first corrections $x_3^1, y_3^1, z_3^1(\theta)$. These functions are determined by the equations

$$\begin{aligned} 2Axx_3^0(\theta) + x(By_3^0(\theta) + Cz_3^0(\theta)) + x_3^0(\theta)(By + Cz) &= -d_\theta x_3^0(\theta) \\ d_\theta y &= -Dx_3^0(\theta)y - Dxy_3^0(\theta) \\ Ey - z &= 0 \end{aligned} \tag{4.12}$$

The variables x, z are eliminated and the problem is reduced to the single differential equation

$$d_\theta y + \beta D[x_3^0/(x_3^0 + \beta y_3^0)]y = f(\theta) \tag{4.13}$$

where the right side decays rapidly at infinity

$$f(\theta) = (D/2A)[y_3^0/(x_3^0 + \beta y_3^0)]d_\theta x_3^0 = \mathcal{O}(\exp(-2D\gamma\theta)) \quad \theta \to \infty$$

The solution of the corresponding homogeneous equation can be used to solve the nonhomogeneous equation. This solution is just the derivative $d_\theta y_3^0$ of the solution of the nonlinear equation from the first step. Hence, the solution reads

$$y_3^1(\theta) = d_\theta y_3^0(\theta)\left[y_3^1(0) + \int_0^\theta f(\zeta)/d_\zeta y_3^0(\zeta)\, d\zeta \right]$$

Because the integrand is bounded by the exponential $M \exp(-\gamma D\zeta)$ ($M = \text{const}$), the integral is bounded for large values of θ. Therefore the decreasing factor

$$d_\theta y_3^0(\theta) = \mathcal{O}(\exp(-D\gamma\theta))$$

provides the decay of the function $y_3^1(\theta)$ at infinity

$$y_3^1(\theta) = \mathcal{O} \exp(-D\gamma\theta))$$

The other components $x_3^1(\theta), z_3^1(\theta)$ have similar asymptotics as one can see from the algebraic equations (4.12).

So there are no secular terms on the third scale $\theta = \varepsilon^2 t$. The asymptotic solution (4.7) is available for all times to infinity. Of course, this approximation is not suitable for very small times $\theta \cong \varepsilon$, where other asymptotics apply.

D. The Results for Problem I

Let us consider the leading order terms of the various asymptotic solutions. On the first (fast) scale the third component

$$z_1^0(t) = E - \mathcal{O}(\exp(-t))$$

is increasing and stabilizing to the constant E at infinity. The first and second components are stable

$$x_1^0(t) \equiv 1 \qquad y_1^0(t) \equiv 1$$

On the second scale $\tau = \varepsilon t$, the first component varies from 1 to X^+

$$x_2^0(\tau) = X^+ + \mathcal{O}(\exp(-\lambda\tau)) \qquad \tau \to \infty$$

whereas the components x, z are stable

$$y_2^0(\tau) \equiv 1 \qquad z_2^0(\tau) \equiv E$$

Finally, on the slow scale, $\theta = \varepsilon^2 t$, all three components are varying, and they are stabilizing to the vector $(\gamma, 0, 0)$, $\gamma = (W/A)^{1/2}$.

The limit state

$$(x, y, z) = ((W/A)^{1/2}, 0, 0)$$

is a unique and stable equilibrium of the input system (4.0). Therefore, the process, described by Eqs. (4.0), stabilizes to the state $((W/A)^{1/2}, 0, 0)$, which occurs on the slow time scale θ.

V. ASYMPTOTIC SOLUTION OF PROBLEM II

The problem, given by Eqs. (II), is very similar to the one investigated above. The difference is only some other arrangement of the small factors. The task is as before: To find the asymptotic approximation of the solution as $\varepsilon \to 0$ uniformly over a long time interval on a scale where the stable equilibrium occurs. The determination of the leading term of $\mathcal{O}(1)$ is the main goal. The higher order corrections of $\mathcal{O}(\varepsilon)$ are defined to identify both the secular terms and the time intervals of asymptotic fitness.

At first sight, the situation is analogous to the former case. Looking for the small parameters in the equations

$$
\begin{aligned}
d_t x &= \varepsilon^4(W - Ax^2) - \varepsilon^3 x(By + Cz) \qquad x(0) = 1 \\
d_t y &= -\varepsilon^4 Dxy \qquad y(0) = 1 \\
d_t z &= x(Ey - z) \qquad z(0) = 0
\end{aligned}
\tag{5.0}
$$

one can perceive three time scales, $t, \varepsilon^3 t, \varepsilon^4 t$. It seems natural, by analogy, to presuppose the ordering of the scales so that the slow scale follows the fast one. However, in the case under consideration, the situation is more complicated because the final time scale, which is similar to $\varepsilon^4 t$, occurs after a very slow scale $\varepsilon^5 t$. It is very difficult to explain this result without an analysis of the asymptotic solution. Experience of asymptotic analysis is needed to grasp it in advance.

The input constants affect the leading terms of the asymptotics through just three combinations

$$
\mu = B + CE \qquad \nu = (B + CE)/DW \qquad X_2^+ = W/(B + CE)
$$

A. The Fast Scale

At first, we seek the asymptotic solution in the form of a power series in ε, with coefficients depending on the fast time t, as follows:

$$
\begin{aligned}
x(t; \varepsilon) &= x_1^0(t) + \varepsilon x_1^1(t) + \varepsilon^2 x_1^2(t) + \cdots \\
y(t; \varepsilon) &= y_1^0(t) + \varepsilon y_1^1(t) + \varepsilon^2 y_1^2(t) + \cdots \\
z(t; \varepsilon) &= z_1^0(t) + \varepsilon z_1^1(t) + \varepsilon^2 z_1^2(t) + \cdots
\end{aligned}
\tag{5.1}
$$

The coefficients are defined from a system of recurrence equations.

At the first step, the equations

$$
\begin{gathered}
d_t x = 0 \qquad d_t y = 0 \qquad d_t z + xz = Exy \\
x(0) = 1 \qquad y(0) = 1 \qquad z(0) = 0
\end{gathered}
$$

are solved in terms of elementary functions

$$
x_1^0(t) \equiv 1 \qquad y_1^0(t) \equiv 1 \qquad z_1^0(t) = E[1 - \exp(-t)] \tag{5.2}
$$

At the next two steps, homogeneous equations are obtained.

$$d_t x = 0 \qquad d_t y = 0 \qquad d_t z + z = E(x + y)$$

$$x(0) = 0 \qquad y(0) = 0 \qquad z(0) = 0$$

Hence, both the first and second correction are zero.

$$x_1^1(t) \equiv 0 \qquad y_1^1(t) \equiv 0 \qquad z_1^1(t) \equiv 0 \qquad x_1^2(t) \equiv 0$$

$$y_1^2(t) \equiv 0 \qquad z_1^2(t) \equiv 0$$

Nonzero corrections occur of $\mathcal{O}(\varepsilon^3)$. The coefficients x_1^3, y_1^3, $z_1^3(t)$ of the third power of ε are determined from the inhomogeneous equations

$$d_t x = -[B + C z_1^0(t)] \qquad x(0) = 0$$

$$d_t y = 0 \qquad y(0) = 0$$

$$d_t z = (Ey - z) + x[E - z_1^0(t)] \qquad z(0) = 0$$

One can write out the solution

$$x_1^3(t) = CE - \mu t - CE \exp(-t) \qquad (\mu = B + CE)$$

$$y_1^3(t) \equiv 0$$

$$z_1^3(t) = E[CE + \mu - \mu t] + [CE - CE^2 - E\mu - (E-1)CEt$$
$$+ \mu t^2/2] \exp(-t)] - CE \exp(-2t)$$

We see that the first and third components have the term μt, which is unbounded. Hence, for long time of $\mathcal{O}(1/\varepsilon^3)$, the first nonzero correction in the components x, z has the same order as the leading one.

$$\varepsilon^3 x_1^3(t),\ \varepsilon^3 z_1^3(t) = \mathcal{O}(1) \qquad \text{as} \qquad t = \mathcal{O}(1/\varepsilon^3)$$

At the subsequent two steps, secular terms of only the same order may occur.

$$x_1^n,\ y_1^n,\ z_1^n(t) = \mathcal{O}(t) \qquad t \to 0 \qquad (n = 4, 5)$$

Secular terms of the next order emerge just for $x_1^6(t) = \mathcal{O}(t^2)$. Thus, the formulas (5.1) give the approximate solution only if the time is not too long $t \ll \varepsilon^{-3}$.

A new typical time variable is needed to construct the asymptotic solution suitable for $t \cong \varepsilon^{-3}$. The new independent variable is $\tau = \varepsilon^3 t$ as one can guess by looking at the secular terms.

B. The First Slow Scale

After the change of variable $\tau = \varepsilon^3 t$ the equations (5.0) for the new unknown functions $x_2(\tau; \varepsilon) = x(t; \varepsilon)$, $y_2(\tau; \varepsilon) = y(t; \varepsilon)$, and $z_2(\tau; \varepsilon) = z(t; \varepsilon)$ are

$$\begin{aligned} d_\tau x_2 &= \varepsilon[W - Ax_2^2] - x_2(By_2 + Cz_2) \\ d_\tau y_2 &= -\varepsilon D x_2 y_2 \\ \varepsilon^3 d_\tau z_2 &= x_2(Ey_2 - z_2) \end{aligned} \tag{5.3}$$

The initial data are taken from the matching of the asymptotic series. Indeed, the nondecreasing terms (consts) of the asymptotics at infinity on the fast scale ($t \to \infty$) give the nonzero terms (consts) of the asymptotics at zero on the slow scale ($\tau \to 0$). This requirement results in initial data that naturally differ from the original ones (5.0). There are asymptotic series valid in the general case. If we restrict ourselves to two terms in the asymptotics, the initial data read

$$\{x_2, y_2, z_2\}(\tau; \varepsilon)|_{\tau=0} = \{1, 1, E\} + \varepsilon^3\{CE, 0, E(2CE + B)\} + \mathcal{O}(\varepsilon^4)$$

We seek an asymptotic solution similar to Eq. (5.1) with coefficients depending on the new time τ.

$$\begin{aligned} x_2(\tau; \varepsilon) &= \overset{0}{x}_2(\tau) + \varepsilon \overset{1}{x}_2(\tau) + \varepsilon^2 \overset{2}{x}_2(\tau) + \cdots \\ y_2(\tau; \varepsilon) &= \overset{0}{y}_2(\tau) + \varepsilon \overset{1}{y}_2(\tau) + \varepsilon^2 \overset{2}{y}_2(\tau) + \cdots \\ z_2(\tau; \varepsilon) &= \overset{0}{z}_2(\tau) + \varepsilon \overset{1}{z}_2(\tau) + \varepsilon^2 \overset{2}{z}_2(\tau) + \cdots \end{aligned} \tag{5.4}$$

The problems determining the coefficients of the asymptotic solution are modified. In particular, one algebraic and two differential equations give the leading order terms.

$$\begin{gathered} d_\tau x = -x(By + Cz) \qquad x(0) = 1 \\ d_\tau y = 0 \qquad y(0) = 1 \\ x[Ey - z] = 0 \end{gathered}$$

Because the solutions for y, z are trivial, the system reduces to a problem for the first component

$$d_\tau x = -\mu x \qquad x(0) = 1 \qquad (\mu = B + CE)$$

The single differential equation is easy to solve and the first approximation is

$$x_2^0(\tau) = \exp(-\mu\tau) \qquad y_2^0(\tau) \equiv 1 \qquad z_2^0(\tau) \equiv E \tag{5.5}$$

We see that the function $x_2^0(\tau)$ tends to zero at infinity.

At the next step, the coefficients are determined from the inhomogeneous equations

$$d_\tau x + \mu x = W - Ax_2^0 - x_2^0(By + Cz) \qquad x(0) = 0$$
$$d_\tau y = -Dx_2^0 \qquad y(0) = 0$$
$$Ey - z = 0$$

where the right sides are determined as before. The second equation is easily solved because x_2^0 is a known function. Hence,

$$y_2^1(\tau) = -(D/\mu)[1 - x_2^0(\tau)] \qquad z_2^1(\tau) = Ey_2^1(\tau)$$

If we now eliminate both y and z from the first equality, a single differential equation for x_2^1 is obtained

$$d_\tau x + \mu x = W + Dx_2^0 - (A + D)(x_2^0)^2 \qquad x(0) = 0$$

This equation can easily be solved, and so the first correction can also be written out in the explicit form

$$\begin{gathered} x_2^1(\tau) = X_2^+ - [X_2^+ + (A + D)/2 + Dt] \exp(-\mu\tau) \\ + ((A + D)/2) \exp(-2\mu\tau) \\ y_2^1(\tau) = -(D/\mu)[1 - \exp(-\mu\tau)] \\ z_2^1(\tau) = -(DE/\mu)[1 - \exp(-\mu\tau)] \\ (X_2^+ = W/(B + CE),\ \mu = B + CE) \end{gathered} \tag{5.6}$$

One can see here that all components are stabilized at infinity. The secular terms on this scale occur just in the second corrections

$x_2^2, y_2^2, z_2^2(\tau)$, which are determined by the equations

$$d_\tau x + \mu x = -2Axx_2^0 - x_2^0(By + Cz) - x_2^1(By_2^1 + Cz_2^1)$$
$$d_\tau y = -D[x_2^0 y_2^1 + x_2^1]$$
$$Ey - z = 0$$

Since the right side of the y equation stabilizes to a nonzero constant, both the second and third components grow at infinity

$$y_2^2(\tau) = -DX_2^+\tau + \mathcal{O}(1) \qquad z_2^2(\tau) = -DEX_2^+\tau + \mathcal{O}(1) \qquad \tau \to \infty$$

The first component remains bounded for long times

$$x_2^2(\tau) = \mathcal{O}(1)$$

Subsequent analysis shows that the degree of the secularities increase over just two steps

$$y_2^n(\tau), z_2^n(\tau) = \mathcal{O}(\tau^k) \qquad x_2^n(\tau) = \mathcal{O}(\tau^{k-1}) \qquad \tau \to \infty \qquad (k = [n/2])$$

Thus, the asymptotic solution on the slow scale $\tau = \varepsilon t$ is suitable as long as $\tau \ll \varepsilon^{-2}$ and it fails for long times $\tau \cong \varepsilon^{-2}$. This reasoning is based on the structure of the secularities $\varepsilon^2\tau$, which shows also that the next scale is just $\varepsilon^2\tau = \varepsilon^5 t$ and not $\varepsilon\tau = \varepsilon^4 t$. At first sight, this last fact does not agree with the structure of the original equations, because there are no terms of order $\mathcal{O}(\varepsilon^5)$ on the right sides. To understand this situation, we have to take into account the new normalization of the variable x. Namely, in accordance with the asymptotics (5.4)–(5.6) the component $x(\tau; \varepsilon)$ has order $\mathcal{O}(\varepsilon)$, if the time is long, $\tau \gg |\ln \varepsilon|$. Hence, if we are normalizing x to unity, the additional factor ε from the variable x appears in the equations. After that one can see terms of $\mathcal{O}(\varepsilon^5)$ in the equations and it is no wonder that the scale $\varepsilon^2\tau = \varepsilon^5 t$ occurs.

Concerning the new normalization of the component x, there is no need to solve it at once. It will inevitably emerge in the structure of the asymptotic solution beginning with the $\mathcal{O}(\varepsilon)$ terms.

The last remark concerns the general problem of identifying the typical magnitudes of dependent variables, x, y, z, especially in the case when some of the variables are zero at the initial moment. To make an error here is not fatal and it does not crash the asymptotics. The structure of the asymptotic solution compensates for any mistakes in the normalization, although a vagueness may occur when the time scales are defined.

C. The Second Slow Scale

After changing the independent variable $\theta = \varepsilon^2 \tau = \varepsilon^5 t$ the equations for the new unknown functions

$$x_3(\theta; \varepsilon) = x(t; \varepsilon) \qquad y_3(\theta; \varepsilon) = y(t; \varepsilon) \qquad z_3(\theta; \varepsilon) = z(t; \varepsilon)$$

take the form

$$\begin{aligned} \varepsilon^2 d_\theta x_3 &= \varepsilon(W - Ax_3^2) - x_3(By_3 + Cz_3) \\ \varepsilon d_\theta y_3 &= -Dx_3 y_3 \\ \varepsilon^5 d_\theta z_3 &= x_3(Ey_3 - z_3) \end{aligned} \tag{5.7}$$

The initial data are taken from the matching condition and have the asymptotic form

$$\{x_3, y_3, z_3\}(\theta; \varepsilon)|_{\theta=0} = \{0, 1, E\} + \varepsilon\{X_2^+, -D/\mu, -DE/\mu\} + \mathcal{O}(\varepsilon^2)$$

We seek an asymptotic solution similar to Eq. (5.4) with coefficients depending on the new slow time. It is natural that the term of $\mathcal{O}(1)$ is absent in the first component x, because it is zero in the initial data

$$\begin{aligned} x_3(\theta; \varepsilon) &= \varepsilon x_3^1(0) + \varepsilon^2 x_3^2(\theta) + \cdots \\ y_3(\theta; \varepsilon) &= y_3^0(\theta) + \varepsilon y_3^1(\theta) + \varepsilon^2 y_3^2(\theta) + \cdots \\ z_3(\theta; \varepsilon) &= z_3^0(\theta) + \varepsilon z_3^1(\theta) + \varepsilon^2 z_3^2(\theta) + \cdots \end{aligned} \tag{5.8}$$

Two algebraic and one differential equations are obtained for the leading terms $x_3^1, y_3^0, z_3^0(\theta)$ of the asymptotics

$$\begin{aligned} 0 &= W - x(By + Cz) \\ d_\theta y &= -Dxy \qquad y(0) = 1 \\ 0 &= x(Ey - z) \end{aligned}$$

The algebraic equations allow us to eliminate both the first and the third components

$$z = Ey \qquad x = 1/D\nu y \qquad \text{where } \nu = (B + CE)/WD$$

The differential equation for y is trivial

$$d_\theta y = -1\nu \qquad y(0) = 1$$

and we get

$$x_2^1(\theta) = (1/D)(\nu - \theta)^{-1} \qquad y_2^0(\theta) = (\nu - \theta)/\nu \qquad z_2^0(\theta) = E(\nu - \theta)/\nu \tag{5.9}$$

No more complicated problems are encountered at subsequent steps. For example, x_3^2, y_3^1, $z_3^1(\theta)$ are found from the equations

$$x[By_2^0 + Cz_2^0] + x_2^1[By + Cz] = 0$$

$$d_\theta y = -D[x_2^1 y + xy_2^0] \qquad y(0) = -D/\mu$$

$$y - Ez = 0$$

The solution is trivial.

$$x_3^2(\theta) = -(\nu/\mu)(\nu - \theta)^{-2} \qquad y_3^1(\theta) = -D/\mu \qquad z_3^1(\theta) = -DE/\mu$$

It is not hard to see that the asymptotic solution, obtained in such a manner, has a singularity at the finite moment $\theta = \nu$ because of the zero in the denominator. The singularities become stronger at subsequent steps.

$$x_3^n(\theta) = \mathcal{O}((\nu - \theta)^{-n}) \qquad y_3^n, z_3^n(\theta) = \mathcal{O}((\nu - \theta)^{-n+2})$$

Thus, the asymptotic solution, given by Eqs. (5.8), fails near the moment $\theta = \nu$.

One has to understand that there are no singularities in the exact solution of problem (5.7). There are singularities in the asymptotic solution due only to the nonapplicability of the anzatz (5.8) near the time point $\theta = \nu$.

The structure of the secular terms suggests the next scale as $\rho = (\theta - \nu)/\varepsilon$. Note, this new scale $(\varepsilon^5 t - \nu)/\varepsilon = \varepsilon^4 t - \nu/\varepsilon$ corresponds to the obvious scale $\varepsilon^4 t$, but this occurs when the slow time θ is near ν and not zero. This new scale is fast with respect to the slow one $\theta = \varepsilon^5 t$. Therefore, it seems a good idea to call it an explosive scale.

D. Explosive Scale

After the change of the independent variable

$$\rho = (\theta - \nu)/\varepsilon = (\varepsilon^5 t - \nu)/\varepsilon$$

the original Eqs. (5.0) change slightly. The new equations for the unknown functions

$$x_4(\rho;\varepsilon)=x(t;\varepsilon) \qquad y_4(\rho;\varepsilon)=y(t;\varepsilon) \qquad z_4(\rho;\varepsilon)=z(t;\varepsilon)$$

have the form

$$\begin{gathered}\varepsilon d_\rho x_4=\varepsilon[W-Ax_4^2]-x_4[By_4+Cz_4]\\ d_\rho y_4=-Dx_4y_4\\ \varepsilon^4 d_\rho z_4=x_4(Ey_4-z_4)\end{gathered}\tag{5.10}$$

It is only the boundary conditions that are significantly different. The problem is considered now on the unbounded interval $-\infty<\rho<\infty$, which corresponds to a neighborhood of the time in which explosive growth occurs $|\theta-\nu|\ll 1$. For the functions x_4, y_4, $z_4(\rho;\varepsilon)$, the asymptotics at minus infinity (as $\rho\to-\infty$) are determined. This determination is done by matching them with the slow asymptotic solution (5.8).

In fact, the question under consideration is to find an asymptotic solution that describes the fast passage across the instant of explosive growth on the slow time scale $\theta=\nu$. From a formal point of view, the boundary asymptotic conditions are obtained in the same way, as was described in sections 4.2 and 4.3. Namely, the asymptotic expansions of the coefficients x_3^n, y_3^n, $z_3^n(\theta)$ as $\theta\to\nu-0$ are substituted into the series (5.8). After that, the change of independent variable is performed $\theta-\nu=\varepsilon\rho$ and the expressions involving like powers of ε are gathered together. Consequently, double asymptotic series in powers of ε, ρ are obtained. The expressions for the powers ε^n are now considered as the asymptotic data of the solution on the new (fast) scale.

$$\begin{gathered}x_4(\rho;\varepsilon)=[-(1/D)\rho^{-1}+(\nu/\mu)\rho^{-2}+\mathcal{O}(\rho^{-3})]+\mathcal{O}(\varepsilon)\\ y_4(\rho;\varepsilon)=\varepsilon[-(1/\nu)\rho-D/\mu+\mathcal{O}(\rho^{-1})]+\mathcal{O}(\varepsilon^2)\\ z_4(\rho;\varepsilon)=\varepsilon[-E\rho-DE/\mu+\mathcal{O}(\rho^{-1})]+\mathcal{O}(\varepsilon^2)\\ \rho\to-\infty \qquad \varepsilon\to 0\end{gathered}$$

These are the correct matching requirements. The asymptotic conditions, obtained above, suggest the following anzatz on this scale

$$
\begin{aligned}
x_4(\rho;\varepsilon) &= x_4^0(\rho) + \varepsilon x_4^1(\rho) + \varepsilon^2 x_4^2(\rho) + \cdots \\
y_4(\rho;\varepsilon) &= \varepsilon[y_4^0(\rho) + \varepsilon y_4^1(\rho) + \varepsilon^2 y_4^2(\tau) + \cdots \\
z_4(\rho;\varepsilon) &= \varepsilon[z_4^0(\rho) + \varepsilon z_4^1(\rho) + \varepsilon^2 z_4^2(\rho) + \cdots
\end{aligned}
\tag{5.11}
$$

One can now derive the equations for the coefficients. At the first step, a system of two differential equations for the functions x_4^0, $y_4^0(\rho)$ is obtained.

$$
\begin{aligned}
d_\rho x &= W - Ax^2 - \mu xy \\
d_\rho y &= -Dxy
\end{aligned}
\tag{5.12}
$$

The third component $z_4^0(\rho)$ can be found from

$$Ey - z = 0$$

To complete the formulation of the problem, two additional asymptotic conditions are assigned.

$$
\begin{aligned}
x(\rho) &= (1/D)[-\rho^{-1} + (1/W)\rho^{-2} + \mathcal{O}(\rho^{-3})] \\
y(\rho) &= (D/\mu)[-W\rho - 1 + \mathcal{O}(\rho^{-1})] \qquad \rho \to -\infty
\end{aligned}
\tag{5.13}
$$

We are unable to write out the solution of problems (5.12) and (5.13) in explicit form, in contrast to the previous stages. The study of the mathematical problems (5.12) and (5.13) is a separate question.

However, it is not hard to draw the phase portrait of the two-dimensional system (5.12). If we look at Fig. 2, we can see that there is a single stable equilibrium $((W/A)^{1/2}, 0, 0)$. Exactly the same one occurs in the original system (5.0). Therefore, the process described by (5.0) ends in this stable equilibrium and occurs on the time scale ρ. The exponents, which appear in the solution of the linearized equations, give the rate of asymptotic approximation to the stable state

$$
\begin{aligned}
x_4^0(\rho) &= \gamma + \mathcal{O}(\rho \exp(-2A\gamma\rho)) \qquad \gamma = (W/A)^{1/2} \\
y_4^0(\rho) &= \mathcal{O}(\rho \exp(-D\gamma\rho)) \qquad \rho \to \infty
\end{aligned}
$$

Thus, on the explosive scale, we have the asymptotic approximations of x_4^0, $y_4^0(\rho)$ both at plus and at minus infinity (as $\rho \to \pm\infty$). To obtain the

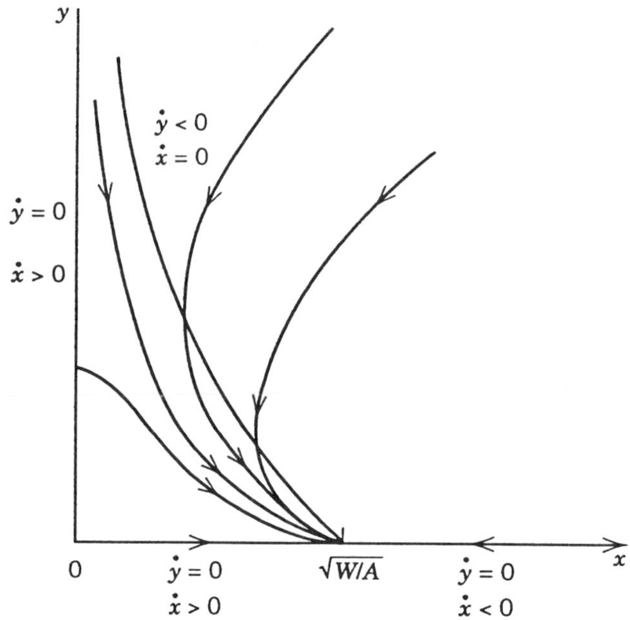

Figure 2. Phase portrait of the system (5.12).

kinetic curves on the finite interval $-M < \rho < +M$ ($M = \text{const} > 0$) other methods have to be applied, for example, numerical simulation.

E. The Results for Problem II

Let us consider the leading order terms of the various asymptotic solutions. On the first (fast) scale the third component is increasing and it is stabilizing to a constant at infinity.

$$z_1^0(t) = E - \mathcal{O}(\exp(-t))$$

The first and second components are stable.

$$x_1^0(t) \equiv 1 \qquad y_1^0(t) \equiv 1$$

On the second scale, the first component is decreasing to zero.

$$x_2^0(\tau) = \exp(-\mu\tau)$$

The components y, z do not vary

$$y_2^0(\tau) \equiv 1 \qquad z_2^0(\tau) \equiv E$$

Hence, for time $\tau \gg 1$, a change in the typical magnitude of the variable x occurs. Because x takes on values of order $\mathcal{O}(\varepsilon)$, the next time scale turns out to be $\varepsilon^2\tau$.

All three components x, y, z vary on the slow scale $\theta = \varepsilon^2\tau = \varepsilon^5 t$. Singularities occur in the asymptotic solution at the finite time $\theta = \nu$.

$$x_2^1(\theta) = (\nu - \theta)^{-1}/D \qquad y_2^0(\theta) = (\nu - \theta)/\nu \qquad z_2^0(\theta) = E(\nu - \theta)/\nu$$

In fact, this is the time at which the slow scale θ breaks down.

Near the time $\theta = \nu$ a new (faster) scale occurs. At this fourth stage, all three components vary and tend to the vector $(\gamma, 0, 0)$, $\gamma = (W/A)^{1/2}$.

The limit state

$$(x, y, z) = ((W/A)^{1/2}, 0, 0)$$

is a unique and stable equilibrium of the original system (5.0). Therefore, the process described by Eqs. (5.0) stabilizes to the state $(\gamma, 0, 0)$ on the last time scale $\rho = (\varepsilon^4 t - \nu/\varepsilon)$.

VI. PRACTICAL APPLICATIONS

The formulas, obtained in Section 3.4, solve a direct problem of chemical kinetics. That is, the dynamics of the concentrations are computed for given rate constants K_j.

The same formulas can be used to solve inverse problems, when the rate constants are determined from kinetic curves obtained experimentally. The simplest way is to compare the experimental and theoretical curves with each other for the correct values of K_j.

If one considers either Problem (I) or (II), the question is to find some coefficients of the differential equations by comparing the solutions and the given (experimental) curves with each other. Since we use the explicit form of the asymptotic solutions, only those coefficients (or their combinations) can be found that are present in the formulas. In practice, the leading terms of the asymptotics provide an accuracy corresponding to the experimental accuracy. In Figures 3–5 both the experimental and theoretical curves for Problem (I) are presented on various time scales.

If the formulas (4.5) are exploited, one finds either the relation $\lambda = \sqrt{(B + CE)^2 + 4AW}$ or $X^+ = (1/2A)[\lambda - (B + CE)]$. To find λ the dynamics of the substance x is required on the second time scale $\tau \cong 1$. To find X^+ it is sufficient to measure the concentration of x when the time is long for the second scale ($\tau \gg 1$) but small for the third scale ($\theta \ll 1$).

Thus, the theoretical formulas provide the solution of some inverse

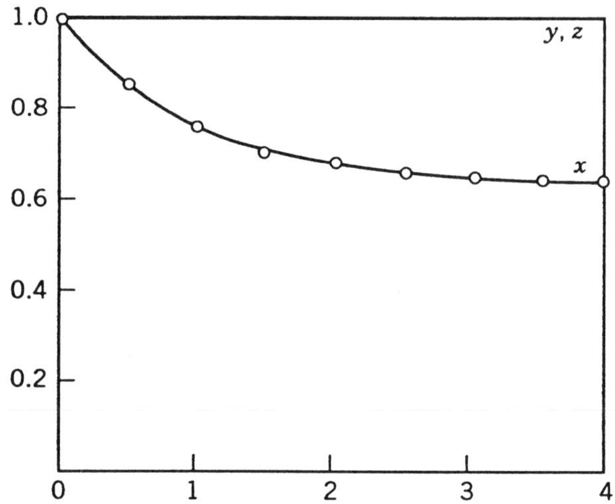

Figure 3. Time scale II. CL method. Comparison between asymptotic solution ---- and experiment ⊙ for $[RO_2]_0 = 1.3 \times 10^{-7}\, M\, L^{-1}$; $[InH]_0 = 9.1 \times 10^{-6}\, M\, L^{-1}$; $[In]_0 = 0$; $2k_6 = 3.2 \times 10^6\, L\, M^{-1}\, s^{-1}$; $k_7 = 2.2 \times 10^4\, L\, M^{-1}\, s^{-1}$; $k_8 = 5 \times 10^8\, L\, M^{-1}\, s^{-1}$; $T = 323$ K; $[RO_2] = x$, $[InH] = y$, $[In] = z$.

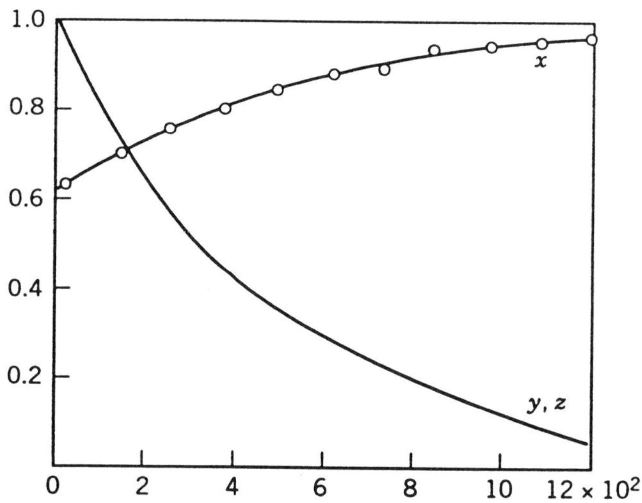

Figure 4. Time scale III. CL method. Comparison between asymptotic solution ---- and experiment ⊙ for $[RO_2]_0 = 1.3 \times 10^{-7}\, M\, L^{-1}$, $[InH]_0 = 9.1 \times 10^{-6}\, M\, L^{-1}$; $[In]_0 = 0$; $2k_6 = 3.2 \times 10^6\, L\, M^{-1}\, s^{-1}$; $k_7 = 2.2 \times 10^4\, L\, M^{-1}\, s^{-1}$; $k_8 = 5 \times 10^8\, L\, M^{-1}\, s^{-1}$; $T = 323$ K; $[RO_2] = x$, $[InH] = y$, $[In] = z$.

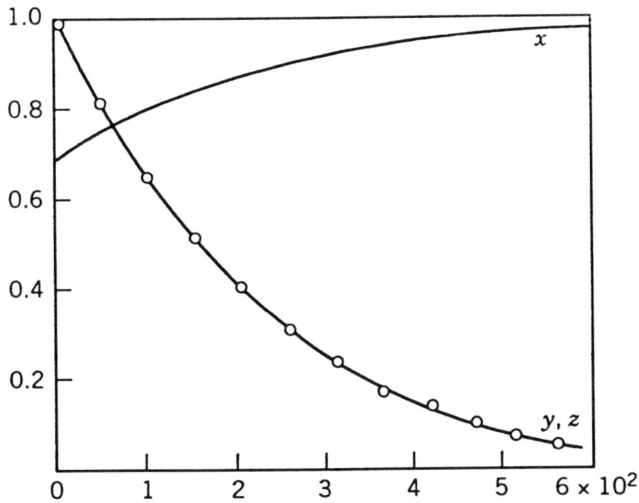

Figure 5. Time scale III. Spectrophotometry method. Comparison between asymptotic solution ---- and experiment ⊙ for $[RO_2]_0 = 2.0 \times 10^{-7}\ M\ L^{-1}$; $[InH]_0 = 8.1 \times 10^{-6}\ M\ L^{-1}$; $[In]_0 = 0$; $2k_6 = 3.2 \times 10^6\ L\ M^{-1}\ s^{-1}$; $k_7 = 2.8 \times 10^4\ L\ M^{-1}\ s^{-1}$; $k_8 = 5 \times 10^8\ L\ M^{-1}\ s^{-1}$; $T = 323$ K; $[RO_2] = x$, $[InH] = y$, $[In] = z$.

kinetic problems. It should be emphasized that the various formulas describe the process on different time scales. This fact imposes different restrictions on both the apparatus and the experimental methods.

For example, the relation $D\gamma = D\sqrt{W/A}$ can be found from either Eq. (4.8) or (4.9). To do this, the experimental dynamics of either x or y is needed on the slow time $\theta = \varepsilon^2 t$.

Note that, if a small factor ε has the value of 10^{-1}, the process under discussion runs 10 times more slowly than the processes on the scale $\tau = \varepsilon t$. Generally, the experimental curves, obtained on the slow time scale $\theta = \varepsilon^2 t$, give no information on the fast time scale $\tau = \varepsilon t$. Therefore, the relation $\lambda = \sqrt{(B + CE)^2 + 4AW}$, that is present in another x-formula (4.2), cannot be found without a new additional (fast) experiment. Roughly speaking, the fast processes are not recorded by slow experimental tools.

To evaluate the capability of the given formulas in practice, one has to rewrite them in dimension variables.

A. The Fast Time Scale, I. (t)

In the dimension variables $X = [RO_2]/[RO_2]_0$, $Y = [InH]/[InH]_0$, $Z =$

[In] the solution (4.2) takes the form

$$[RO_2] = [RO_2]_0 = \sqrt{W_i/2K_6} \tag{6.1}$$

$$[InH] = [InH]_0 \tag{6.2}$$

$$[In] = (K_7/K_8)[InH]_0(1 - \exp(-K_8(W_i/2K_6)T)) \tag{6.3}$$

One can see that $[RO_2]$ and [InH] are stable and equal to their initial values. The concentration of [InH] increases and tends to the steady-state concentration at large times.

$$[In]_{st} = (K_7/K_8)[InH]_0 \tag{6.4}$$

The constant $[In]_{st}$ is taken as the initial data on the next (slow) time scale. The approximate solutions (6.1)–(6.3) is valid as long as

$$0 \le t \ll t^I_{\max} = 10\sqrt{2K_6/W_i}/K_8 \tag{6.5}$$

Expressions (6.3) and (6.5) contain the relations K_7/K_8, and $K_8/\sqrt{2K_6}$. So, if $2K_6$ is known, we can find the values of K_7 and K_8.

The use of (6.4) to evaluate the ratio of the rate constant K_8 to the known K_7 was described in the literature [16]. Equation (6.2) has not been applied in practice.

B. The First Slow Time Scale, II. ($\tau = \varepsilon t$)

Returning to the dimension variables y, z, the solution (4.5) takes the form

$$\ln\frac{x_2 + \alpha_0 + \sqrt{1+\alpha_0^2}}{x_2 + \alpha_0 - \sqrt{1+\alpha_0^2}} = 2\sqrt{(1+\alpha_0^2)2K_6W_iT} + \ln\frac{1+\alpha_0 - \sqrt{1+\alpha_0^2}}{1+\alpha_0+\sqrt{1+\alpha_0}} \tag{6.6}$$

$$[InH] = [InH] \tag{6.7}$$

$$[In] = K_7[InH]_0 \tag{6.8}$$

Here $\alpha_0 = K_7[InH]_0/\sqrt{2K_6W_i} = \beta[InH]_0$. The inhibitor concentration remains fixed and equal to $[In]_{st}$. The concentration of $x_2 = [RO_2]/[RO_2]_0$ decreases (as time increases) and tends to the quantity X^+, which is the final value of x on this time scale

$$X^+ \equiv [RO_2]/[RO_2]_0 = \sqrt{1+\alpha_0^2} - \alpha_0 \tag{6.9}$$

As was pointed out, the constant X^+ is the initial value of the variable x on the next time scale.

To measure the concentration of $[RO_2]$ both the electron spin resonance (ESR) method and the chemiluminescence (CL) method can be usefully applied. Equation (6.6) is used in the original form, if the ESR method is applied. When the CL method is applied, it is expedient to transform Eq. (6.6) to a more convenient form, containing the parameters $\sqrt{i}$, $\sqrt{i_m}$, which can be measured experimentally

$$x = \sqrt{I/I_0} = \sqrt{i} \qquad X^+ = \sqrt{I_m/I_0} = \sqrt{i_m}$$

Here I_0 is the initial intensity, measured before the inhibitor was added to the reaction, I is the current intensity, and I_m is the minimal intensity of the emitted light.

By using the relations

$$\sqrt{1+\alpha_0^2} - \alpha_0 = \sqrt{i_m} \qquad (1+\alpha_0)^2 + \alpha_0 = 1/\sqrt{i_m} \tag{6.10}$$

we obtain from (6.6), (6.10)

$$\ln\frac{\sqrt{i} + 1/\sqrt{i_m}}{\sqrt{i} - \sqrt{i_m}} - \ln\frac{1 + 1/\sqrt{i_m}}{1 - \sqrt{i_m}} = 2\sqrt{(1+\alpha_0)2k_6W_i}T \tag{6.11}$$

$$1/\sqrt{i_m} - \sqrt{i_m} = 2\alpha_0 \tag{6.12}$$

Formulas (6.6), (6.9), and (6.11) and (6.12) can also be used to evaluate both K_7 and $2K_6W_i$; see Fig. 3. If the stationary concentration of radicals In is measured under similar conditions, then the rate constant K_8 can be found from (6.4).

The kinetic parameter

$$\alpha_0 = K_7[\text{InH}]_0/\sqrt{2K_6W_i} = \beta[\text{InH}]_0 \tag{6.13}$$

corresponds to the experimentally measured quantity $\sqrt{i_m}$; see the relations (6.10) and (6.12).

The experimental results can be treated in two ways.

1. The kinetic curve, plotted for changing concentration of peroxide radicals, following addition of the inhibitor, is a straight line with respect to the coordinates of Eq. (6.12).

$$[\ln(\sqrt{i} + 1/\sqrt{i_m}) - \ln(\sqrt{i} - \sqrt{i_m})] \div T$$

The effective rate constant is determined by the slope of the curve

$$K_{\text{eff}} = 2\sqrt{(1+\alpha_0^2)2K_6W_i} \tag{6.14}$$

and according to Eq. (6.4) we obtain

$$2K_6W_i = K_{\text{eff}}^2/4(1+\alpha_0^2) \tag{6.15}$$

From the known value of $2K_6$ we find the value

$$W_i = K_{\text{eff}}^2/4(1+\alpha_0^2)2K_6 \tag{6.16}$$

To find K_7, the following relation can be derived from Eqs. (6.12), (6.13), and (6.14).

$$K_7 = K_{\text{eff}}(\alpha_0/2[\text{InH}]_0)\sqrt{1+\alpha_0^2} \tag{6.17}$$

Thus, Eqs. (6.16) and (6.17) allow the determination of both W_i and K_7 from a single experimental curve. It is expedient to carry out a series of experiments with similar inhibitor concentrations in order to obtain more precise values of the parameters W_i, K_7.

2. Another expression can be derived for evaluating K_7 from a series of kinetic curves, obtained for various inhibitor concentrations. By following Eqs. (6.12) and (6.13), we can plot the kinetic parameter α_0 against $[\text{InH}]_0$. The slope of the resulting curve is

$$\beta = K_7/\sqrt{2K_6W_i} \tag{6.18}$$

If we substitute (6.15) into (6.18) we find

$$K_7 = K_{\text{eff}}\beta/2(1+\alpha_0^2) \tag{6.19}$$

Here a straight line of experimental data with respect to the coordinates $\alpha_0 \div [\text{InH}]_0$ is required in order to find β and after that K_7. The formulas (5.11)–(5.13) are valid as long as

$$T_{\max}^{I} < T \ll T_{\max}^{II} = 10/\sqrt{2k_6W_i} \tag{6.20}$$

The first inequality of this condition ensures, according to Eq. (6.3), that [In] has the quasistationary value $[\text{In}]_{st}$ and that it remains stable.

The inequality $T \ll T_{\max}^{II} = 10/\sqrt{2k_6W_i}$ determines the upper limit of the slow scale times and it is equivalent to condition (6.5). Then $T_{\max}^{II} \cong 50\,\text{s}$, if W_i and $2K_6$ are taken as above.

Formulas (6.9) and (6.12) can also be applied only for the stationary mode of $[RO_2]$. The upper time limit for the fitness of these equations is determined by (6.20), while the lower time limit is given by the value $T_{st}^{RO_2}$. To estimate approximately $T_{st}^{RO_2}$, the formulas (6.6) and (6.9) can be used in the form

$$\gamma = \frac{x_0 - x}{x_0 - X^+} > 0,99 \qquad x_0 = 1$$

This result means that the variable x differs from the constant X^+ no more than 1%. Thus, from Eqs, (6.6) and (6.9), we obtain

$$\ln \frac{0,01(1+\alpha_0) + 1,99\sqrt{1+\alpha_0^2}}{0,01(1+\alpha_0) - 0,01\sqrt{1+\alpha_0^2}} - \ln \frac{1+\alpha_0+\sqrt{1+\alpha_0^2}}{1+\alpha_0-\sqrt{1+\alpha_0}}$$

$$\leq 2\sqrt{(1+\alpha_0)2k_6W_i}T_{st}^{RO_2} \qquad (6.21)$$

Substituting for W_i, $2K_6$, and α_0, we find that $T_{st}^{RO_2} \geq 10$ s. Therefore, both of the quantities X^+ and $\sqrt{i_m}$ have to be measured within a time interval from 10 to 50 s, see Fig. 3.

Formula (6.12), giving the ratio $K_7/\sqrt{2K_6}$, was reported in [17], where the method of steady-state concentrations was applied. This equation is widely used in practice for quantitative studies of liquid-phase inhibited oxidation. Relation (6.11) was obtained in [18] by the same method of steady-state concentrations under severe constraints.

C. The Second Slow Time Scale, III. ($\theta = \varepsilon^2 t = \varepsilon\tau$)

In this case the dimensionless variables x, y for the current concentrations of the inhibitor and the peroxide radicals can be used. The time T and [In] are taken in dimensional form. We have from (4.10)

$$x_3 = \sqrt{1 + (\alpha_0 y_3)^2} - \alpha_0 y_3 \qquad (6.22)$$

$$\ln\{[(1+\sqrt{\alpha_0 y_3})^2 - \alpha_0 y_3]/y_3\} - [\alpha_0 y_3 + \sqrt{1+(\alpha_0 y_3)^2}]$$

$$= K_7\sqrt{W_i/2k_6}t + \ln[1+\sqrt{1+\alpha_0^2}] - [\alpha_0 + \sqrt{1+\alpha_0^2}] \qquad (6.23)$$

$$[\text{In}] = K_7[\text{InH}]_0 y_3/K_8 \qquad (6.24)$$

Another form of the solution is obtained from (4.9).

$$\ln\frac{1+x_3}{1+x_3}-\frac{1}{x_3}=K_7\sqrt{W_i/2K_6}T+\ln\frac{1+\sqrt{1+\alpha_0^2}-\alpha_0}{1-\sqrt{1+\alpha_0^2}+\alpha_0}-\frac{1}{\sqrt{1+\alpha_0^2}-\alpha_0} \tag{6.25}$$

$$y_3=\frac{1}{2\alpha_0}(1/x_3-x_3) \tag{6.26}$$

$$[\mathrm{In}]=\frac{\sqrt{2K_6W_i}}{K_8}(1/x_3-x_3) \tag{6.27}$$

In the case under consideration, on the slowest time scale, the concentrations of the radicals In and RO_2 tend to zero and $[RO_2]_0$, respectively, under the inhibitor consumption. Equations (6.25)–(6.27) are applied for the ESR analysis of the $[RO_2]$ measurements. For the CL method, it is expedient to use the same relations in another form

$$\ln\frac{1+\sqrt{i}}{1-\sqrt{i}}-\frac{1}{\sqrt{i}}=k_7\sqrt{W_i/2k_6}T+\ln\frac{1+\sqrt{i_m}}{1-\sqrt{i_m}}-\frac{1}{\sqrt{i_m}} \tag{6.28}$$

$$[\mathrm{InH}]=\frac{\sqrt{2K_6W_i}}{2K_7}(1/\sqrt{i}-\sqrt{i}) \tag{6.29}$$

$$[\dot{\mathrm{In}}]=\frac{\sqrt{2K_6W_i}}{2K_8}(1/\sqrt{i}-\sqrt{i}) \tag{6.30}$$

Equations (6.22)–(6.24), (6.25)–(6.27), and (5.8), (5.9) are suitable for the times (see Fig. 4)

$$T_{\max}^{\mathrm{II}}\ll T\ll T_{\max}^{\mathrm{III}}=10\sqrt{2K_6W_i}/K_7 \tag{6.31}$$

The lower time limit $T_{\max}^{\mathrm{II}}$, corresponds to the upper time limit of scale II, when RO_2 has the stationary concentration $RO_2=X^+$.

The upper time limit $T_{\max}^{\mathrm{III}}=10\sqrt{2K_6W_i}/K_7$, found from the inequality $\varepsilon\ll\theta\ll\varepsilon^{-1}$, corresponds to the time of complete inhibitor consumption.

Similar solutions (6.28) and (6.29) were reported in [19, 20] by using the steady-state concentration method.

As $\alpha_0\to 0$, that is, at rather low inhibitor concentration, formula (6.23) reduces to the form

$$\ln \frac{[\mathrm{InH}]}{[\mathrm{InH}]_0} \approx K_7 \sqrt{W_i/2K_6} T$$

This particular expression was described elsewhere [21].

D. Determination of K_7 and W_i by the CL Method

Equations (6.28) and (6.29) are applied to find K_7 and W_i. Formula (6.29), describing consumption of InH on the time scale III, as the time θ tends to zero, turns out to be formula (6.12) on the time scale II due to the matching condition. Taking into account (6.13), one can rewrite it as follows:

$$\alpha_0 = \beta[\mathrm{InH}]_0 = 0,5(1/\sqrt{i} - \sqrt{i}) \tag{6.32}$$

Experimental results can be treated in two ways.

1. The kinetic curve, plotted for varying concentrations of the peroxide radical, is a straight line with respect to the coordinates (see Fig. 4)

$$\ln[(1+\sqrt{i})/(1-\sqrt{i})] - 1/\sqrt{i} \div T$$

The effective rate constant is determined by the slope of the resulting straight line.

$$K_{\mathrm{eff}} = K_7 \sqrt{W_i/2K_6} \tag{6.33}$$

By following Eqs. (6.32) and (6.33), we obtain

$$K_7 = \sqrt{\alpha_0 K_{\mathrm{eff}} 2K_6/[\mathrm{InH}]_0} \tag{6.34}$$

$$W_i = K_{\mathrm{eff}}[\mathrm{InH}]_0/\alpha_0 \tag{6.35}$$

Thus, starting from the value $[\mathrm{InH}]_0$ and the known value $2K_6$, we can find K_7 and W_i. To do this, the experimentally determined parameters K_{eff} and $\alpha_0(\sqrt{i_m})$ are used. They are obtained from a single experimental curve. See Fig. 4.

2. From the series of kinetic curves, obtained for various $[\mathrm{InH}]_0$, α_0 is plotted against $[\mathrm{InH}]_0$. The parameter β is found from the slope of the resulting straight line.

$$\beta = K_7/\sqrt{2K_6 W_i} \tag{6.36}$$

From Eqs. (6.32) and (6.36), we derive the following expressions for K_7

and W_i

$$K_7 = \sqrt{\beta K_{\text{eff}} 2K_6} \tag{6.37}$$

$$W_i = K_{\text{eff}}/\beta \tag{6.38}$$

E. Determination of K_7 and K_i by the Spectrophotometry Method

Equation (4.10) and formula (6.23) are applied to find K_7 and W_i from the kinetics of either the InH consumption or the P_8 accumulation.The method of spectrophotometry (SPM) can be used to this end. See Fig. 5. For the original dimensional time T, these formulas take the form

$$W = -\frac{dy_3}{dT} = K_{\text{eff}}\sqrt{1 + (\alpha_0 y_3)^2} - \alpha_0 y_3 y_3 \tag{6.39}$$

$$\ln|[(1 + \sqrt{\alpha_0 y_3)^2} - \alpha_0 y_3]/y_3| - [\alpha_0 y_3 + \sqrt{1 + (\alpha_0 y_3)^2}]$$
$$= K_{\text{eff}} T + \ln[1 + \sqrt{1 + \alpha_0^2}] - [\alpha_0 + \sqrt{1 + \alpha_0^2}] \tag{6.40}$$

As before $K_{\text{eff}} = K_7\sqrt{W_i/2K_6}$, $\alpha_0 = \beta[\text{InH}]_0$.

As has already been shown in this work, detectable inhibitor consumption occurs only on the time scale III, and no additional information concerning the inhibitor behavior on other time scales is available. See Figs. 3 and 4.

Effective formulas for the determination of K_7 and W_i can be derived from Eq. (6.39) and (6.40) with some simplifying assumptions.

If $y \to 1$, which occurs as $T \to 0$, an approximate formula is derived for the initial consumption rate y

$$W \approx K_{\text{eff}}[\sqrt{1 + \alpha_0^2} - \alpha_0] \tag{6.41}$$

For small values $\alpha_0 \ll 1$ (e.g., $\alpha_0 \leq 0.4$) the formula (6.41) can be reduced to the simpler version

$$W' \approx K_{\text{eff}}(1 - \alpha_0) \tag{6.42}$$

For large $\alpha_0 \gg 1$ (e.g., $\alpha_0 \geq 1.5$) an approximation of Eq. (6.41) is as follows:

$$W'' \approx K_{\text{eff}}/2\alpha_0 \tag{6.43}$$

Of course, the above simplifications cause additional errors of up to 10%.

Substitution of expressions (6.13) and (6.36) into the formulas (6.42) and (6.43) gives

$$W'' \approx k_7[\sqrt{W_i/2k_6} - k_7[\text{InH}]_0/2k_6] \qquad W' \approx W_i/2 \tag{6.44}$$

Relation (6.44) is applied to find K_7 and W_i from the experimental curves of inhibitor consumption for various values of $[\text{InH}]_0$. Formula (6.44) is the equation of a straight line in the coordinates $W' \div [\text{InH}]_0$. The parameter $K_7\sqrt{W_i/2K_6}$ is found from the intersection of the resulting straight line with the ordinate axis, while $K_7^2/2K_6$ is calculated from the slope of the tangent ($\text{tg}\,\varphi_i$). Then,

$$K_7 \approx \sqrt{2K_6\,\text{tg}\,\varphi_i} \qquad W_i \approx K_{\text{eff}}^2/\text{tg}\,\varphi_i \tag{6.45}$$

ACKNOWLEDGMENTS

This research was made possible in part by Grant NMB000 from the International Science Foundation and by Grant 94-01-00193a from the Russian Foundation of Fundamental Research.

REFERENCES

1. L. S. Polyak, M. Ya. Gol'denberg, and A. A. Levitckii, *Numerical methods for chemical kinetics*, Nauka, Novosibirsk, Russia, 1984 (Russian).
2. E. T. Denisov, *The kinetics of homogeneous chemical reactions*, Vyschaya Shkola, Moscow, Russia, 1978 (Russian).
3. E. A. Novikov, *Numerical methods for solving differential equations of chemical kinetics*, Mathematical methods for chemical kinetics, Nauka, Novosibirsk, Russia, 1990, pp. 53–68 (Russian).
4. A. Ya. Dubowitckii and V. A. Dubowitckii, *J. Vychislit. Math. Math. Phys.*, **23**, 1060 (1983) (Russian).
5. A. M. Il'in, *Mat. Zametki*, **6**, 237 (1969).
6. P. Berge, Y. Pomeau, and C. Vidal, *L'ordre dans le chaos*, Hermann, Paris, 1988.
7. A. N. Ivanova, *J. Vychislit. Math. Math. Phys.*, **34**, 1206 (1994) (Russian).
8. A. M. Il'in, *Matching of Asymptotic Expansions of Solutions of Boundary Value Problems*, AMS, Providence, RI, 1992.
9. L. A. Kalyakin, S. I. Maslennikov, and V. D. Komissarov, *Int. J. Chem. Kinet.*, **25**, 681 (1993).
10. A. H. Nayfeh, *Perturbation Methods*, Wiley-Interscience, New York, 1981.
11. A. B. Vasil'eva and V. F. Butuzov, *Asymptotic Expansions of Solutions of Singularly Perturbed Equations*, Nauka, Moscow, Russia, 1973 (Russian).
12. D. Summers, J.M. W. Scott, and S. K. Ralph, *Int. J. Chem. Kinet.*, **19**, 553 (1987).
13. V. G. Gorskii, E. A. Katcman, and T. N. Shwetcova-Shilovckaya, *Mathematical*

Aspects of the Quasiequilibrium of Reactions in Chemical Kinetics, Mathematical methods for chemical kinetics, Nauka, Novosibirsk, Russia, 1990, pp. 136–152 (Russian).

14. N. M. Emanuel, E. T. Denisov, and Z. K. Maizus, *Chain Reactions of the Oxidation of Hydrocarbons in the Liquid Phase*, Nauka, Moscow, Russia, 1980 (Russian).
15. E. T. Denisov, *The reactions of radical inhibitors and the mechanism of inhibited oxidation of hydrocarbons*, Itogi Nauki i Techniki, Kinetika & kataliz, Vol. 17, Moscow, Russia, 1987, pp. 3–115 (Russian).
16. V. I. Rubtsov, V. A. Roginsky, V. Z. Dubinsky, and V. B. Miller, *Kinet. Kataliz*, **19**, 1140 (1978) (Russian).
17. V. Ya. Shlyapintokh, O. N. Karpukhin, and L. M. Postnikov, *Chemiluminescence Methods for the Study of Slow Chemical Processes*, Nauka, Moscow, Russia, 1966 (Russian).
18. I. F. Rusina, A. B. Gagarina, and N. M. Emanuel, *Dokl. Akad. Nauk SSSR*, **285**, 665 (1985) (Russian).
19. V. M. Goltberg and L. K. Obukhova, *Dokl. Akad. Nauk SSSR*, **165**, 860 (1965) (Russian).
20. V. A. Belyakov, E. L. Shanina, V. A. Roginskiy, and V. B. Miller, *Izv. Akad. Nauk SSSR*, *Ser.* Khimicheskaya., 2685 (1975) (Russian).
21. E. T. Denisov and A. L. Aleksandrov, Zhurnal Fizich. Khimii., **38**, 491 (1964) (Russian).

SINGULARLY PERTURBED PROBLEMS WITH BOUNDARY AND INTERIOR LAYERS: THEORY AND APPLICATION*

V. F. BUTUZOV and A. B. VASILIEVA

Department of Physics, Moscow State University, Moscow, Russia

CONTENTS

* Portions of this chapter were adapted with permission from A. B. Vasil'eva, V. F. Butuzov, and L. V. Kalachev, "The Boundary Function Method for Singular Perturbation Problems." © 1995 by the Society for Industrial and Applied Mathematics.

Singular Perturbation Problems in Chemical Physics: Analytic and Computational Methods, Edited by John J. H. Miller, Advances in Chemical Physics Series, Vol. XCVII.
ISBN 0-471-11531-2

I. INTRODUCTION

A. Singularly Perturbed Differential Equations

Differential equations are often used as mathematical models describing processes in physics, chemistry, and biology. In the investigation of a number of applied problems, an important role is played by differential equations that contain small parameters at the highest derivatives. Such equations are called *singularly perturbed differential equations*. These equations describe various processes that are characterized by boundary and/or interior layers. Consider some simple examples:

Example 1.1

$$\varepsilon \frac{du}{dt} = -u + t \qquad 0 \le t \le 1 \qquad u(0) = 1 \tag{1.1}$$

where ε is a small positive parameter: $0 < \varepsilon \ll 1$. The solution of this problem is

$$u(t) = (1+\varepsilon)\exp\{-t/\varepsilon\} + t - \varepsilon$$

The graph of $u_\varepsilon(t)$ for small $\varepsilon > 0$ is presented in Fig. 1.

Note two characteristic features of this problem: (1) in the subinterval $[\delta, 1]$ (where δ is a small number) the solution $u_\varepsilon(t)$ is close to $\bar{u}_0(t) = t$, that is, to the solution of the equation that we obtain from (1.1) for $\varepsilon = 0$. We will call such an equation the *reduced* equation. Thus, the solution $\bar{u}_0(t) = t$ of the reduced equation gives an approximation for the solution $u_\varepsilon(t)$ of Problem (1.1) in the subinterval $[\delta, 1]$ for small $\varepsilon > 0$; (2) in the subinterval $[0, \delta]$ the solution $u_\varepsilon(t)$ changes rapidly from the initial value $u_\varepsilon(0) = 1$ to values close to $\bar{u}_0(t)$. In this subinterval, $\bar{u}_0(t)$ does not approximate $u_\varepsilon(t)$. The subinterval $[0, \delta]$ is called a *boundary layer*.

A generalization of this example is *Tikhonov's system* (z and y are vector functions) [1]

$$\varepsilon \frac{dz}{dt} = F(z, y, t) \qquad \frac{dy}{dt} = f(z, y, t) \qquad 0 \le t \le T$$

$$z(0) = z^0 \qquad y(0) = y^0$$

This system and other problems for singularly perturbed ordinary differential equations will be investigated in Sections II–V. Solutions with boundary and/or interior layers will be considered. Our main goal will be the construction of an approximation to the solution valid outside the boundary (interior) layer as well as within the boundary (interior) layer, that is, so-called *uniform approximation* in the entire t domain. This approximation will have an asymptotic character. The definition of an asymptotic approximation with respect to a small parameter will be introduced in Section I.B.

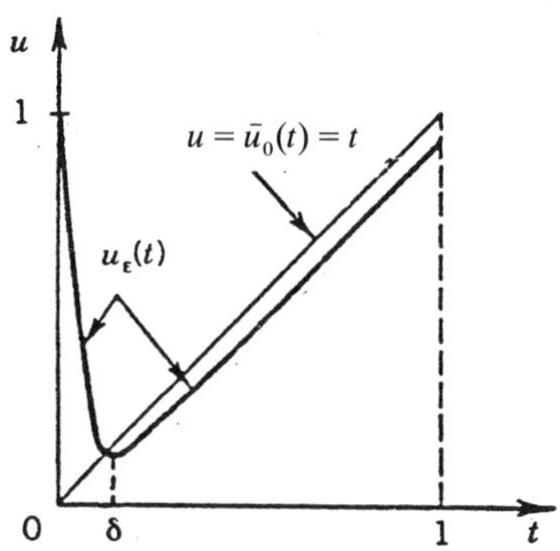

Figure 1. The exact solution $u_\varepsilon(t)$ and the solution $\bar{u}_0(t)$ of the reduced equation.

Example 1.2

$$\varepsilon^2 \Delta u - k^2(x, y)u = f(x, y, \varepsilon) \qquad (x, y) \in \Omega \qquad u|_{\partial\Omega} = 0 \qquad (1.2)$$

Here $\varepsilon > 0$ is a small parameter, $\Delta = (\partial^2/\partial x^2) + (\partial^2/\partial y^2)$ is the Laplace operator and Ω is a bounded domain. As we will see later, the solution $u_\varepsilon(x, y)$ of problem (1.2) will be close to the solution $\bar{u}_0 = -f(x, y, 0)/k^2(x, y)$ of the reduced problem (which we obtain from (1.2) for $\varepsilon = 0$) outside some small vicinity of the boundary $\partial\Omega$. However, in this small vicinity, $\bar{u}_0$ will not generally be an approximation for $u_\varepsilon(x, y)$, since $u_\varepsilon|_{\partial\Omega} = 0$, but $\bar{u}_0$ does not necessarily satisfy this condition. Hence, a boundary layer is formed in the vicinity of $\partial\Omega$.

Problem (1.2) and more complicated problems (with corner boundary layers, with interior layers) for partial differential equations will be considered in Sections VI–X. The method of constructing of the uniform approximations for the solutions with boundary and interior layers will be presented and different applied problems will be investigated by this method.

B. Asymptotic Approximations

Let $u_\varepsilon(x)$ be a solution of some problem with a small parameter ε defined for $x \in D$, $0 < \varepsilon \leq \varepsilon_0$. Let D_1 be a subdomain of D (in particular, D_1 might coincide with D) and some function $U(x, \varepsilon)$ defined in $\{D_1 \times (0 < \varepsilon \leq \varepsilon_0)\}$.

The function $U(x, \varepsilon)$ is called an *asymptotic approximation* of the solution $u_\varepsilon(x)$ with respect to the parameter ε in the subdomain D_1 if

$$\sup_{D_1} \|u_\varepsilon(x) - U(x, \varepsilon)\| \to 0 \qquad \text{as } \varepsilon \to 0$$

Moreover, if

$$\sup_{D_1} \|u_\varepsilon(x) - U(x, \varepsilon)\| = O(\varepsilon^k)$$

then we say that $U(x, \varepsilon)$ is an asymptotic approximation of $u_\varepsilon(x)$ in D_1 with accuracy of the order ε^k.

Here $\|u(x)\| = |u(x)|$ if u is scalar and $\|u(x)\| = \max_{1 \leq k \leq m} |u_k(x)|$ if $u(x) = \mathrm{col}(u_1(x, \ldots, u_m(x))$; the symbolic notation $\nu(\varepsilon) = O(\varepsilon^k)$ means that there exist numbers $C > 0$ and $\varepsilon_0 > 0$ such that for $0 < \varepsilon \leq \varepsilon_0$ the inequality $\|\nu(\varepsilon)\| \leq C\varepsilon^k$ holds.

To illustrate this definition, let us return to the Examples 1.1 and 1.2 of Section I. In Example 1.1 the function $U = t + \exp\{-t/\varepsilon\}$ is the asymptotic approximation for the solution $u_\varepsilon(t)$ in the whole interval $D = \{0 \le t \le 1\}$ with accuracy of order ε, but the solution $\bar{u}_0(t) = t$ of the reduced equation gives an asymptotic approximation for $u_\varepsilon(t)$ with accuracy of order ε only in the subinterval $D_1 = \{\delta \le x \le 1\}$, that is, outside the boundary layer. In addition, as we will see later in Example 1.2, the solution $\bar{u}_0(x, y)$ of the reduced problem will be an asymptotic approximation for the solution $u_\varepsilon(x, y)$ of the singularly perturbated problem (1.2) only outside the boundary layer. In order to construct the asymptotic approximation for $u_\varepsilon(x, y)$ in the whole domain D, it is necessary to add so-called *boundary layer functions* to the solution $\bar{u}_0(x, y)$ of the reduced problem.

By an *asymptotic method* we understand some method of construction of an asymptotic approximation $U(x, \varepsilon)$ for the solution $u_\varepsilon(x)$. Generally, the construction of $U(x, \varepsilon)$ is reduced to solving a set of simpler problems than the problem for $u_\varepsilon(x)$. The practical value of an asymptotic method is determined by the possibility of effectively finding $U(x, \varepsilon)$ with the help of simpler problems.

In this chapter, we will consider the *boundary layer function method* (or, in short, boundary function method), which gives the possibility of constructing a uniform asymptotic approximation (or, briefly, uniform asymptotics) for the solution $u_\varepsilon(\varepsilon)$ of the singularly perturbed problem (i.e., an asymptotic approximation in the whole domain D).

For wide classes of singularly perturbed problems, we will present algorithms for constructing series in powers of ε of the form

$$\sum_{k=0}^{\infty} \varepsilon^k (\bar{u}_k(x) + \Pi_k(\xi) + \cdots + P_k(\zeta)) \tag{1.3}$$

where $\bar{u}_k, \Pi_k, \ldots, P_k$ will be found as solutions of simpler problems, $\bar{u}_k$ will be the so-called *regular* terms of the asymptotics, $\Pi_k(\xi), \ldots, P_k(\xi)$ will be different boundary layer and/or interior layer functions that describe (together with $\bar{u}_k(x)$) the solution $u_\varepsilon(x)$ in regions of the boundary and/or interior layers, $\xi = \xi(x, \varepsilon), \ldots, \zeta = \zeta(x, \varepsilon)$ will be the so-called boundary layer (interior layer) variables. The series (1.3) has the following property: The truncated series (partial sum)

$$U_n(x, \varepsilon) = \sum_{k=0}^{n} \varepsilon^k [\bar{u}_k(x) + \Pi_k(\xi) + \cdots + P_k(\zeta)]$$

will give a uniform approximation to the solution $u_\varepsilon(x)$ with an accuracy of order ε^{n+1}, that is,

$$\sup_D \|u_\varepsilon(x) - U_n(x, \varepsilon)\| = O(\varepsilon^{n+1}) \tag{1.4}$$

A series (1.3) satisfying condition (1.4) is called an *asymptotic series* for the function $u_\varepsilon(x)$ (or an *asymptotic expansion* of $u_\varepsilon(x)$) in the domain D as $\varepsilon \to 0$.

It is important to note that the asymptotic series (1.3) might not converge to $u_\varepsilon(x)$ and might even diverge.

II. INITIAL VALUE PROBLEM

A. The Passage to the Limit

This section is based on the fundamental work done by Tikhonov [1] on the dependence of solutions of differential equations on small parameters.

Consider the initial value problem (in following with this paper [1], we will use the notation μ for the small parameter).

$$\mu \frac{dz}{dt} = F(z, y, t) \qquad \frac{dy}{dt} = f(z, y, t) \quad 0 \le t \le T \tag{2.1}$$

$$z(0, \mu) = z^0 \qquad y(0, \mu) = y^0 \tag{2.2}$$

Here $\mu > 0$ is the small parameter, z and y are vector functions of arbitrary dimensions M and m, respectively. We use the notation $z(t, \mu)$ and $y(t, \mu)$ for the solution of (2.1), (2.2). In general, it is not possible to find an exact solution of this problem. Our goal is to define an approximate solution using the fact that the parameter μ is small. Setting $\mu = 0$ in (2.1), we obtain

$$0 = F(z, y, t) \qquad \frac{dy}{dt} = f(z, y, t) \tag{2.3}$$

The order of this system is lower than that of the original system since the first equation of (2.3) is no longer a differential equation. We will call (2.3) the *degenerate* or *reduced* system. For this system, we should retain only the initial condition for y.

To solve (2.3) we have to start by expressing z from the first equation of (2.3) as a function of y and t. It is worth mentioning that this operation is not unique, that is, the equation $F(z, y, t) = 0$ might have several solutions (roots) with respect to z. In this case, the question of how to choose the appropriate root arises. Suppose, we somehow choose one root $z = \varphi(y, t)$, isolated with respect to z, when $(y, t) \in D = \{\|y\| \leq a, 0 \leq t \leq T\}$. The next two conditions 1^0, 2^0 show which root one must choose. Consider the so-called *associated system*

$$\frac{d\tilde{z}}{d\tau} = F(\tilde{z}, y, t) \qquad \tau \geq 0 \tag{2.4}$$

The independent variable is the new variable τ, and y and t enter the system (2.4) as parameters. The system (2.4) is considered for any values of (y, t) in the domain D. It is obvious that $\tilde{z} = \varphi(y, t)$ is a rest point of system (2.4).

Condition 1^0. *Let the stationary point $\tilde{z} = \varphi(y, t)$ of the associated system be asymptotically stable in the sense of Lyapunov, uniformly in $(y, t) \in D$ as $\tau \to \infty$.*

In the particular case when the system (2.1) does not contain y and $M = 1$, the following sufficient stability condition may be used.

$$F_z(\varphi(t), t) < 0 \qquad 0 \leq t \leq T \tag{2.5}$$

One can see this by considering the direction field on the (z, t) plane (see [2]).

There might exist several roots of the equation $F(z, y, t) = 0$ that satisfy condition 1^0. To make the final choice of the root, consider the associated system (2.4) for the initial parameters $y = y^0$ and $t = 0$

$$\frac{d\tilde{z}}{d\tau} = F(\tilde{z}, y^0, 0) \tag{2.6}$$

with the initial condition

$$\tilde{z}(0) = z^0 \tag{2.7}$$

Here z^0 is the same vector that enters the initial condition (2.2). This vector, generally speaking, is not close to the rest point $\tilde{z} = \varphi(y^0, 0)$ of system (2.7). Therefore, the solution $\tilde{z}(\tau)$ of (2.6), (2.7) might not necessarily tend to the rest point $\varphi(y^0, 0)$ as $\tau \to \infty$. We demand that $\tilde{z}(\tau)$ approach the rest point.

Condition 2^0. *Let the solution $\tilde{z}(\tau)$ of the problem (2.6), (2.7) exist for $\tau \geq 0$ and tend to the stationary point $\varphi(y^0, 0)$ as $\tau \to \infty$.*

In such a case, we will say that z^0 belongs to the *domain* (*or basin*) *of attraction* of the rest point $\varphi(y^0, 0)$.

In the particular case when $M = 1$, the system (2.1) does not contain y, the equation $F(z, t) = 0$ has three roots $\varphi_1(t) < \varphi(t) < \varphi_2(t)$ and condition (2.5) is fulfilled; the domain of attraction of $\varphi(0)$ is the interval $\varphi_1(0) <$

$z^0 < \varphi_2(0)$. One may see this by considering the direction field on the (z, τ) plane (see [2]).

Substituting the root $\varphi(y, t)$ that satisfies the conditions $1^0, 2^0$ into the second equation of (2.3) we obtain

$$\frac{dy}{dt} = f(\varphi(y, t), y, t) \tag{2.8}$$

Solving this equation with the initial condition

$$y(0) = y^0 \tag{2.9}$$

we obtain $y = \bar{y}(t)$ in the interval $0 \leq t \leq T$. Then we put $\bar{z}(t) = \varphi(\bar{y}(t), t)$. It is seen that determining $\bar{z}(t)$ and $\bar{y}(t)$ happens to be an easier problem than obtaining the solution $z(t, \mu)$, $y(t, \mu)$ of the original problem (2.1), (2.2).

Under conditions $1^0, 2^0$ *the following equalities hold:*

$$\lim_{\mu \to 0} y(t, \mu) = \bar{y}(t) \quad \text{for} \quad 0 \leq t \leq T$$

$$\lim_{\mu \to 0} z(t, \mu) = \bar{z}(t) \quad \text{for} \quad 0 \leq t \leq T$$

This statement is known as **Tihonov's theorem**. The equalities show that in the limit (as $\mu \to 0$) the solution $z(t, \mu)$, $y(t, \mu)$ of (2.1), (2.2) tends to the solution of the reduced problem (2.3), (2.9). That is why the above theorem is called the *theorem on passage to the limit*. Notice that the transition to the limit for y takes place for all t in the interval $0 \leq t \leq T$. Moreover, this limiting process is uniform. Transition to the limit for z takes place for any t except $t = 0$. This is clear since $z(0, \mu) = z^0 \neq \bar{z}(0)$. The limiting process for z will be uniform outside a small neighborhood of the initial point. We call this neighborhood the *boundary (initial) layer*.

Thus, $\bar{y}$ is an asymptotic approximation of $y(t, \mu)$ in the entire interval $0 \leq t \leq T$, and $\bar{z}$ is an asymptotic approximation of $z(t, \mu)$ for $\delta \leq t \leq T$, where δ is any arbitrary small but fixed number. Possible behavior of the solution $z(t, \mu)$, $y(t, \mu)$ for small μ is shown in Fig. 2.

Example 2.1. Let us consider the kinetic system [3].

$$\frac{dy}{dt} = -kyz \qquad \frac{dz}{dt} = -kyz + \alpha m \qquad \frac{dm}{dt} = kyz - \alpha m \tag{2.10}$$

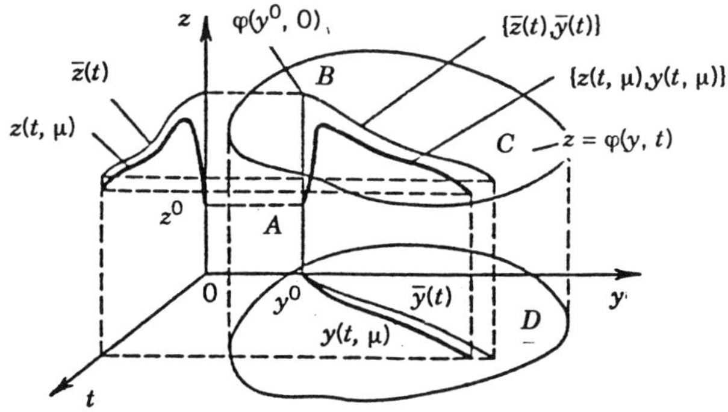

Figure 2. The exact solution $z(t, \mu)$, $y(t, \mu)$ and its asymptotic approximation $\bar{z}(t)$, $\bar{y}(t)$. C: the surface $z = \varphi(y, t)$; D: projection of this surface into the plane $z = 0$; A: the initial point with coordinates z^0, y^0, 0; B: the point with coordinates $\varphi(y^0, 0)$, y^0, 0.

under initial condition

$$y(0) = y^0 \qquad z(0) = z^0 \qquad m(0) = 0 \tag{2.11}$$

Here $y(t)$, $z(t)$, $m(t)$ are the concentrations of substances Y, Z, M in chemical reactions $Y + Z \rightarrow M + N_1\uparrow$, $M \rightarrow Z + N_2\uparrow$. We can treat M as a catalyst.

From (2.10), (2.11) we have

$$z + m = \text{const} = z^0 \qquad \frac{dz}{dy} = 1 - \frac{\alpha}{k}\frac{z^0 - z}{yz}$$

By introducing dimensionless variables $\zeta = z/z^0$, $\eta = y/y^0$ we obtain

$$\mu\frac{d\zeta}{d\eta} = 1 - \lambda\frac{1-\zeta}{\zeta\eta} \qquad 0 \le \eta \le 1 \qquad \zeta(1) = 1 \tag{2.12}$$

where $\mu = z^0/y^0$ (assumed to be small), $\lambda = \alpha/(ky^0)$.

The degenerate equation corresponding to the (2.12) is $1 - \lambda(1 - \zeta/\zeta\eta) = 0$. It has the unique root $\zeta = \varphi(\eta) = \lambda/(\lambda + \eta)$, that is asymptotically stable since

$$\left.\frac{\partial F}{\partial \zeta}\right|_{\zeta = \lambda(\lambda+\eta)} = \frac{(\lambda+\eta)^2}{\lambda\eta} > 0$$

(we must take into account that the initial value of ζ is given in (2.12) at

$\eta = 1$, i.e., at the right boundary point of the interval $(0, 1)$, and therefore the stability condition (2.5) has the opposite (positive) sign).

The point $\zeta = 1$ belongs to the domain of attraction of $\varphi(1)$ ($\varphi(\eta)$ is the unique root of the degenerate equation, therefore the domain of attraction is $-\infty < \zeta^0 < +\infty$).

Thus, Tikhonov's theorem is valid and $\bar{\zeta} = \lambda/(\lambda + \eta)$ is the asymptotic approximation to $\zeta(\eta, \mu)$ for $0 \leq \eta \leq 1 - \delta$ (see Fig. 3). The uniform asymptotic approximation to $\zeta(\eta, \mu)$ for $0 \leq \eta \leq 1$ (the dashed curve) will be constructed below (Example 2.2).

B. Asymptotic Algorithm

Since $\bar{z}$ does not approximate $z(t, \mu)$ for small μ in the boundary layer, we need to construct an asymptotic approximation for $z(t, \mu)$ that is valid in the entire interval $0 \leq t \leq T$ (*a uniform approximation*). Moreover, Tikhonov's theorem does not give the order of accuracy of the asymptotic approximation $\bar{y}(t)$ for the solution $y(t, \mu)$ in $0 \leq t \leq T$, and that of $\bar{z}(t)$ for $z(t, \mu)$ outside the boundary layer.

Our goal is to construct an approximation with higher accuracy and a

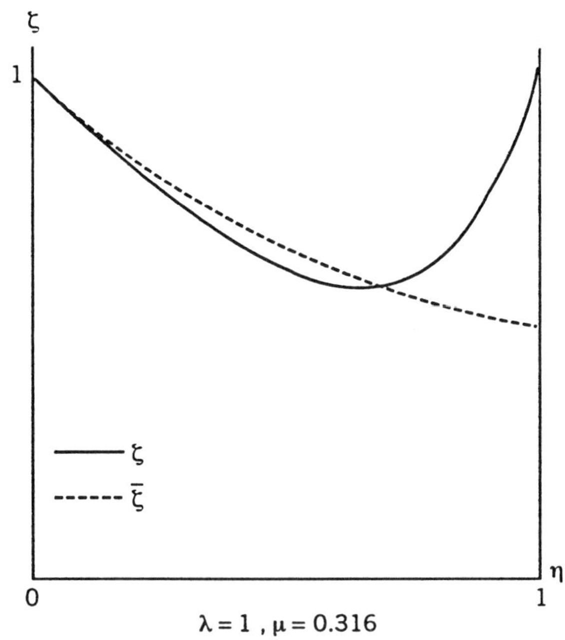

Figure 3. The graphs of $\zeta(\eta, \mu)$ and $\bar{\zeta}(t) = \lambda/(\lambda + \eta)$.

complete asymptotic expansion for the solution of the problem (2.1), (2.2). We suppose that $F(z, y, t)$ and $f(z, y, t)$ are sufficiently smooth.

Let $\lambda_i(t)$ $(i = 1, \ldots, M)$ denote the eigenvalues of the matrix $F_z(t) = F_z(\bar{z}(t), \bar{y}(t), t)$.

Condition 3^0. *Let*

$$Re\ \lambda_i(t) < 0 \quad \text{for} \quad 0 \le t \le T \qquad i = 1, \ldots, M \tag{2.13}$$

Notice that condition 1^0 in Section I.A follows for the domain

$$D_1 = \{\|y - \bar{y}(t)\| \le \eta, 0 \le t \le T\}$$

from 3^0. Notice also that (2.5) is a particular case of (2.13).

We will seek an asymptotic expansion for the solution of (2.1), (2.2) in a form that is typical for singularly perturbed problems

$$x(t, \mu) = \bar{x}(t, \mu) + \Pi x(\tau, \mu) \tag{2.14}$$

Here $\tau = t/\mu$ and $x = \{z, y\}$,

$$\bar{x}(t, \mu) = \bar{x}_0(t) + \mu \bar{x}_1(t) + \cdots + \mu^k \bar{x}_k(t) + \cdots \tag{2.15}$$

is the regular part of the expansion,

$$\Pi x(\tau, \mu) = \Pi_0 x(\tau) + \mu \Pi_1 x(\tau) + \cdots + \mu^k \Pi_k x(\tau) + \cdots \tag{2.16}$$

is the boundary layer part. By substituting the series (2.14) into system (2.1) we obtain the equalities

$$\begin{aligned} \mu \frac{d\bar{z}}{dt} + \frac{d\Pi z}{d\tau} &= F(\bar{z} + \Pi z, \bar{y} + \Pi y, t) \\ \frac{d\bar{y}}{dt} + \frac{1}{\mu} \frac{d\Pi y}{d\tau} &= f(\bar{z} + \Pi z, \bar{y} + \Pi y, t) \end{aligned} \tag{2.17}$$

Next, we represent the right-hand sides of these equations in a form similar to (2.14): $F = \bar{F} + \Pi F$, $f = \bar{f} + \Pi f$. Let us illustrate how this can be done for the function F

$$\begin{aligned} F(\bar{z} + \Pi z, \bar{y} + \Pi y, t) &= F(\bar{z}(t, \mu), \bar{y}(t, \mu), t) \\ &+ [F(\bar{z}(\tau\mu, \mu) + \Pi z(\tau, \mu), \bar{y}(\tau\mu, \mu) + \Pi y(\tau, \mu), \tau\mu) \\ &- F(\bar{z}(\tau\mu, \mu), \bar{y}(\tau\mu, \mu), \tau\mu)] \equiv \bar{F} + \Pi F \end{aligned}$$

Equalities (2.17) can now be written as

$$\mu \frac{d\bar{z}}{dt} + \frac{d\Pi z}{d\tau} = \bar{F} + \Pi F \qquad \frac{d\bar{y}}{dt} + \frac{1}{\mu}\frac{d\Pi y}{d\tau} = \bar{f} + \Pi f \tag{2.18}$$

Let us substitute the series (2.15) and (2.16) into (2.18) and represent $\bar{F}$, ΠF, $\bar{f}$, and Πf in the form of a power series in μ

$$\begin{aligned}\bar{F} &= F(\bar{z}_0(t) + \mu\bar{z}_1(t) + \cdots, \bar{y}_0(t) + \mu\bar{y}_1(t) + \cdots, t)\\ &= F(\bar{z}_0(t), \bar{y}_0(t), t) + \mu[\bar{F}_z(t)\bar{z}_1(t) + \bar{F}_y(t)\bar{y}_1(t)] + \cdots\\ &\quad + \mu^k[\bar{F}_z(t)\bar{z}_k(t) + \bar{F}_y(t)\bar{y}_k(t) + F_k(t)] + \cdots\\ &\equiv \bar{F}_0 + \mu\bar{F}_1 + \mu^k\bar{F}_k + \cdots\end{aligned}$$

where the elements of the matrices $\bar{F}_z(t)$ and $\bar{F}_y(t)$ are calculated at the point $(\bar{z}_0(t), \bar{y}_0(t), t)$, and the functions $F_k(t)$ are expressed recursively through $\bar{z}_i(t)$ and $\bar{y}_i(t)$ with $i<k$;

$$\begin{aligned}\Pi F &= F(\bar{z}_0(\tau\mu) + \mu\bar{z}_1(\tau\mu) + \cdots + \Pi_0 z(\tau) + \mu\Pi_1 z(\tau) + \cdots,\\ &\qquad \bar{y}_0(\tau\mu) + \mu\bar{y}_1(\tau\mu) + \cdots + \Pi_0 y(\tau) + \mu\Pi_1 y(\tau) + \cdots, \tau\mu)\\ &\qquad - F(\bar{z}_0(\tau\mu) + \mu\bar{z}_1(\tau\mu) + \cdots, \bar{y}_0(\tau\mu) + \mu\bar{y}_1(\tau\mu) + \cdots, \tau\mu)\\ &= [F(\bar{z}_0(0) + \Pi_0 z(\tau), \bar{y}_0(0) + \Pi_0 y(\tau), 0) - F(\bar{z}_0(0), \bar{y}_0(0), 0)]\\ &\qquad + \mu[F_z(\tau)\Pi_1 z(\tau) + F_y(\tau)\Pi_1 y(\tau) + G_1(\tau)] + \cdots\\ &\qquad + \mu^k[F_z(\tau)\Pi_k z(\tau) + F_y(\tau)\Pi_k y(\tau) + G_k(\tau)]\\ &\equiv \Pi_0 F + \mu\Pi_1 F + \cdots + \mu^k\Pi_k F + \cdots\end{aligned}$$

where the elements of $F_z(\tau)$ and $F_y(\tau)$ are calculated at the point $(\bar{z}_0(0) + \Pi_0 z(\tau), \bar{y}_0(0) + \Pi_0 y(\tau), 0)$, and the functions $G_k(t)$ are expressed recursively through $\Pi_i z(\tau)$ and $\Pi_i y(\tau)$ with $i<k$. Similar expansions hold for $\bar{f}$ and Πf.

Let us now equate coefficients of like powers of μ on both sides of (2.18) (separately, for the coefficients depending on t and on τ). This will provide equations for the terms of the series (2.15) and (2.16).

For the leading term $\bar{x}_0(t) = \{\bar{z}_0(t), \bar{y}_0(t)\}$ of the regular part of the asymptotics, we obtain the system of equations

$$0 = \bar{F}_0 \equiv F(\bar{z}_0(t), \bar{y}_0(t), t) \qquad \frac{d\bar{y}_0}{dt} = \bar{f}_0 \equiv f(\bar{z}_0(t), \bar{y}_0(t), t) \tag{2.19}$$

which evidently coincide with the reduced system (2.3).

For the leading term $\Pi_0 x(\tau) = \{\Pi_0 z(\tau), \Pi_0 y(\tau)\}$ of the boundary layer part of the asymptotics, we obtain

$$\frac{d\Pi_0 z}{d\tau} = \Pi_0 F \equiv F(\bar{z}_0(0) + \Pi_0 z, \bar{y}_0(0) + \Pi_0 y, 0) - F(\bar{z}_0(0), \bar{y}_0(0), 0)$$

$$= F(\bar{z}_0(0) + \Pi_0 z, \bar{y}_0(0) + \Pi_0 y, 0) \qquad \frac{d\Pi_0 y}{d\tau} = 0 \tag{2.20}$$

(Notice that $F(\bar{z}_0(0), \bar{y}_0(0), 0) = 0$ by virtue of the first equation of (2.19).)

For the terms $\bar{x}_k(t)$ and $\Pi_k x(\tau)$ $(k \geq 1)$ we have the equations

$$\begin{aligned} \frac{d\bar{z}_{k-1}}{dt} &= \bar{F}_k \equiv \bar{F}_z(t)\bar{z}_k + \bar{F}_y(t)\bar{y}_k + F_k(t) \\ \frac{d\bar{y}_k}{dt} &= \bar{f}_k \equiv \bar{f}_z(t)\bar{z}_k + \bar{f}_y(t)\bar{y}_k + f_k(t) \end{aligned} \tag{2.21}$$

$$\begin{aligned} \frac{d\Pi_k z}{d\tau} &= \Pi_k F \equiv F_z(\tau)\Pi_k z + F_y(\tau)\Pi_k y + G_k(\tau) \\ \frac{d\Pi_k y}{d\tau} &= \Pi_{k-1} f \end{aligned} \tag{2.22}$$

To define the solutions of (2.21) and (2.22), we have to impose initial conditions. Substituting (2.14) into (2.2), we obtain

$$\bar{z}_0(0) + \mu\bar{z}_1(0) + \cdots + \Pi_0 z(0) + \mu\Pi_1 z(0) + \cdots = z^0$$

$$\bar{y}_0(0) + \mu\bar{y}_1(0) + \cdots + \Pi_0 y(0) + \mu\Pi_1 y(0) + \cdots = y^0$$

Hence, equating the coefficients of like powers of μ in these relations implies that

$$\bar{z}_0(0) + \Pi_0 z(0) = z^0 \qquad \bar{y}_0(0) + \Pi_0 y(0) = y^0 \tag{2.23}$$

$$\bar{z}_k(0) + \Pi_k z(0) = 0 \qquad \bar{y}_k(0) + \Pi_k y(0) = 0 \qquad k = 1, 2, \ldots \tag{2.24}$$

Equations (2.23) contain four unknowns: $\bar{z}_0(0)$, $\Pi_0 z(0)$, $\bar{y}_0(0)$, and $\Pi_0 y(0)$ (a similar situation takes place in (2.24)). It is clear that by using only (2.23) it is impossible to define these four unknowns. We need to take into consideration some additional ideas. First, it is important to mention that the first equation of (2.19) is not differential, and therefore we need not impose any initial condition on $\bar{z}_0(t)$. Thus, we need not treat $\bar{z}_0(0)$ in (2.23) as an unknown. The same is true for the quantities $\bar{z}_k(0)$ in the systems (2.24). Second, recall that the functions $\Pi_k x(\tau)$ should be boundary functions, that is, they should approach zero as $\tau \to \infty$. It turns out that it is sufficient to impose this condition only on the functions $\Pi_k y(\tau)$

$$\Pi_k y(\infty) = 0 \qquad k = 0, 1, 2, \dots \tag{2.25}$$

Let us now show that Eqs. (2.19)–(2.22) together with conditions (2.23)–(2.25) allow us to determine successively all the terms of the series (2.15), (2.16). From the second equation of (2.20) we obtain $\Pi_0 y(\tau) = \text{const}$, and since $\Pi_0 y(\infty) = 0$ by virtue of (2.25) we have $\Pi_0 y(\tau) \equiv 0$. Thus, the function $y(t, \mu)$ does not have a boundary layer in the zeroth-order approximation.

Since $\Pi_0 y(0) = 0$, from the second equation of (2.23) we can find $\bar{y}_0(0) = y^0$. Thus, for $\bar{z}_0(t)$ and $\bar{y}_0(t)$ we have the system (2.19), which coincides with the reduced system (2.3), and the initial condition, which coincides with (2.9). For the solution $\bar{z}_0(t)$, $\bar{y}_0(t)$ of the problem (2.3), (2.9), let us take the solution that was mentioned in Tikhonov's theorem: $\bar{z}_0 = \bar{z}(t) = \varphi(\bar{y}, t)$, $\bar{y}_0 = \bar{y}(t)$.

Since $\bar{z}_0(t)$ is defined, the value $\bar{z}_0(0)$ is known. Therefore, the first equation of (2.23) yields

$$\Pi_0 z(0) = z^0 - \bar{z}_0(0) = z^0 - \varphi(y^0, 0) \tag{2.26}$$

To find $\Pi_0 z(\tau)$ we have to solve the first equation of (2.20) with this initial condition (also take into account that $\Pi_0 y(\tau) = 0$). The change of variable $\tilde{z}(\tau) = \Pi_0 z(\tau) + \varphi(y^0, 0)$ transforms the problem for $\Pi_0 z$ to the associated system (2.6) with the initial condition (2.7). By virtue of the condition 2^0 of Tikhonov's theorem,

$$\Pi_0 z(\tau) \to 0 \qquad \text{for} \qquad \tau \to \infty$$

It can be proved (using condition 3^0) that $\Pi_0 z(\tau)$ satisfies the exponential estimate (see [4])

$$\|\Pi_0 z(\tau)\| \le C \exp\{-k\tau\} \qquad \tau \ge 0 \tag{2.27}$$

Here and below we will use C and k to represent appropriate positive numbers which are, generally speaking, different in different inequalities.

Thus, all terms of the zeroth-order approximation are defined.

Let us consider the equations for $\bar{x}_1(t)$ and $\Pi_1 x(\tau)$. We have

$$G_1(\tau) = +(F_z(\tau) - \bar{F}_z(0))(\bar{z}_0'(0)\tau + \bar{z}_1(0)) + (F_y(\tau) - \bar{F}_y(0))(\bar{y}'(0)\tau + \bar{y}_1(0)) + (F_t(\tau) - \bar{F}_t(0))\tau$$

$$\Pi_0 f = f(\bar{z}_0(0) + \Pi_0 z(\tau), y^0, 0) - f(\bar{z}_0(0), y^0, 0)$$

The last equation (2.22) for $k = 1$ with condition (2.25) for $k = 1$ yields

$$\Pi_1 y(\tau) = -\int_r^\infty \Pi_0 f(s)\, ds$$

Hence

$$\Pi_1 y(0) = -\int_0^\infty \Pi_0 f(s)\, ds$$

From the second equation of (2.24) for $k = 1$ we obtain

$$y_1(0) = \int_0^\infty \Pi_0 f(s)\, ds$$

This condition specifies the unique solution $\bar{z}_1(t), \bar{y}_1(t)$ of the linear system (2.21) for $k = 1$. Then we solve the first equation (2.22) with the condition $\Pi_1 z(0) = -\bar{z}_1(0)$ (see (2.24)) and obtain $\Pi_1 z(\tau)$. It can be shown that $\Pi_1 z(\tau)$ satisfies an exponential estimate of type (2.27) [4].

The higher order terms of the asymptotic expansion can be constructed by the same method for any $k \geq 1$: [4, 5].

Let $X_n(t, \mu)$ denote the n-th partial sum of the expansion (2.14)

$$X_n(t, \mu) = \sum_{k=0}^{n} \mu^k [\bar{x}_k(t) + \Pi_k x(\tau)]$$

Under conditions 1^0–3^0, the series (2.14) is an asymptotic series as $\mu \to 0$ for the solution $x(t, \mu) = \{z(t, \mu), y(t, \mu)\}$ of the problem (2.1), (2.2) in the interval $0 \leq t \leq T$, that is, the following estimate holds:

$$\max_{0 \leq t \leq T} \|x(t, \mu) - X_n(t, \mu)\| = O(\mu^{n+1}) \tag{2.28}$$

The proof of this statement (Vasil'eva) can be found in [4, 5].

Example 2.2. Let us construct an asymptotic approximation to $\zeta(\eta, \mu)$ (see Example 2.1) with accuracy $O(\mu)$ for $0 \leq \eta \leq 1$. For this purpose we must add to $\bar{\zeta}_0 = \lambda/(\lambda + \eta)$ the boundary function $\Pi_0 \zeta$. In this case, Eqs.

(2.20), (2.23) have the form

$$\frac{d\Pi_0\zeta}{d\tau} = 1 - \lambda\left(1 - \frac{\lambda}{\lambda+1} - \Pi_0\zeta\right)\Big/\left(\frac{\lambda}{\lambda+1} + \Pi_0\zeta\right)$$
$$\Pi_0\zeta(0) = \frac{1}{\lambda+1} \qquad \tau = \frac{\eta-1}{\mu} \tag{2.29}$$

By solving problem (2.29) we obtain

$$\eta = 1 + \frac{\mu}{(\lambda+1)^2}\{\lambda \ln[(\lambda+1)\Pi_0\zeta] + (\lambda+1)\Pi_0\zeta - 1\}$$

The parameter $\Pi_0\xi$ can be obtained from here as an implicit function.

According to (2.28) we have $\zeta(\eta, \mu) - [\lambda/(\lambda+\eta) + \Pi_0\zeta] = O(\mu)$, uniformly for $0 \le \eta \le 1$.

Example 2.3. If we return to the system (2.10) we may remove the small parameter in another way (by the use of another condition). By using the relation $z = z^0 - m$, we obtain

$$\frac{dy}{dt} = -kz^0y + kym \qquad \frac{dm}{dt} = kz^0y - kym - \alpha m \tag{2.30}$$

By the change of variables $y = s/(kz^0)$, $m = Cz^0$, $t = \tilde{t}/(kz^0)$, $(kz^0 \sim 1)$ the system (2.30) can be transformed to

$$\mu\frac{dC}{dt} = +s - sC - \lambda C \qquad C(0) = 0$$
$$\frac{ds}{dt} = -s + sC \qquad s(0) = s^0$$

where $\tilde{t}$ is denoted again as t, $\lambda = \alpha z^0 \sim 1$ and $\mu = k(z^0)^2$ is small.

To construct the zeroth-order terms $\bar{C}_0 + \Pi_0 C$, $\bar{s}_0$ $(\Pi_0 s = 0)$ of the asymptotic solution of this problem (i.e., the asymptotic approximation with accuracy $O(\mu)$) we have

$$\frac{d\bar{s}_0}{dt} = -\frac{\lambda\bar{s}_0}{\bar{s}+\lambda} \qquad \bar{s}_0(0) = s^0 \qquad \bar{C}_0 = \frac{\bar{s}_0}{\bar{s}_0+\lambda}$$

By solving this problem we obtain $t(\bar{s}_0)$. The function $\bar{s}_0(t)$ is defined implicitly. The boundary function $\Pi_0 C$ can be defined explicitly

$$\frac{d\Pi_0 C}{d\tau} = -(s_0 + \lambda)\Pi_0 C \qquad \Pi_0 C(0) = -\frac{s^0}{s^0 + \lambda} \qquad \tau = \frac{t}{\mu}$$

Hence,

$$\Pi_0 C = -\frac{s^0}{s^0 + \lambda} \exp\left\{-(s^0 + \lambda)\frac{t}{\mu}\right\}$$

Thus

$$C = \frac{\bar{s}_0(t)}{\bar{s}_0(t) + \lambda} - \frac{s_0}{s^0 + \lambda} \exp\left\{-(s^0 + \lambda)\frac{t}{\mu}\right\} + O(\mu)$$

$$s = \bar{s}_0(t) + O(\mu)$$

The system (2.30), and a more complicated one, can be found in [5].

III. BOUNDARY VALUE PROBLEM

A. Conditionally Stable Case

Consider the system

$$\mu \frac{dz}{dt} = F(z, y, t) \qquad \frac{dy}{dt} = f(z, y, t) \qquad 0 \le t \le T \tag{3.1}$$

in the case when y is an m vector and z contains two scalar components z_1 and z_2. Let the eigenvalues $\lambda_{1,2}(t)$ of matrix $\bar{F}_z(t) = F_z(\bar{z}_0(t), \bar{y}_0(t), t)$ (here $\bar{z}_0(t), t)$ is the solution of the reduced problem (2.8), (2.9)) satisfy instead of condition (2.13) another condition.

Condition 4^0. $\lambda_1(t) < 0$, $\lambda_2(t) > 0$ for $0 \le t \le T$

Simple examples show that the solution of such *initial value* problems are not bounded for $\mu \to 0$. Therefore the theorem on the passage to the limit does not hold.

For (3.1) in the case when condition 4^0 holds, we must replace initial conditions for z by boundary conditions (assume that $T = 1$)

$$z_1(0, \mu) = z_1^0 \qquad z_2(1, \mu) = z_2^0 \tag{3.2}$$

and retain an initial condition for y

$$y(0, \mu) = y^0 \tag{3.3}$$

The study of this singularly perturbed problem shows that its solution tends to the solution of the reduced problem that retains only the initial condition for y. The asymptotic expansion of the solution now contains boundary functions that appear not only in the vicinity of the point $t = 0$, but also in the vicinity of $t = 1$.

Let us describe the algorithm to construct the asymptotics of the solution of problem (3.1)–(3.3) assuming sufficient smoothness of the functions $F(z, y, t)$ and $f(x, y, t)$. We seek an asymptotic expansion of the solution in the form (where x stands for both variables z and y, i.e., $x = \{z, y\}$)

$$x(t, \mu) = \bar{x}(t, \mu) + \Pi x(\tau_0, \mu) + Qx(\tau_1, \mu) \tag{3.4}$$

Here $\bar{x}(t, \mu)$ is the regular power series

$$\bar{x}(t, \mu) = \bar{x}_0(t) + \mu\bar{x}_1(t) + \cdots + \mu^k\bar{x}_k(t) + \cdots$$

$\Pi x(\tau_0, \mu)$ $(\tau_0 = t/\mu)$ is a boundary layer series in the vicinity of $t = 0$:

$$\Pi x(\tau_0, \mu) = \Pi_0 x(\tau_0) + \mu\Pi_1 x(\tau_0) + \cdots + \mu^k\Pi_k x(\tau_0) + \cdots$$

and $Qx(\tau_1, \mu)[\tau_1 = (t-1)/\mu]$ is a boundary layer series in the vicinity of $t = 1$:

$$Qx(\tau_1, \mu) = Q_0 x(\tau_1) + \mu Q_1 x(\tau_1) + \cdots + \mu^k Q_k x(\tau_1) + \cdots$$

Substituting (3.4) into (3.1), we obtain

$$\begin{aligned} \mu\frac{d\bar{z}}{dt} + \frac{d\Pi z}{d\tau_0} + \frac{dQz}{d\tau_1} &= \bar{F} + \Pi F + QF \\ \frac{d\bar{y}}{dt} + \frac{1}{\mu}\frac{d\Pi y}{d\tau_0} + \frac{1}{\mu}\frac{dQy}{d\tau_1} &= \bar{f} + \Pi f + Qf \end{aligned} \tag{3.5}$$

where

$$\begin{aligned} \bar{F} &= F(\bar{z}(t, \mu), \bar{y}(t, \mu), t) \\ \Pi F &= F(\bar{z}(\tau_0\mu, \mu) + \Pi z(\tau_0, \mu), \bar{y}(\tau_0\mu, \mu) + \Pi y(\tau_0), \tau_0\mu) \\ &\quad - F(\bar{z}(\tau_0\mu, \mu), \bar{y}(\tau_0\mu, \mu), \tau_0\mu) \\ QF &= F(\bar{z}(1 + \tau_1\mu, \mu) + Qz(\tau_1, \mu), \bar{y}(1 + \tau_1\mu, \mu) + Qy(\tau_1\mu), 1 + \tau_1\mu) \\ &\quad - F(\bar{z}(1 + \tau_1\mu, \mu), \bar{y}(1 + \tau_1\mu, \mu), 1 + \tau_1\mu) \end{aligned}$$

For the coefficients of the expansion ΠF, a representation similar to that discussed in Section II.B holds (with τ replaced by τ_0); for QF, we have an analogous representation with $t=0$ changed to $t=1$ and τ changed to τ_1. Similar representations hold for Πf and Qf.

Substituting (3.4) into the boundary conditions (3.2), (3.3), and assuming that $\Pi_k x(1/\mu)$ and $Q_k x(-1/\mu)$ are less than any power of μ, we obtain

$$
\begin{aligned}
z_1^0 &= \bar{z}_{10}(0) + \mu\bar{z}_{11}(0) + \cdots + \Pi_0 z_1(0) + \mu\Pi_1 z_1(0) + \cdots \\
z_2^0 &= \bar{z}_{20}(1) + \mu\bar{z}_{21}(1) + \cdots + Q_0 z_2(0) + \mu Q_1 z_2(0) + \cdots \\
y^0 &= \bar{y}_0(0) + \mu\bar{y}_1(0) + \cdots + \Pi_0 y(0) + \mu\Pi_1 y(0) + \cdots
\end{aligned}
\tag{3.6}
$$

(the first subindex of $\bar{z}_{1k}(0)$ and of $\bar{z}_{2k}(0)$ denotes the component of the vector $\bar{z}$).

In (3.5), let us equate separately terms depending on t, τ_0, and τ_1, respectively. For the zeroth order, we have

$$
0 = \bar{F}_0 \equiv F(\bar{z}_0(t), \bar{y}_0(t), t) \qquad \frac{d\bar{y}_0}{dt} = \bar{f}_0 \equiv f(\bar{z}_0(t), \bar{y}_0(t), t) \tag{3.7}
$$

(this is, evidently, the reduced system for the original system (3.1)).

$$
\begin{aligned}
&\frac{d\Pi_0 z}{d\tau_0} = \Pi_0 F \equiv F(\bar{z}_0(0) + \Pi_0 z, \bar{y}_0(0) + \Pi_0 y, 0) \\
&\frac{d\Pi_0 y}{d\tau_0} = 0 \qquad (\tau_0 \geq 0) \\
&\frac{dQ_0 z}{d\tau_1} = Q_0 F \equiv F(\bar{z}_0(1) + Q_0 z, \bar{y}_0(1) + Q_0 y, 1) \\
&\frac{dQ_0 y}{d\tau_1} = 0 \qquad (\tau_1 \leq 0)
\end{aligned}
\tag{3.8}
$$

Additional conditions are obtained from (3.6), equating the coefficients of μ^0.

$$
\bar{z}_{10}(0) + \Pi_0 z_1(0) = z_1^0 \qquad \bar{z}_{20}(1) + Q_0 z_2(0) = z_0^2 \tag{3.9}
$$

$$
\bar{y}_0(0) + \Pi_0 y(0) = y^0 \tag{3.10}
$$

In addition, we impose on the boundary functions the decay conditions similar to those of Section II.B. For $\Pi_k y(\tau_0)$ and $Q_k y(\tau_1)$, we ask that

$$\Pi_k y(\infty) = 0 \qquad Q_k y(-\infty) = 0 \qquad k = 0, 1, 2, \ldots$$

Then $\Pi_0 y(\tau_0) \equiv 0$, $Q_0 y(\tau_1) \equiv 0$, and from (3.10) we obtain $\bar{y}_0(0) = y^0$. System (3.7) with this initial condition defines the reduced solution $\bar{z}_0(t)$, $\bar{y}_0(t)$. We assume that for this solution condition 4^0 is satisfied.

Let us now consider problem (3.8), (3.9). If we take into account the equality $\Pi_0 y(\tau_0) \equiv 0$, we have

$$\frac{d\Pi_0 z}{d\tau_0} = F(\bar{z}_0(0) + \Pi_0 z, y^0, 0) \tag{3.11}$$

$$\Pi_0 z_1(0) = z_1^0 - \bar{z}_{10}(0) \tag{3.12}$$

Thus, for the two-dimensional vector function $\Pi_0 z$ only one condition (3.12) is given. This condition does not specify the solution uniquely. To find $\Pi_0 z$ and other Π terms, we demand that they decay to zero as $\tau_0 \to \infty$.

$$\Pi_k z(\infty) = 0 \qquad k = 0, 1, 2, \ldots$$

By virtue of condition 4^0 the rest point $\Pi_0 z = 0$ of (3.11) is not asymptotically stable: It is a saddle point. In Fig. 4 the saddle point O is represented along with the two separatrices (1 and 2) passing through it. Arrows indicate the directions corresponding to increasing τ_0. In order for the solution $\Pi_0 z$ of (3.11), (3.12) to satisfy the condition $\Pi_0 z(\infty) = 0$, separatrix 1 should cross the straight line $\Pi_0 z_1 = z_1^0 - \bar{z}_{10}(0)$ (point M on the Fig. 4). This crossing will define the value $\Pi_0 z_2(0)$ and thereby define the entire solution $\Pi_0 z(\tau_0)$.

We impose the following condition:

Condition 5^0. *Let the straight line* $\Pi_0 z_1 = z_1^0 - \bar{z}_{10}(0)$ *intersect the*

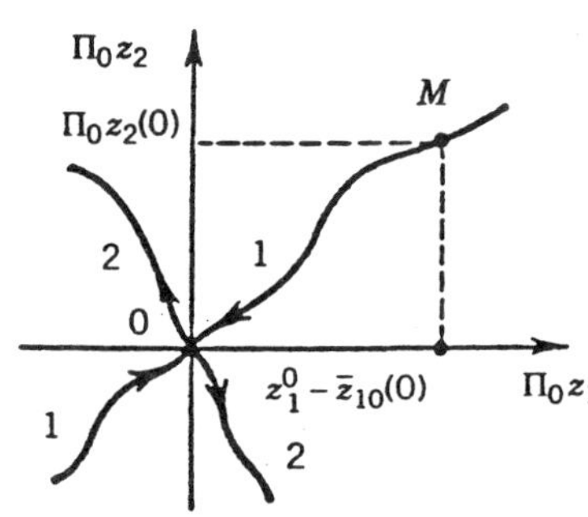

Figure 4. The saddle point $\Pi_0 z_1 = \Pi_0 z_2 = 0$ and two separatrices 1 and 2 passing through it.

separatrix that enters the rest point (saddle) $\Pi_0 z = 0$ *of the system (3.11) as* $\tau_0 \to \infty$.

Remark 1: We call this case *conditionally stable* since the rest point $\Pi_0 z = 0$ is now not asymptotically stable, and "attracts" only those solutions whose initial points lie on the separatrix 1.

Remark 2: It might happen that the separatrix 1 does not contain a point with abscissa $z_1^0 - \bar{z}_{10}(0)$ (Fig. 5(*a*)) or, on the contrary, contains more than one such point (points M_1 and M_2 in Fig. 5(b)). In the first case, a solution with the assumed asymptotic behavior (3.4) does not exist. In the second case, two asymptotic solutions of type (3.4) exist.

It can be shown that $\Pi_0 z(\tau_0)$ satisfies an exponential estimate

$$\|\Pi_0 z(\tau_0)\| \le C \exp(-k\tau_0) \qquad \tau_0 \ge 0 \tag{3.13}$$

The boundary function $Q_0 z(\tau_1)$ is defined analogously to $\Pi_0 z(\tau_0)$. In the case of $Q_0 z(\tau_1)$, the separatrix entering the saddle point as $\tau_1 \to -\infty$ is used. A condition similar to 5^0 (we denote it $5^{0\prime}$) should be imposed. Function $Q_0 z(\tau_1)$ then has an exponential estimate of type (3.13)

$$\|Q_0 z(\tau_1)\| \le C \exp(k\tau_1) \qquad \tau_1 \le 0 \tag{3.14}$$

Now let us consider the equations for $\bar{x}_1(t)$ and $\Pi_1 x(\tau_0)$

$$\frac{d\bar{z}_0}{dt} = \bar{F}_z(t)\bar{z}_1 + \bar{F}_y(t)\bar{y}_1 \qquad \frac{d\bar{y}_1}{dt} = \bar{f}_z(t)\bar{z}_1 + \bar{f}_y(t)y_1 \tag{3.15}$$

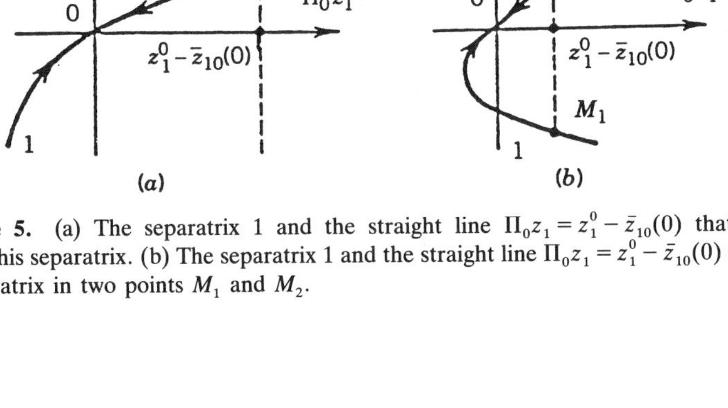

Figure 5. (a) The separatrix 1 and the straight line $\Pi_0 z_1 = z_1^0 - \bar{z}_{10}(0)$ that is not crossing this separatrix. (b) The separatrix 1 and the straight line $\Pi_0 z_1 = z_1^0 - \bar{z}_{10}(0)$ crossing this separatrix in two points M_1 and M_2.

where $\bar{F}_z(t) = F_z(\bar{z}_0(t), \bar{y}_0(t), t)$, and $\bar{F}_y(t)$, $\bar{f}_z(t)$, $\bar{f}_y(t)$ have analogous meanings;

$$\frac{d\Pi_1 z}{d\tau_0} = \Pi_1 F \equiv F_z(\tau_0)\Pi_1 z + F_y(\tau_0)\Pi_1 y + G_1(\tau_0)$$
$$\frac{d\Pi_1 y}{d\tau_0} = \Pi_0 f \tag{3.16}$$

where $F_z(\tau_0) = F_z(\bar{z}_0(0) + \Pi_0 z(\tau_0), y^0, 0)$, and $F_y(\tau_0)$ has a similar meaning; $G_1(\tau_0)$ is given by formula

$$G_1(\tau_0) = [F_z(\tau_0) - \bar{F}_z(0)][\bar{z}_0'(0)\tau_0 + \bar{z}_1(0)]$$
$$+ [F_y(\tau_0) - \bar{F}_y(0)][\bar{y}_0'(0)\tau_0 + \bar{y}_1(0)] + [F_t(\tau_0) - \bar{F}_t(0)]\tau_0$$

The additional conditions for (3.15), (3.16) are

$$\bar{z}_{11}(0) + \Pi_1 z_1(0) = 0 \qquad \bar{y}_1(0) + \Pi_1 y(0) = 0 \tag{3.17}$$

$$\Pi_1 z(\infty) = 0 \qquad \Pi_1 y(\infty) = 0 \tag{3.18}$$

The second equation of (3.16) with condition (3.18) yields

$$\Pi_1 y(\tau_0) = -\int_{\tau_0}^{\infty} \Pi_0 f(s)\, ds$$

Hence,

$$\Pi_1 y(0) = -\int_0^{\infty} \Pi_0 f(s)\, ds$$

and from the second equation of (3.17) we obtain

$$\bar{y}_1(0) = \int_0^{\infty} \Pi_0 f(s)\, ds$$

This initial condition specifies the unique solution $\bar{z}_1(t), \bar{y}_1(t)$ of the linear system (3.15). There now remains the first equation of (3.16) with boundary conditions $\Pi_1 z_1(0) = -\bar{z}_{11}(0)$, $\Pi_1 z(\infty) = 0$ to define $\Pi_1 z(\tau_0)$. It can be shown that these conditions specify $\Pi_1 z(\tau_0)$ uniquely (see [4]).

The problem for $Q_1x(\tau_0)$ has a form analogous to that for $\Pi_1x(\tau_0)$. We obtain

$$Q_1y(\tau_1) = -\int_{\tau_1}^{-\infty} Q_0f(s)\,ds$$

The function $\bar{z}_{21}(t)$ is already determined, therefore $\bar{z}_{21}(1)$ is known, and for $Q_1z(\tau_1)$ we have the additional conditions $Q_1z_2(0) = -\bar{z}_{21}(1)$ and $Q_1z(-\infty) = 0$. This allows us to determine $Q_1z(\tau_1)$ uniquely.

Analogously, $\bar{x}_k(t)$, $\Pi_kx(\tau_0)$, $Q_kx(\tau_1)$ can be obtained termwise for any arbitrary k. For the Π and Q functions exponential estimates of type (3.13), (3.14) hold.

Remark: The algorithm described undergoes minor changes when instead of (3.2), the boundary conditions $z_1(0) = z_1^0$, $z_1(1) = z_1^1$ or $z_2(0) = z_2^0$, $z_2(1) = z_2^1$ are given (values of only one of the components of the vector z are prescribed at each endpoint of the interval $[0, 1]$), or when boundary conditions of the type $(dz_1/dt)(0) = z_1^0$, $(dz_2/dt)(1) = z_2^1$ are given. For y one may also impose boundary conditions.

Let $X_n(t, \mu)$ denote the n-th partial sum of the expansion (3.4).

$$X_n(t, \mu) = \sum_{k=0}^{n} \mu^k[\bar{x}_k(t) + \Pi_kx(\tau_0) + Q_kx(\tau_1)]$$

Under conditions 4^0, 5^0, *and* $5^{0\prime}$, *for sufficiently small* μ *there exists a unique solution* $x(t, \mu)$ *of the problem (3.1)–(3.3) in a vicinity of the leading term of the asymptotic expansion* $\bar{x}_0(t) + \Pi_0x(t/\mu) + Q_0x((t-1)/\mu)$. *The series (3.4) is the asymptotic expansion of this solution in the interval* $0 \le t \le 1$ *for* $\mu \to 0$, *that is,*

$$\max_{0 \le t \le 1} \|x(t, \mu) - X_n(t, \mu)\| = O(\mu^{n+1}) \tag{3.19}$$

The proof of this statement is presented in [4].

Example 3.1. Consider the problem

$$\mu \frac{dz_1}{dt} = z_2 - 1 \qquad \mu \frac{dz_2}{dt} = z_1 + 1 \qquad z_1(0, \mu) = 0 \qquad z_2(1, \mu) = 0$$

The exact solution of this problem has the form

$$z_1 = -1 + \frac{1+\exp\{-1/\mu\}}{1+\exp\{-2/\mu\}} \exp\left\{-\frac{t}{\mu}\right\} + \frac{-1+\exp\{-1/\mu\}}{+1+\exp\{-2/\mu\}} \exp\left\{\frac{t-1}{\mu}\right\}$$

$$z_2 = +1 - \frac{1+\exp\{-1/\mu\}}{1+\exp\{-2/\mu\}} \exp\left\{-\frac{t}{\mu}\right\} + \frac{-1+\exp\{-1/\mu\}}{+1+\exp\{-2/\mu\}} \exp\left\{\frac{t-1}{\mu}\right\}$$

(3.20)

Let us construct the zeroth-order terms of the asymptotics of the solution $z_{i0} = \bar{z}_{i0} + \Pi_0 z_i + Q_0 z_i$ $(i = 1, 2)$. We easily obtain $\bar{z}_{10} = -1$, $\bar{z}_{20} = 1$. The eigenvalues of the matrix F_z are $\lambda_{1,2} = \pm 1$ and, thus, condition 4^0 is satisfied. The system (3.11) with condition (3.12) and the decay condition at infinity has the form

$$\frac{d}{d\tau_0}\Pi_0 z_1 = \Pi_0 z_2 \qquad \frac{d}{d\tau_0}\Pi_0 z_2 = \Pi_0 z_1 \qquad \Pi_0 z_1(0) = 1 \qquad \Pi_0 z_i(\infty) = 0$$

From this system we obtain $\Pi_0 z_1 = -\Pi_0 z_2$. Now by solving

$$\frac{d}{d\tau_0}\Pi_0 z_1 = -\Pi_0 z_1 \qquad \Pi_0 z_1(0) = 1$$

we find $\Pi_0 z_1 = \exp\{-\tau_0\}$, $\Pi_0 z_2 = -\exp\{-\tau_0\}$. For $Q_0 z_1$, $Q_0 z_2$ we have

$$\frac{d}{d\tau_1}Q_0 z_1 = Q_0 z_2 \qquad \frac{d}{d\tau_1}Q_0 z_2 = Q_0 z_1 \qquad Q_0 z_2(0) = -1 \qquad Q_0 z_i(-\infty)$$

Analogous to the case of Π functions, we obtain $Q_0 z_1 = -\exp\{\tau_1\}$, $Q_0 z_2 = -\exp\{\tau_1\}$. Thus we have: $Z_{10} = -1 + \exp\{-t/\mu\} - \exp\{(t-1)/\mu\}$, $z_{20} = 1 - \exp\{-t/\mu\} - \exp\{(t-1)/\mu\}$.

We may immediately observe that these formulas and the exact solution (3.20) differ in value by at most $O(\mu)$.

Example 3.2. Consider the problem

$$\mu\frac{dz_1}{dt} = yz_2 - 3 \qquad \mu\frac{dz_2}{dt} = yz_1 \qquad \frac{dy}{dt} = z_1 + 1 \qquad t \in (0, 1)$$

$$z_1(0) = -1 \qquad z_2(1) = 1 \qquad y(0) = 1$$

This problem is nonlinear and we cannot obtain the exact solution. Let us construct the zeroth-order terms of the asymptotics of the solution. The regular terms of the zeroth order are

$$\bar{y}_0(t) = 1 + t \qquad \bar{z}_{10}(t) = 0 \qquad \bar{z}_{20}(t) = \frac{3}{1+t}$$

The eigenvalues of the matrix F_z are $\lambda_{1,2} = \pm \bar{y}_0(t)$ and, thus, condition 4^0 is satisfied. Functions $\Pi_0 y$ and $Q_0 y$ are identically zero. The system (3.11) with condition (3.12) and the decay condition at infinity has the form

$$\frac{d\Pi_0 z_1}{d\tau_0} = \Pi_0 z_2 \qquad \frac{d\Pi_0 z_2}{d\tau_0} = \Pi_0 z_1$$

$$\Pi_0 z_1(0) = -1 - \bar{z}_{10}(0) = -1 \qquad \Pi_0 z_i(\infty) = 0$$

From this system we obtain $\Pi_0 z_1(\tau_0) = -\Pi_0 z_2(\tau_0)$. Now by solving

$$\frac{d\Pi_0 z_1}{d\tau_0} = -\Pi_0 z_1 \qquad \Pi_0 z_1(0) = -1$$

we obtain $\Pi_0 z_1(\tau_0) = -\exp\{-\tau_0\}$, $\Pi_0 z_2(\tau_0) = \exp\{-\tau_0\}$.

Similarly, for $Q_0 z_1$ and $Q_0 z_2$ we have

$$\frac{dQ_0 z_1}{d\tau_1} = 2Q_0 z_2 \qquad \frac{dQ_0 z_2}{d\tau_1} = 2Q_0 z_1$$

$$Q_0 z_2(0) = 1 - \bar{z}_{20}(1) = -\tfrac{1}{2} \qquad Q_0 z_i(-\infty) = 0$$

Analogous to the case of Π functions, we obtain $Q_0 z_2 = Q_0 z_1 = -(\tfrac{1}{2})\exp\{2\tau_1\}$.

All the zeroth-order terms have now been determined. From (3.19) it follows that

$$Z_{10} = -\exp\left\{-\frac{t}{\mu}\right\} - \tfrac{1}{2}\exp\left\{\frac{t-1}{\mu}\right\} \qquad z_1(t, \mu) = Z_{10} + O(\mu)$$

$$Z_{20} = \frac{3}{1+t} + \exp\left\{-\frac{t}{\mu}\right\} - \tfrac{1}{2}\exp\left\{\frac{t-1}{\mu}\right\} \qquad z_2(t, \mu) = Z_{20} + O(\mu)$$

$$Y_0 = 1 + t \qquad y(t, \mu) = Y_0 + O(\mu)$$

Example 3.3. Let us consider the chemical reaction in Example 2.1. We assume now that some diffusion is present in the system. Then we have three partial differential equations of parabolic type

$$-D_1 \frac{\partial^2 y}{\partial x^2} + \frac{\partial y}{\partial t} = -kyz \qquad -D_2 \frac{\partial^2 z}{\partial x^2} + \frac{\partial z}{\partial t} = -kyz + \alpha m$$

$$-D_3 \frac{\partial^2 m}{\partial x^2} + \frac{\partial m}{\partial t} = kyz - \alpha m$$

We must impose some boundary conditions in addition to the initial conditions, for instance,

$$y(0, t) = \gamma \qquad \frac{\partial y}{\partial x}(1, t) = 0$$

$$z(0, t) = z^0 \qquad \frac{\partial z}{\partial x}(1, t) = 0$$

$$m(0, t) = 0 \qquad \frac{\partial m}{\partial x}(1, t) = 0$$

Taking $D_1 = 1$, $D_2 = D_3 = \varepsilon^2$ (ε is small) consider the *stationary* solution, that satisfies the ordinary differential equations

$$\frac{d^2 y}{dx^2} = kyz \qquad \varepsilon^2 \frac{d^2 z}{dx^2} = kyz - \alpha m \qquad \varepsilon^2 \frac{d^2 m}{dx^2} = -kyz + \alpha m$$

$$y(0) = \gamma \qquad \frac{dy}{dx}(1) = 0 \qquad z(0) = z^2 \qquad \frac{dz}{dx}(1) = 0 \tag{3.21}$$

$$m(0) = \frac{dm}{dx}(1) = 0$$

These equations describe the regime, to which the process comes when t is sufficiently large.

From second and third equations of (3.21) we obtain $(d^2 z/dx^2) + (d^2 m/dx^2) = 0$. Hence, $(dz/dx) + (dm/dx) = a$, $z + m = ax + b$. From the boundary conditions we have $a = 0$, $b = z^0$. Hence,

$$m = z^0 - z \qquad \frac{d^2 y}{dx^2} = kyz$$

$$\varepsilon \frac{d^2 z}{dx^2} = kyz - \alpha(z^0 - z) = z(ky + \alpha) - \alpha z^0$$

Introducing the variables $y = y_1$, $(dy/dx) = y_2$, $z = z_1$, $\varepsilon(dz/dx) = z_2$, we obtain

$$\varepsilon\frac{dz_1}{dx}=z_2 \qquad \varepsilon\frac{dz_2}{dx}=z_1[ky_1+\alpha]-\alpha z^0 \qquad \frac{dy_1}{dx}=y_2 \qquad \frac{dy_2}{dx}=ky_1z_1$$

$$y_1(0)=\gamma \qquad y_2(1)=0 \qquad z_1(0)=z^0 \qquad z_2(1)=0$$

Thus, we have equations of type (3.1). We must take into account the fact that in (3.9) the independent variable is denoted by t, and here by x. The boundary conditions are different from (3.2) but our algorithm is valid (see the remark before statement (3.19)).

To construct the zeroth-order approximation we have

$$\bar{z}_0=\bar{z}_{10}=\frac{\alpha z^0}{k\bar{y}_0+\alpha} \qquad \bar{z}_{20}=0$$

$$\frac{d^2\bar{y}_0}{dx^2}=k\bar{y}_0\frac{\alpha z^0}{k\bar{y}_0+\alpha} \qquad \bar{y}_0(0)=\gamma \qquad \frac{d\bar{y}_0}{dx}(1)=0$$

We cannot obtain an explicit representation for $\bar{y}_0$. But it is easy to obtain an approximate solution when γ (or α) is small (it is a regular perturbation). We will not solve this problem here. Notice that $\bar{y}_0(x)\geq 0$ in $[0,1]$.

For the boundary function $\Pi_0 z$ we have

$$\frac{d^2}{d\tau_0^2}\Pi_0 z=[k\bar{y}_0(0)+\alpha]\Pi_0 z \quad \Pi_0 z(0)=z^0-\frac{\alpha z^0}{k\bar{y}_0(0)+\alpha} \tag{3.22}$$

If we take into account the condition at $\tau_0=\infty$ we find that

$$\Pi_0 z=\frac{z^0 k\gamma}{k\gamma+\alpha}\exp\left\{\frac{-\sqrt{k\gamma+\alpha}}{\mu}x\right\}$$

For $Q_0 z$, the differential equation in (3.22) is valid [$\bar{y}_0(0)$ and τ_0 must be replaced by $\bar{y}_0(1)$ and τ_1], but the boundary condition is different: $(d/d\tau_1)Q_0 z(0)=0$ (because according to (3.9) $Q_0 z_2(0)=0$). Hence, $Q_0 z(\tau_1)\equiv 0$. Thus, the zeroth-order approximation is constructed.

B. Some Generalizations

Let us mention some generalizations of the problem discussed above.

1. The boundary conditions might be of the more complicated form

$$R(z_1(0), z_2(0), y(0), z_1(1), z_2(1), y(1))=0 \tag{3.23}$$

where R is an $(m+2)$ dimensional vector. In this case, a theorem on the existence of a solution of the boundary value problem can also be proved. An asymptotic expansion of type (3.4) can be constructed, but now we have to consider z_1^0, z_2^0, y^0 as functions of μ for which the following asymptotic representations hold:

$$z_i^0 = z_{i0}^0 + \mu z_{i1}^0 + \cdots \quad i = 1, 2 \qquad y^0 = y_0^0 + \mu y_1^0 + \cdots$$

The algorithm described earlier can be extended to this case [6].

2. The results of Section III.A are generalized to the case when the vector function z has arbitrary dimension ℓ. In this case, instead of condition 4^0 the following condition must hold:

$$\begin{aligned} &Re\ \lambda_i(t) < 0 \qquad (i = 1, 2, \ldots, r) \\ &Re\ \lambda_i(t) > 0 \qquad (i = r+1, \ldots, \ell) \end{aligned} \tag{3.24}$$

where the $\lambda_i(t)$ are eigenvalues of the matrix $F_z(\bar{z}_0(t), \bar{y}_0(t), t)$; $\bar{z}_0(t)$ and $\bar{y}_0(t)$ is the solution of the corresponding reduced problem, and $t \in [0, 1]$. This general case is discussed in detail in [4].

Remark: There are many boundary value problems where solutions with boundary layer asymptotics exist when the $\lambda_i(t)$ satisfy condition 3^0 of Section II.B. In that case, the boundary layer appears only in the vicinity of the point $t = 0$ (see [4], §13 for details).

IV. THE CRITICAL CASE

A. The Initial Value Problem for a System with a Small Nonlinearity

One of the main conditions in the theorem on the passage to the limit is the condition for the existence of an isolated root $z = \varphi(y, t)$ of the reduced equation $F(z, y, t) = 0$. In a variety of applied problems leading to singularly perturbed equations, and particularly in most problems of chemical kinetics, this condition is not satisfied because the reduced equation does not have an isolated root. Instead, it has a family of solutions depending on one or several parameters. This case will be called the *critical case*.

It turns out that under suitable conditions, the asymptotics for both the initial value problem and the boundary value problem in the critical case have the same form as in Sections II.B and III.A. In particular, the solution of the initial value problem approaches one of the solutions of the reduced equation in the limit as $\mu \to 0$. But the algorithm for constructing the asymptotic expansion undergoes some changes.

Let us consider this algorithm for a system of equations with a small nonlinearity:

$$\mu \frac{dx}{dt} = A(t) + \mu f(x, t, \mu) \qquad 0 \le t \le T \tag{4.1}$$

$$x(0, \mu) = x^0 \tag{4.2}$$

Here x and f are m-dimensional vector functions, $A(t)$ is an $m \times m$ matrix, and $\mu > 0$ is a small parameter. We assume that $A(t)$ and $f(x, t, \mu)$ are sufficiently smooth.

Condition 1^0. *Let the matrix $A(t)$ have zero as an eigenvalue of multiplicity k, $0 < k < m$, for each $t \in [0, T]$:*

(1) $$\lambda_i(t) = 0 \qquad i = 0, 1, 2, \ldots, k$$

while other eigenvalues of $A(t)$ satisfy the inequality,

(2) $$Re\ \lambda_i(t) < 0 \qquad i = k+1, \ldots, m$$

From condition 1^0 (1) it follows that $det\, A(t) = 0$, and therefore the reduced equation $A(t)x = 0$ obtained from (4.1), when $\mu = 0$, has a family of solutions represented by a linear combination (with arbitrary coefficients) of the eigenvectors e_i of the matrix $A(t)$ corresponding to $\lambda = 0$.

Condition 2^0. *Let the k vectors $e_i(t)$, $i = 1, 2, \ldots, k$ be linearly independent for each $t \in [0, T]$.*

As in Section II.B, we will seek the asymptotic expansion of the solution of (4.1), (4.2) as a sum of regular and boundary function series

$$x(t, \mu) = \bar{x}(t, \mu) + \Pi x(\tau, \mu) = \sum_{i=0}^{\infty} \mu^i (\bar{x}_i(t) + \Pi_i x(\tau)) \tag{4.3}$$

where $\tau = t/\mu$. Substituting (4.3) into (4.1), (4.2) and representing f in the form $f = \bar{f} + \Pi f$ as in Section II.B, we arrive at the equalities

$$\mu \frac{d\bar{x}}{dt} + \frac{d\Pi x}{d\tau} = A(t)\bar{x} + A(\tau\mu)\Pi x + \mu \bar{f} + \mu \Pi f \tag{4.4}$$

$$\bar{x}(0, \mu) + \Pi x(0, \mu) = x^0 \tag{4.5}$$

Expanding $A(\tau\mu)$, $\bar{f}$, and Πf into power series in μ, and equating coefficients of like powers of μ on both sides of (4.4) and (4.5) (separately for the coefficients depending on t and on τ), we obtain problems defining the terms of the series (4.3).

For $\bar{x}_0(t)$ we have the reduced equation

$$A(t)\bar{x}_0 = 0$$

The general solution of this equation, by virtue of condition 2^0, can be written in the form

$$\bar{x}_0 = \sum_{i=1}^{k} \alpha_i(t)e_i(t) \tag{4.6}$$

where the $\alpha_i(t)$ are some arbitrary scalar functions. Let us introduce the $m \times k$ matrix $e(t)$ whose columns are the eigenvectors $e_i(t)$, $(i = 1, \ldots, k)$ and the k-dimensional vector function $\alpha(t)$ with elements $\alpha_i(t)$. Then (4.6) can be written in the more compact form

$$\bar{x}_0 = e(t)\alpha(t) \tag{4.7}$$

For $\Pi_0 x(\tau)$, we obtain the problem

$$\frac{d\Pi_0 x}{d\tau} = A(0)\Pi_0 x \qquad \tau \geq 0 \tag{4.8}$$

$$\Pi_0 x(0) = x^0 - \bar{x}_0(0) = x^0 - \sum_{i=1}^{k} \alpha_i(0)e_i(0) \tag{4.9}$$

The general solution to (4.8) can be represented in the form

$$\Pi_0 x(\tau) = \sum_{i=1}^{k} C_i e_i(0) + \sum_{i=k+1}^{m} C_i w_i(\tau)\exp\{\lambda_i(0)\tau\} \tag{4.10}$$

where C_i are arbitrary constants, and $w_i(\tau)$ are known vector functions whose elements are polynomials in τ. In particular, if $\lambda_i(0)$ $(i = k + 1, \ldots, m)$ arc simplc cigcnvalucs, thcn thc w_i arc thc τ indcpendent eigenvectors of $A(0)$ corresponding to $\lambda_i(0)$.

By virtue of condition 1^0 (2), the second sum in (4.10) approaches zero as $\tau \to \infty$. Let us require that the entire boundary function $\Pi_i x(\tau)$ approaches zero as $\tau \to \infty$

$$\Pi_i x(\infty) = 0$$

For this condition to hold for $\Pi_0 x$, we need to take $C_i = 0$, $i = 1, \ldots, k$. Thus,

$$\Pi_0 x(\tau) = \sum_{i=k+1}^{m} C_i w_i(\tau)\exp\{\lambda_i(0)\tau\} \tag{4.11}$$

Substituting this expression into the initial conditions (4.9), we obtain

$$\sum_{i=1}^{k} \alpha_i(0)e_i(0) + \sum_{i=k+1}^{m} C_i w_i(0) = x^0 \tag{4.12}$$

This system is of m linear algebraic equations for the m unknowns $\alpha_i(0)$ $(i = 1, \ldots, k)$ and C_i $(i = k+1, \ldots, m)$. It is known (see, e.g., [7]) that the vectors $e_i(0)$ $(i = 1, \ldots, k)$ and $w_i(0)$ $(i = k+1, \ldots, m)$ are linearly independent. Therefore the system (4.12) has a unique solution.

Thus, the function $\Pi_0 x(\tau)$ is completely determined by formula (4.11) and by virtue of the condition 1^0 (2), it satisfies the exponential estimate

$$\|\Pi_0 x(\tau)\| \leq C \exp\{-k\tau\} \tag{4.13}$$

For the as yet unknown functions $\alpha_i(t)$ in the expression (4.6) for $\bar{x}_0$, we obtained the initial values $\alpha_i(0)$. Let us introduce the notation: $\alpha_i(0) = \alpha_i^0$, $\alpha(0) = \alpha^0$.

The functions $\alpha_i(t)$ are defined completely in the next step of the algorithm, that is, during the solution of the equation for $\bar{x}_1(t)$:

$$A(t)\bar{x}_1 = -f(\bar{x}_0(t), t, 0) + \frac{d\bar{x}_0}{dt} = -f(e(t)\alpha(t), t, 0) + \frac{d}{dt}[e(t)\alpha(t)] \tag{4.14}$$

Since $\det A(t) = 0$, this system is solvable if and only if the right-hand side of (4.14) is orthogonal to each of the eigenvectors $g_j(0)$ $(j = 1, \ldots, k)$ of the adjoint matrix $A^*(t)$ corresponding to the possibly multiple eigenvalue $\lambda = 0$. Let us introduce the $k \times m$ matrix $g(t)$ whose rows are the vectors $g_j(t)$. The orthogonality condition can be written as

$$g(t)\left[-f(e(t)\alpha(t), t, 0) + e(t)\frac{d\alpha}{dt} + \frac{de}{dt}(t)\alpha(t)\right] = 0 \tag{4.15}$$

The eigenvectors $g_j(t)$ can be chosen in such a way that $g(t)e(t) = I_k$, where I_k is the $k \times k$ identity matrix. Therefore, from (4.15), we obtain

$$\frac{d\alpha}{dt} = F(\alpha, t) \tag{4.16}$$

The explicit form of F is clear from the comparison of (4.15) and (4.16).

Condition 3^0. *Let Eq. (4.16) with the initial condition* $\alpha(0) = \alpha^0$ *have a solution* $\alpha(t)$ *for* $0 \leq t \leq T$.

Now $\bar{x}_0(t)$ is completely determined, and the solution of (4.14) can be written in the form

$$\bar{x}_1 = \tilde{x}_1 + \sum_{i=1}^{k} \beta_i(t)e_i(t) = \tilde{x}_1 + e(t)\beta(t)$$

where $\tilde{x}_1(t)$ is a known function (a particular solution of (4.14)), but $\beta(t)$ remains an arbitrary k-dimensional vector function.

For $\Pi_1 x(\tau)$, we obtain the problem

$$\frac{d\Pi_1 x}{d\tau} = A(0)\Pi_1 x + \varphi_1(\tau) \qquad \tau \geq 0 \tag{4.17}$$

$$\Pi_1 x(0) = -\bar{x}_1(0) = -\tilde{x}_1(0) - \sum_{i=1}^{k} \beta_i(0)e_i(0) \tag{4.18}$$

where $\varphi_1(\tau) = A'(0)\tau\Pi_0 x(\tau) + [f(\bar{x}_0(0) + \Pi_0 x(\tau), 0, 0) - f(\bar{x}_0(0), 0, 0)]$. Evidently, $\varphi_1(\tau)$ satisfies an estimate of type (4.13).

The general solution of (4.17) can be written in the form

$$\Pi_1 x = \tilde{\Pi}_1 x(\tau) + \sum_{i=1}^{k} d_i e_i(0) + \sum_{i=k+1}^{m} d_i w_i(\tau) \exp\{\lambda_i(0)\tau\}$$

Here $\tilde{\Pi}_1 x(\tau)$ is a known function [a particular solution of (4.17)], which can be chosen in such a way that it satisfies an estimate of type (4.13), d_i are arbitrary constants, and $e_i(0)$ and $w_i(\tau)$ are the same vectors as in (4.10).

From the condition $\Pi_1 x(\infty) = 0$ we obtain $d_i = 0$ $(i = 1, \ldots, k)$. Next, substituting the expression for $\Pi_1 x(\tau)$ into the initial condition (4.18), we arrive at the linear algebraic system

$$\sum_{i=k} \beta_i(0)e_i(0) + \sum_{i=k+1}^{m} d_i w_i(0) = -\tilde{x}_1(0) - \tilde{\Pi}_1 x(0)$$

We can uniquely define $\beta_i(0)$ $(i = 1, \ldots, k)$ and d_i $(i = k+1, \ldots, m)$ from this system. Thus, $\Pi_1 x(\tau)$ will be completely determined, and for the unspecified $\beta(t)$ the initial condition will be defined. This function is completely determined in the next step of the construction of the asymptotics during the solution of the equation for $\bar{x}_2(t)$. The solvability condition for this equation provides a linear differential equation for $\beta(t)$

$$\frac{d\beta}{dt} = B(t)\beta + g_1(t) \tag{4.19}$$

where $B(t) = F_\alpha(\alpha(t), t)$ is a known matrix, and $g_1(t)$ is a known vector function. Solving this equation with the initial condition obtained above, determines $\beta(t)$, and thus the function $\bar{x}_1(t)$ is completely defined.

Construction of higher order terms of the series (4.3) can be done analogously. The main result about the series (4.3) can be formulated in the form:

Under conditions 1^0–3^0 *for sufficiently small* μ, *the problem (4.1), (4.2) has a unique solution* $x(t, \mu)$, *and the series (4.3) is the asymptotic series for this solution in the interval* $0 \le t \le T$, *that is, for any* n, *the following relation holds:*

$$\max_{[0,T]} \|x(t, \mu) - X_n(t, \mu)\| = O(\mu^{n+1}) \tag{4.20}$$

where $X_n(t, \mu)$ *is the n-th partial sum of the series (4.3).*

The proof of this statement is standard (see [8]).

B. Other Problems in the Critical Case

1. The nonlinear problem. Consider the problem (x is m-dimensional vector function)

$$\begin{aligned} \mu \frac{dx}{dt} &= F(x, t, \mu) \\ x(0, \mu) &= x^0 \end{aligned} \tag{4.21}$$

Suppose that the reduced equation $F(x, t, 0) = 0$ has for each t in $[0, T]$ a family of solutions

$$x = \varphi(t, \alpha_1, \ldots, \alpha_n) = \varphi(t, \alpha) \tag{4.22}$$

Let the rank of the matrix $(\partial\varphi/\partial\alpha)(t, \alpha)$ be equal to k.

By differentiating the identity $F(\varphi(t, \alpha), t, 0) \equiv 0$ with respect to α we obtain $F_x(\varphi(t, \alpha), t, 0)\varphi_\alpha(t, \alpha) \equiv 0$, which implies that the matrix $F_x(\varphi(t, \alpha), t, 0)$ has the eigenvalue $\lambda(t, \alpha) \equiv 0$ and that the columns of the matrix φ_α are eigenvectors corresponding to $\lambda \equiv 0$.

Suppose that the multiplicity of the eigenvalue $\lambda \equiv 0$ is exactly equal to k and that the remaining eigenvalues $\lambda_i(t, \alpha)$ of the matrix $F_x(\varphi(t, \alpha), t, 0)$ satisfy the conditions

$$Re\, \lambda_i(t, \alpha) < 0 \qquad i = 1, \ldots, m - k \tag{4.23}$$

Under these conditions, we can develop an asymptotic algorithm generalizing that in Section IV.A. We will not do it here as it is done in detail in [8].

Some applications immediately demonstrating this generalization will be considered below, in Section IV.C.

2. The boundary value problem. It may happen that the remaining eigenvalues $\lambda_i(t, \alpha)$ of the matrix $F_x(\varphi(t, \alpha), t, 0)$ satisfy the conditions

$$Re\ \lambda_i(t, \alpha) < 0 \quad i = 1, \dots, p \qquad Re\ \lambda_i(t, \alpha) > 0 \quad i = p+1, \dots, m-k$$

Then we have the *critical conditionally stable* case. In this case, we must impose the following boundary conditions: $p + q$ $(q \leq k)$ conditions at $t = 0$, and $m - p - q$ conditions at $t = 1$.

The theory of such a problem is developed in [8] for some special cases

$$\begin{aligned} \mu \frac{dz}{dt} &= A(u, t)y + \mu B(u, t) \\ \mu \frac{dy}{dt} &= z \\ \mu \frac{du}{dt} &= C(u, t)y + \mu D(u, t) \qquad 0 \leq t \leq 1 \end{aligned} \tag{4.24}$$

where z and y are scalar functions, and u is a k-dimensional vector function. We prescribe the following boundary conditions for (4.24).

$$z(0, t) = z^0 \qquad z(1, \mu) = z^1 \qquad u(0, \mu) = u^0 \tag{4.25}$$

Suppose that $A(u, t) > 0$ in some domain $G = \{(\|u\| < d) \times (0 \leq t \leq 1)\}$. It is clear that the reduced system

$$A(u, t)y = 0 \qquad z = 0 \qquad C(u, t)y = 0$$

has the family of solutions

$$\bar{z} = 0 \qquad \bar{y} = 0 \qquad \bar{u} = \alpha$$

where α is an arbitrary k-dimensional vector. Consider the matrix F_x.

$$F_x = \begin{pmatrix} 0 & A(\alpha, t) & 0 \\ 1 & 0 & 0 \\ 0 & C(\alpha, t) & 0 \end{pmatrix}$$

It is easy to see that F_x has $\lambda = 0$ as an eigenvalue of multiplicity k as well as two eigenvalues $\lambda_{1,2}(\alpha, t) = \pm\sqrt{A(\alpha, t)}$. Thus we have indeed the critical conditionally stable case.

Problems (4.24), (4.25) may occur in the theory of semiconductor devices (see [5, 8]). The problem in Example 3.3 belongs to the critical case too. There we reduced it to the noncritical case, using the integral $z + m = ax + b$.

C. The Equations of Chemical Kinetics

Let us discuss the statement of chemical kinetics problems in general. Consider n chemical reactions involving m substances, namely,

$$a_{i1}X_1 + \cdots + a_{im}X_m \underset{k_i^-}{\overset{k_i^+}{\rightleftharpoons}} b_{i1} + \cdots + b_{im}X_m \qquad i = 1, \ldots, n$$

Here X_i denotes the ith substance; a_{ik} are integers denoting the number of molecules of the kth substance, which participate in the ith forward and reverse reactions, respectively; k_i^+ and k_i^- are the rate constants of the reactions.

If we denote the concentration of substance X_j by x_j, then the changes in x_j determined by the ith reactions with rate constants k_i^+ and k_i^- (during time dt) are given, respectively, by

$$dx_j = -k_i^+ x_1^{a_{i1}} \cdots x_m^{a_{im}}(a_{ij} - b_{ij})\, dt$$

for the forward reaction, and

$$dx_j = k_i^- x_1^{b_{i1}} \cdots x_m^{b_{im}}(a_{ij} - b_{ij})\, dt$$

for the reverse reaction. Consequently, the total change in x_j is equal to

$$dx_j = \sum_{i=1}^{n} \gamma_{ij} w_i \, dt$$

where

$$w_i = k_i^+ x_1^{a_{i1}} \cdots x_m^{a_{im}} - k_i^- x_1^{b_{i1}} \cdots y_m^{b_{im}}$$

and $\gamma_{ij} = b_{ij} - a_{ij}$. Thus we obtain the system of differential equations

$$\frac{dx_j}{dt} = \sum_{i=1}^{n} \gamma_{ij} w_i \qquad j = 1, \ldots, m \tag{4.26}$$

Under actual conditions, the rate constants of the various reactions differ from each other. Large rate constants correspond to fast reactions. This property can be expressed by means of a small parameter μ.

Suppose that the first s reactions are fast, so

$$k_i^{\pm} = \frac{1}{\mu} \hat{k}_i^{\pm} \qquad i = 1, \ldots, s$$

with $\hat{k}_i^{\pm}$ moderate in size. Let us introduce the notation

$$\hat{w}_i = \mu w_i = \hat{k}_i^{+} x_1^{a_{i1}} \cdots x_m^{a_{im}} - \hat{k}_i^{-} x^{b_{i1}} \cdots x_m^{b_{im}} \qquad i = 1, \ldots, s$$

Now the system (4.26) can be rewritten as

$$\mu \frac{dx_j}{dt} = \sum_{i=1}^{s} \gamma_{ij} \hat{w}_i + \mu \sum_{i=s+1}^{n} \gamma_{ij} w_i \qquad j = 1, \ldots, m \tag{4.27}$$

Setting $\mu = 0$, we obtain the reduced system

$$0 = \sum_{i=1}^{s} \gamma_{ij} \hat{w}_i \qquad j = 1, \ldots, m \tag{4.28}$$

In practice, it often turns out that the system (4.28) has a family of solutions that depends on one or more arbitrary parameters. Thus, the problem for the singularly perturbed equation (4.27) is the nonlinear critical case.

The well-known method of quasistationary concentration of Semenov–Bodenstein is, from the mathematical point of view, a method from singular perturbation theory (see, e.g., [9]).

Example 4.1. Let us now discuss an example of an actual chemical reaction. The system

$$\frac{dx_1}{dt} = -k_1^+ x_1 + k_1^- x_2$$

$$\frac{dx_2}{dt} = +k_1^+ x_1 - k_1^- x_2 - k_2^+ x_2 x_3 - k_3^+ x_2 x_4$$

$$\frac{dx_3}{dt} = -k_2^+ x_2 x_3$$

$$\frac{dx_4}{dt} = -k_3^+ x_2 x_4$$

with conditions

$$x_i(0) = x_i^0 \qquad i = 1, \ldots, 4$$

occurs in studying the reaction kinetics of organometallic compounds [8]. The rate constants have the order of magnitude

$$k_1^+ \sim 10 \qquad k_1^- \sim 10^9 \qquad k_2^+ \sim 10^{10} \qquad k_3^+ \sim 10^8 \qquad k_2^- = k_3^- = 0$$

By dividing each equation by k_3^+ and making the substitutions

$$\mu = \frac{1}{k_3^+} \qquad A = k_1^+ \qquad B = \frac{k_1^-}{k_3^+} \qquad C = \frac{k_2^+}{k_3^+}$$

we obtain

$$\begin{aligned} \mu \frac{dx_1}{dt} &= -\mu A x_1 + B x_2 \\ \mu \frac{dx_2}{dt} &= +\mu A x_1 - B x_2 - C x_2 x_3 - x_2 x_4 \\ \mu \frac{dx_3}{dt} &= -C x_2 x_3 \\ \mu \frac{dx_4}{dt} &= -x_2 x_4 \end{aligned} \tag{4.29}$$

The reduced system

$$0 = Bx_2 \qquad 0 = -Bx_2 - Cx_2x_3 - x_2x_4 \qquad 0 = -Cx_2x_3 \qquad 0 = -x_2x_4$$

has the family of solutions [see (4.22)]

$$x_1 = \alpha_1 \qquad x_2 = 0 \qquad x_3 = \alpha_2 \qquad x_4 = \alpha_3$$

The matrix $F_x(\varphi(t, \alpha), t, 0)$ is

$$\begin{pmatrix} 0 & B & 0 & 0 \\ 0 & -B - C\alpha_2 - \alpha_3 & 0 & 0 \\ 0 & -C\alpha_2 & 0 & 0 \\ 0 & -\alpha_3 & 0 & 0 \end{pmatrix}$$

It has the eigenvalues $\lambda_1 = \lambda_2 = \lambda_3 = 0$; $\lambda_4 = -B - C\alpha_2 - \alpha_3 < 0$ (because $B > 0$, $C > 0$, and α_2, α_3 are nonnegative).

Analogous to (4.6) we have

$$\bar{x}_{10} = \alpha_1(t) \qquad \bar{x}_{20} = 0 \qquad \bar{x}_{30} = \alpha_2(t) \qquad \bar{x}_{40} = \alpha_3(t)$$

For the Π functions of zeroth order we obtain (according to the general rule in Section II.B)

$$
\begin{aligned}
\frac{d\Pi_0 x_1}{d\tau} &= +B\Pi_0 x_2 \\
\frac{d\Pi_0 x_2}{d\tau} &= -B\Pi_0 x_2 - C\Pi_0 x_2[\alpha_2(0) + \Pi_0 x_3] - \Pi_0 x_2[\alpha_3(0) + \Pi_0 x_4] \\
\frac{d\Pi_0 x_3}{d\tau} &= -C\Pi_0 x_2[\alpha_2(0) + \Pi_0 x_3] \\
\frac{d\Pi_0 x_4}{d\tau} &= -\Pi_0 x_2[\alpha_3(0) + \Pi_0 x_4]
\end{aligned}
\tag{4.30}
$$

$$
\begin{gathered}
\Pi_0 x_1(0) = x_1^0 - \alpha_1(0) \qquad \Pi_0 x_2(0) = x_2^0 \qquad \Pi_0 x_3(0) = x_3^0 - \alpha_2(0) \\
\Pi_0 x_4(0) = x_4^0 - \alpha_3(0) \qquad \Pi_0 x_i(\infty) = 0 \ \ (i = 1, 2, 3, 4)
\end{gathered}
\tag{4.31}
$$

The system (4.30) is sufficiently simple and can be integrated, if $\Pi_0 x_1$ is chosen as an independent variable. From the first and third equations we have

$$\alpha_2(0) + \Pi_0 x_3 = C_1 \exp\left\{-\frac{C}{B}\Pi_0 x_1\right\}$$

The first and fourth equations give

$$\alpha_3(0) + \Pi_0 x_4 = C_2 \exp\left\{-\frac{1}{B}\Pi_0 x_1\right\}$$

Finally, the first and second equations yield

$$\Pi_0 x_2 = -\Pi_0 x_1 + C_1 \exp\left\{-\frac{C}{B}\Pi_0 x_1\right\} + C_2 \exp\left\{-\frac{1}{B}\Pi_0 x_1\right\} + C_3$$

Here C_i $(i = 1, 2, 3)$ are constants of integration. The resulting three expressions are analogous to (4.10). From the conditions (4.31) at $\tau = 0$ we obtain

$$
\begin{aligned}
\Pi_0 x_2 &= -\Pi_0 x_1 + \left[\exp\left\{-\frac{C}{B}\Pi_0 x_1\right\} - 1\right]\alpha_2(0) \\
&\quad + \left[\exp\left\{-\frac{1}{B}\Pi_0 x_1\right\} - 1\right]\alpha_3(0) \\
\Pi_0 x_3 &= \left[\exp\left\{-\frac{C}{B}\Pi_0 x_1\right\} - 1\right]\alpha_2(0)
\end{aligned}
\tag{4.32}
$$

$$\Pi_0 x_4 = \left[\exp\left\{ -\frac{1}{B} \Pi_0 x_1 \right\} - 1 \right] \alpha_3(0)$$

Substituting (4.32) into (4.31), we obtain the following algebraic system defining $\alpha_i(0)$ $(i = 1, 2, 3)$ and $\Pi_0 x_1(0)$

$$\begin{aligned} x_1^0 &= \alpha_1(0) + \Pi_0 x_1(0) \\ x_2^0 &= -\Pi_0 x_1(0) + \left[\exp\left\{ -\frac{C}{B} \Pi_0 x_1 \right\} - 1 \right] \alpha_2(0) \\ &\quad + \left[\exp\left\{ -\frac{\Pi_0 x_1(0)}{B} \right\} - 1 \right] \alpha_3(0) \\ x_3^0 &= \alpha_2(0) + \left[\exp\left\{ -\frac{C}{B} \Pi_0 x_1(0) \right\} - 1 \right] \alpha_2(0) \\ x_4^0 &= \alpha_3(0) + \left[\exp\left\{ -\frac{\Pi_0 x_1(0)}{B} \right\} - 1 \right] \alpha_3(0) \end{aligned} \tag{4.33}$$

This system is solvable (see [8]).

One can also write equations for $\alpha_1(t)$, $\alpha_2(t)$, and $\alpha_3(t)$. As in Section IV.A, these equations can be derived as solvability conditions for the regular terms of first order. In the present case, the solution of the equations for the α_i is a matter of integrating by quadratures. In this way, one can determine $\bar{x}_{10}(t) = \alpha_1(t)$, $\bar{x}_{20}(t) = 0$, $\bar{x}_{30} = \alpha_2(t)$, and $\bar{x}_{40} = \alpha_4(t)$.

The determination of $\Pi_0 x_i$ $(i = 1, 2, 3, 4)$ reduces to the integration of the scalar equation

$$\begin{aligned} \frac{d\Pi_0 x_1}{d\tau} &= B\left[-\Pi_0 x_1 + \left(\exp\left\{ -\frac{C}{B} \Pi_0 x_1 \right\} - 1 \right) \alpha_2(0) \right. \\ &\quad \left. + \left(\exp\left\{ -\frac{\Pi_0 x_1}{B} \right\} - 1 \right) \alpha_3(0) \right] \\ &\qquad \Pi_0 x_1(0) = x_1^0 - \alpha_1(0) \end{aligned}$$

which can be obtained by substituting $\Pi_0 x_2$ from (4.32) into the first Eq. (4.30). After determining $\Pi_0 x_1(\tau)$, the remaining $\Pi_0 x_1$ $(i = 2, 3, 4)$ are found by means of the Eqs. (4.32). Higher order terms of the asymptotic expansion can be constructed as well (see [8]).

V. CONTRAST STRUCTURES

A. Introduction

Let us now return to the system (3.1). Suppose that equation $F(z, y, t) = 0$ has several solutions (roots) $z = \varphi_i(y, t)$. Previously, we considered the case when only one such root was used in the construction of the asymptotic expansion. But some other types of asymptotic behavior, when several such solutions are used in the asymptotic algorithm, are also possible. In these cases, we can obtain a solution with transitions from one root to another or solutions with, so-called, *interior layers* (internal layers).

Let us demonstrate this phenomenon with an example of a scalar second-order equation

$$\mu^2 \frac{d^2 z}{dt^2} = F(z) \qquad 0 \le t \le 1 \tag{5.1a}$$

which is equivalent to the system

$$\mu \frac{dz_1}{dt} = F(z_2) \qquad \mu \frac{dz_2}{dt} = z_1$$

Take

$$z_2(0, \mu) = 0 \qquad z_2(1, \mu) = 0$$

Introducing $\tau = t/\mu$ we obtain

$$\frac{dz_1}{d\tau} = F(z_2) \qquad \frac{dz_2}{d\tau} = z_1 \tag{5.1}$$

Assume that the function $F(z_2)$ has three simple roots $z_2 = \varphi_i$ $(i = 1, 2, 3)$, where $\varphi_1 < \varphi_2 < \varphi_3$ and, moreover, that $F_{z_2}(\varphi_i) > 0$, $i = 1, 3$; $F_{z_2}(\varphi_2) < 0$ (Fig. 6). Let us introduce the phase plane. Since $z_1\, dz_1 = F(z_2)\, dz_2$,

$$\frac{z_1^2}{2} = \int_{\varphi_2}^{z_2} F(z)\, dz + C \equiv \Phi(z_2) + C$$

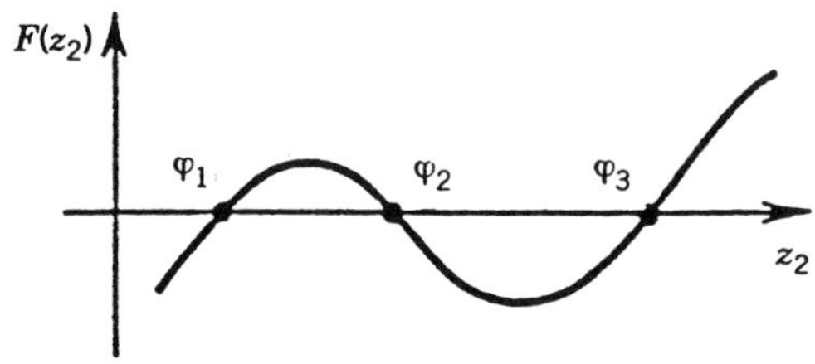

Figure 6. The graph of the function $F(z_2)$ with three zero points $\varphi_1, \varphi_2, \varphi_3$.

and hence,

$$z_1 = \pm\sqrt{2}\sqrt{\Phi(z_2) + C} \tag{5.2}$$

where C is a constant. Depending on the function $\Phi(z_2)$, the family of phase curves (5.2) looks different. One can distinguish three cases. The graphs of $\Phi(z_2)$ and the corresponding families of phase trajectories for each of these cases are presented in Fig. 7. These cases are

$$\begin{aligned} &\text{(a)} \qquad \int_{\varphi_3}^{\varphi_2} F(z)\,dz > \int_{\varphi_1}^{\varphi_2} F(z)\,dz \\ &\text{(b)} \qquad \int_{\varphi_1}^{\varphi_2} F(z)\,dz < \int_{\varphi_3}^{\varphi_2} F(z)\,dz \\ &\text{(c)} \qquad \int_{\varphi_1}^{\varphi_2} F(z)\,dz = \int_{\varphi_3}^{\varphi_2} F(z)\,dz \end{aligned} \tag{5.3}$$

In the last case, the saddle points A_1 and A_2 are connected by separatrices. A so-called *cell* occurs.

The interior of the loop in Fig. 7(*a* and *b*) and the interior of the cell in Fig. 7(*c*) is filled by closed trajectories. On the plane z, t the corresponding solutions are represented graphically (Figs. 8 and 9) (we assume $\varphi_1 < \varphi_2 < 0 < \varphi_3$). There might also exist solutions corresponding to multiple passages along the closed trajectory (Figs. 8 and 9, dashed curves).

Thus we see that these solutions have not only boundary layers, but also *interior layers*. Solutions having such interior layers are called *contrast structures*. A contrast structure of the type represented in Fig. 8 is called a *spike*, and a contrast structure of the type represented in Fig. 9 is called a *step* (or threshold).

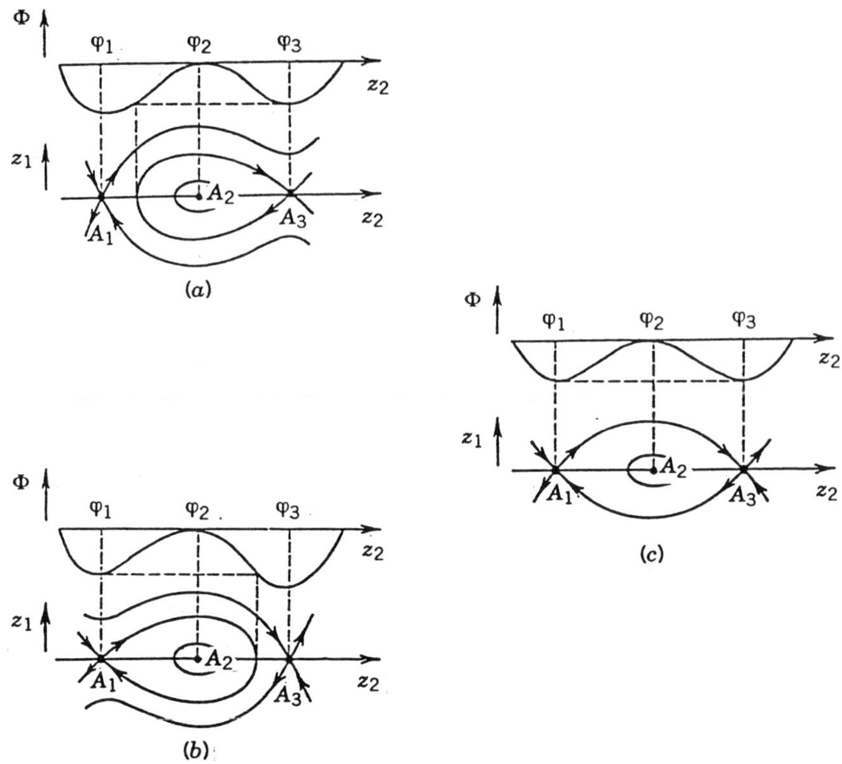

Figure 7. The graphs of the function $\Phi(z_2)$ and corresponding family of phase curves $z_1 = \pm\sqrt{2[\Phi(z_2) + C]}$ for three cases: (*a*) $\Phi(\varphi_1) > \Phi(\varphi_3)$, (*b*) $\Phi(\varphi_1) < \Phi(\varphi_3)$, (*c*) $\Phi(\varphi_1) = \Phi(\varphi_3)$.

Our goal is to investigate contrast structures for more general equations than (5.1), nonautonomous, in particular.

B. Contrast Structures of Step-Type

1. A Second-Order Equation

Consider the problem

$$\mu^2 \frac{d^2 z}{dt^2} = F(z, t) \qquad 0 \le t \le 1$$
$$z(0, \mu) = z(1, \mu) = 0 \tag{5.4a}$$

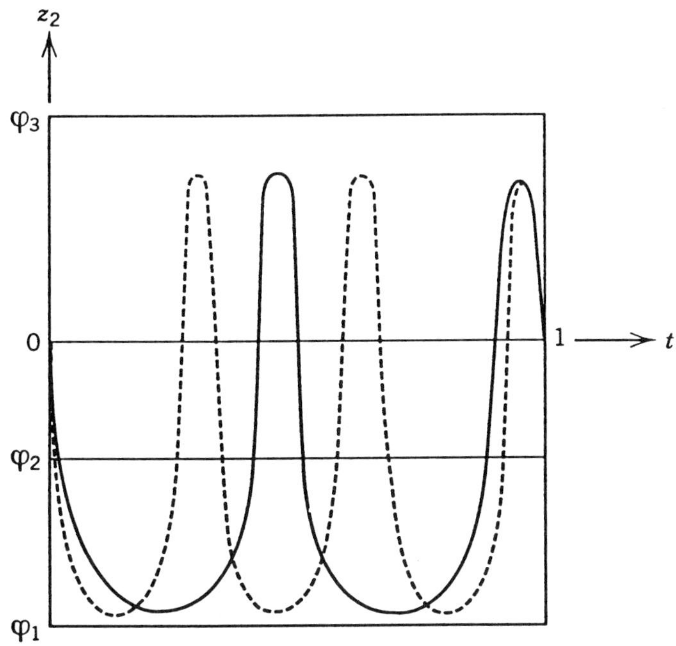

Figure 8. The graphs of the solutions corresponding to the closed phase trajectories filling the interior of the loop in Figure 7(*b*).

which is equivalent to the system

$$\mu \frac{dz_1}{dt} = F(z_2, t)$$

$$\mu \frac{dz_2}{dt} = z_1 \qquad 0 \leq t \leq 1 \tag{5.4b}$$

$$z_2(0, \mu) = z_2(1, \mu) = 0$$

Condition 1^0. *Let the equation* $F(z_2, x) = 0$ *have three simple roots* $z_2 = \varphi_i(t)$ *(i = 1, 2, 3) such that*

1. $\varphi_1(t) < \varphi_2(t) < \varphi_3(t) \quad \text{for} \quad 0 \leq t \leq 1$

2. $F_{z_2}(\varphi_i(t), t) > 0 \quad \text{for} \quad i = 1, 3 \qquad 0 \leq t \leq 1$

and $\qquad F_{z_2}(\varphi_2(t), t) < 0 \quad \text{for} \quad 0 \leq t \leq 1$

It follows from this condition that for each fixed $t \in [0, 1]$ the graph of

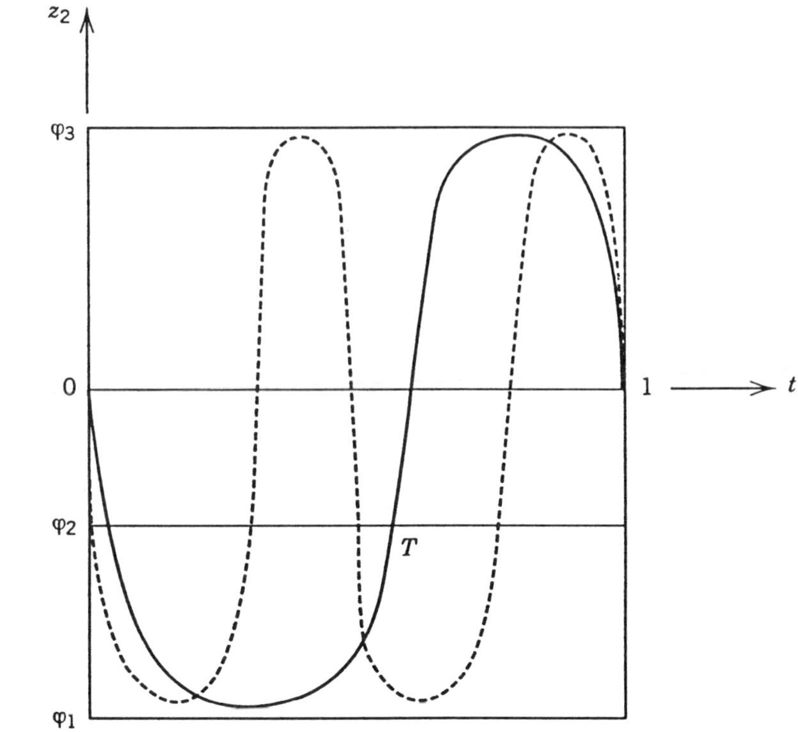

Figure 9. The graphs of the solutions corresponding to the closed phase trajectories filling the interior of the cell in Figure 7(c).

the function $F(z_2, t)$ has the form shown in Fig. 6. As t changes, this graph may deform. It turns out that the phenomenon of a transition from root $\varphi_1(t)$ to root $\varphi_3(t)$ [or from $\varphi_3(t)$ to $\varphi_1(t)$; both of these roots are conditionally stable since the corresponding matrix has eigenvalues $\lambda^i_{1,2} = \pm\sqrt{F_{z_2}(\varphi_i(t), t)}$, $i = 1, 3$] for the solution of (5.4) can take place near some value $t = T$, where the cell [Fig. 7(c)] occurs on the phase plane of the system ($T = \text{const}$)

$$\mu \frac{dz_1}{dt} = F(z_2, T)$$
$$\mu \frac{dz_2}{dt} = z_1 \tag{5.5a}$$

or

$$\frac{dz_1}{d\tau} = F(z_2, T)$$
$$\frac{dz_2}{d\tau} = z_1 \qquad \tau = \frac{t - T}{\mu} \tag{5.5b}$$

One can come to this conclusion using the asymptotic formulas of Section III.A. Let us show that there exists a solution of the type sketched in Fig. 10. We will call the value $t = T(\mu)$, for which $z_2(T(\mu)) = \varphi_2(T(\mu))$, the *point of transition*.

We approach this problem as follows. Assuming that $T(\mu)$ is some, as yet unknown, value between $t = 0$ and $t = 1$, we construct the asymptotic expansion of the solution of (5.4) in the interval $[0, T(\mu)]$, satisfying the condition $z_2(0, \mu) = 0$, $z_2(T(\mu), \mu) = \varphi_2(T(\mu))$ and approaching $\{0, \varphi_3(t)\}$ as $\tau \to 0$. In other words, we construct the asymptotic solution defined in subinterval $[0, T(\mu)]$, or the left part of the solution (see Fig. 10). The algorithm of Section III.A.1 is used for this purpose. We note that at the end points $y = 0$ and $t = T$ the same component z_2 is prescribed (see the remark before statement (3.19)).

Similarly, we construct the asymptotic solution of (5.4) in the interval $[T(\mu), 1]$, satisfying the conditions $z_2(T(\mu), \mu) = \varphi_2(T(\mu))$, $z_2(1, \mu) = 0$

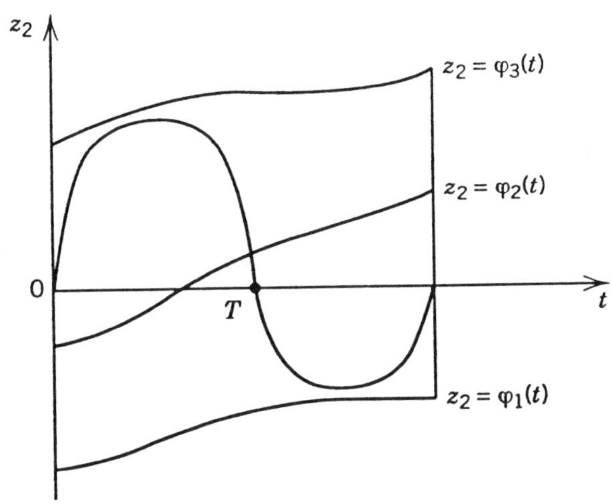

Figure 10. The graph of the solution of step-type with the point of transition $T(\mu)$.

and approaching $\{0, \varphi_1(t)\}$ as $\mu \to 0$, that is, we construct the right part of the asymptotic solution (see Fig. 10).

Equating the values of the expressions for z_1, corresponding to the left and right limits at the point $T(\mu)$, determines $T(\mu)$. Notice that $T(\mu)$ should be sought in the form $T(\mu) = T_0 + \mu T_1 + \mu^2 T_2 + \cdots$. We restrict ourselves to the construction of only the zeroth-order approximation. Therefore, we may omit the subindex 0 in T_0 (below $T_0 = T$).

The asymptotic solution in the interval $[0, T]$ is expressed through the following formulas with accuracy of the order μ (the index ℓ above the boundary functions shows that they are defined on the left subinterval; the index r will be used for the boundary functions defined on the right subinterval):

$$
\begin{aligned}
z_1(t, \mu) &= \overset{(\ell)}{\Pi}_0 z_1(\tau_0) + \overset{(\ell)}{Q}_0 z_1(\tau_T) + O(\mu) \\
z_2(t, \mu) &= \varphi_3(t) + \overset{(\ell)}{\Pi}_0 z_2(\tau_0) + \overset{(\ell)}{Q}_0 z_2(\tau_T) + O(\mu)
\end{aligned}
\tag{5.6}
$$

where $\tau_0 = (t/\mu) \geq 0$, $\tau_T = (t - T)/\mu \leq 0$, and $\overset{(\ell)}{\Pi}_0 z_i$ and $\overset{(\ell)}{Q}_0 z_i$ are defined as the solutions of the problems

$$
\frac{d \overset{(\ell)}{\Pi}_0 z_1}{d\tau_0} = F\left(\varphi_3(0) + \overset{(\ell)}{\Pi}_0 z_2, 0\right) \qquad \frac{d \overset{(\ell)}{\Pi}_0 z_2}{d\tau_0} = \overset{(\ell)}{\Pi}_0 z_1 \tag{5.7a}
$$

$$
\overset{(\ell)}{\Pi}_0 z_2(0) = -\varphi_3(0) \qquad \overset{(\ell)}{\Pi}_0 z_i(\infty) = 0 \qquad i = 1, 2 \tag{5.7b}
$$

$$
\frac{d \overset{(\ell)}{Q}_0 z_1}{d\tau_T} = F\left(\varphi_3(T) + \overset{(\ell)}{Q}_0 z_2, T\right) \qquad \frac{d \overset{(\ell)}{Q}_0 z_2}{d\tau_T} = \overset{(\ell)}{Q}_0 z_1 \tag{5.8a}
$$

$$
\overset{(\ell)}{Q}_0 z_2(0) = \varphi_2(T) - \varphi_3(T) \qquad \overset{(\ell)}{Q}_0 z_i(-\infty) = 0 \qquad i = 1, 2 \tag{5.8b}
$$

Analogously, the asymptotic solution of (5.4) in the subinterval $[T, 1]$ has the form

$$
\begin{aligned}
z_1(t, \mu) &= \overset{(r)}{\Pi}_0 z_1(\tau_T) + \overset{(r)}{Q}_0 z_1(\tau_1) + O(\mu) \\
z_2(t, \mu) &= \varphi_1(t) + \overset{(r)}{\Pi}_0 z_2(\tau_T) + \overset{(r)}{Q}_0 z_2(\tau_1) + O(\mu)
\end{aligned}
\tag{5.9}
$$

where $\tau_T = (t - T)/\mu \geq 0$, $\tau_1 = (t-1)/\mu \leq 0$ and $\overset{(r)}{\Pi}_0 z_i$ and $\overset{(r)}{Q}_0 z_1$ are defined as the solutions of the problems

$$\frac{d\overset{(r)}{\Pi}_0 z_1}{d\tau_T} = F\left(\varphi_1(T) + \overset{(r)}{\Pi}_0 z_2, T\right) \qquad \frac{d\overset{(r)}{\Pi}_0 z_2}{d\tau_T} = \overset{(r)}{\Pi}_0 z_1 \tag{5.10a}$$

$$\overset{(r)}{\Pi}_0 z_2(0) = \varphi_2(T) - \varphi_1(T) \qquad \overset{(r)}{\Pi}_0 z_i(\infty) = 0 \qquad i = 1, 2 \tag{5.10b}$$

$$\frac{d\overset{(r)}{Q}_0 z_1}{d\tau_1} = F\left(\varphi_1(1) + \overset{(r)}{Q}_0 z_2, 1\right) \qquad \frac{d\overset{(r)}{Q}_0 z_2}{d\tau_1} = \overset{(r)}{Q}_0 z_1 \tag{5.11a}$$

$$\overset{(r)}{Q}_0 z_2(0) = -\varphi_1(1) \qquad \overset{(r)}{Q}_0 z_i(-\infty) = \infty \qquad i = 1, 2 \tag{5.11b}$$

Equating the expressions (5.6) and (5.9) for z_1 at $t = T$, and taking into account that at this point $\overset{(\ell)}{\Pi}_0$ and $\overset{(r)}{Q}_0 z_i$ are less than any positive power of μ, we obtain (to the zeroth order)

$$\overset{(\ell)}{Q}_0 z_1(0) = \overset{(r)}{\Pi}_0 z_1(0) \tag{5.12}$$

The above relation is the equation for T. We note that conditions (5.8b) and (5.10b) provide (to the zeroth order) the equality of the expressions (5.6) and (5.9) for z_2 at the point $t = T(\mu)$.

Let us rewrite (5.12) in a different form. We make the change of variables $\tilde{z}_1 = \overset{(\ell)}{Q}_0 z_1$, $\tilde{z}_2 = \varphi_3(T) + \overset{(\ell)}{Q}_0 z_2$. In these new variables the problem (5.8a), (5.8b) becomes

$$\begin{gathered} \frac{d\tilde{z}_1}{d\tau_T} = F(\tilde{z}_2, T) \qquad \frac{d\tilde{z}_2}{d\tau_T} = \tilde{z}_1 \qquad \tau_T \leq 0 \\ \tilde{z}_2(0) = \varphi_2(T) \qquad \tilde{z}_1(-\infty) = 0 \qquad \tilde{z}_2(-\infty) = \varphi_3(T) \end{gathered} \tag{5.13}$$

Analogously, if we make the change of variables $\hat{z}_1 = \overset{(r)}{\Pi}_0 z_1$, $\hat{z}_2 = \varphi_1(T) + \overset{(r)}{\Pi}_0 z_2$ in (5.10a), (5.10b), we obtain

$$\begin{gathered} \frac{d\hat{z}_1}{d\tau_T} = F(\hat{z}_2, T) \qquad \frac{d\hat{z}_2}{d\tau_T} = \hat{z}_1 \qquad \tau_T \geq 0 \\ \hat{z}_2(0) = \varphi_2(T) \qquad \hat{z}_1(\infty) = 0 \qquad \hat{z}_2(\infty) = \varphi_1(T) \end{gathered} \tag{5.14}$$

In the new variables, (5.12) can be written as

$$\tilde{z}_1(0) = \hat{z}_1(0) \tag{5.15}$$

Evidently, both systems of Eqs. (5.13) and (5.14) coincide (up to notation) with the system (5.5a). Assume that the corresponding phase portrait of the system is the one shown in Fig. 7(*a*). Then the point A on the separatrix entering the saddle $(0, \varphi_3(T))$ as $\tau_T \to -\infty$ corresponds to the value $\hat{z}_2(0) = \varphi_2(T)$, and the point B (which does not coincide with A) on the separatrix entering the saddle $(0, \varphi_1(T))$ as $\tau_T \to \infty$ corresponds to the value $\hat{z}_2(0) = \varphi_2(T)$ (see Fig. 11). Therefore condition (5.15) is not satisfied. Similarly, condition (5.15) will not be satisfied in the phase portrait of Fig. 7(*b*). Only the phase portrait of Fig. 7(*c*) provides the necessary equality. So T must be a value of t such that the cell is formed in the phase plane of the associated system. The hypothesis guaranteeing existence of such t will be formulated below (condition 3^0) after the transformation of (5.15) to a special form.

Further, for the existence of the asymptotic solution in each of the subintervals $[0, T]$ and $[T, 1]$, the phase portraits of the associated systems must satisfy for $t = 0$ and $t = 1$ other conditions corresponding to conditions 5^0 and $5^{0\prime}$ of Section III.A.

Condition 2^0. *In the phase plane of the system* $d\tilde{z}_1/d\tau_0 = F(\tilde{z}_2, 0)$, $d\tilde{z}_2/d\tau_0 = \tilde{z}_1$, *let the straight line* $\tilde{z}_2 = 0$ *intersect the separatrix entering the saddle* $(0, \varphi_3(0))$ *as* $\tau_0 \to \infty$.

Condition $2^{0\prime}$. *In the phase plane of the system* $d\hat{z}_1/d\tau_1 = F(\hat{z}_2, 1)$,

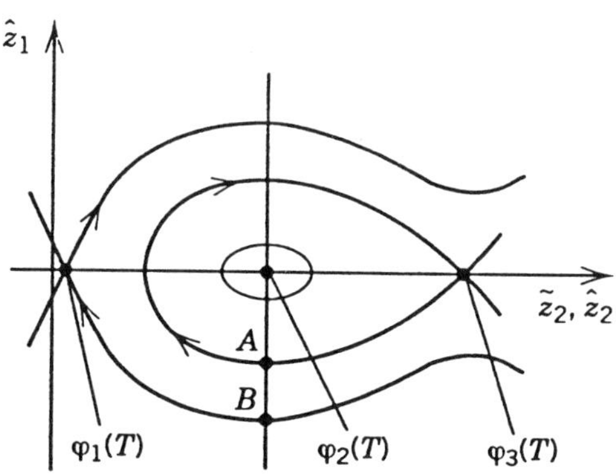

Figure 11. Case (a) in the Figure 7. The point A on the separatrix entering the saddle $(0, \varphi_3(t))$, as $\tau_T \to -\infty$ does not coincide with the point B on the separatrix entering the saddle $(0, \varphi_1(t))$ as $\tau_T \to +\infty$.

$d\hat{z}_2/d\tau_1 = \hat{z}_1$, let the straight line $\hat{z}_2 = 0$ intersect the separatrix entering the saddle $(0, \varphi_1(1))$ as $\tau_1 \to -\infty$.

Let us now rewrite (5.15) in a different form. The separatrix of the system (5.13), which enters the saddle point $(0, \varphi_3(T))$ as $\tau_T \to -\infty$, is described by the equation

$$\tilde{z}_1 = -\sqrt{2}\sqrt{\int_{\varphi_3(T)}^{\tilde{z}_2} F(z, T)\, dz} \tag{5.16a}$$

and the separatrix of the system (5.14), which enters the saddle point $[0, \varphi_1(T)]$ as $\tau_T \to \infty$, is described by

$$\hat{z}_1 = -\sqrt{2}\sqrt{\int_{\varphi_1(T)}^{\hat{z}_2} F(z, T)\, dz} \tag{5.16b}$$

Substituting these expressions into (5.15), we obtain the following equation for T

$$\int_{\varphi_3(T)}^{\varphi_2(T)} F(z, T)\, dz = \int_{\varphi_1(T)}^{\varphi_2(T)} F(z, T)\, dz \quad \text{or} \quad \int_{\varphi_3(T)}^{\varphi_1(T)} F(z, T)\, dz = 0 \tag{5.17}$$

Note that this relation coincides with condition (5.3).

Condition 3^0. *Let Eq. (5.17) have a solution $T = T_0$ with $0 < T_0 < 1$.*

To construct higher order terms of the asymptotics (as well as to prove the existence of such a solution, as shown in Fig. 10), we need one more condition (see [4] for details).

Condition 4^0. *Let*

$$\frac{d}{dT}\int_{\varphi_3(T)}^{\varphi_1(T)} F(z, T)\, dz \bigg|_{T=T_0} = \int_{\varphi_3(T)}^{\varphi_1(T)} F_t(z, T)\, dz \bigg|_{T=T_0} \neq 0$$

(so, the root T_0 of (5.18) is simple).

Under conditions 1^0–4^0, for sufficiently small μ, there exists a solution of the boundary value problem (5.4a) for which

$$\lim_{\mu\to 0} z_2(t, \mu) = \begin{cases} \varphi_3(t) & \text{for} \quad 0 < t < T_0 \\ \varphi_1(t) & \text{for} \quad T_0 < t < 1 \end{cases}$$

$$\lim_{\mu\to 0} z_1(t, \mu) = 0 \quad \text{for} \quad 0 < t < T_0 \quad T_0 < t < 1$$

Remark 1: Along with the constructed solution, there may also exist a solution for which $z_2(t, \mu)$ approaches $\varphi_1(t)$ in $(0, T_0)$ and $\varphi_3(t)$ in $(T_0, 1)$ as $\mu \to 0$.

Remark 2: If a cell in the associated system (5.5a) occurs for several values of T in the interval $(0, 1)$, then solutions with multiple transition layers may exist.

Remark 3: The solutions mentioned above may coexist with each other as well as with solutions without transition layers (such solutions were constructed in Section III.A). Thus, the solution of (5.4a) is not, in general, unique.

Example 5.1. Consider the equation of the pendulum (the parameter μ characterizes the inertia of this physical system)

$$\mu^2 \ddot{\varphi} = -a \sin \varphi + b(t)$$

The forcing power $b(t)$ is a periodic function with respect to t (its period is equal to 1). Let $b(0) = 0$, $b > 0$ $(0 < t < t_0)$, $b(t_0) = 0$, $b < 0$ $(t_0 < t < 1)$, $b(1) = 0$, $|b(t)| < a$.

The degenerate equation $-a \sin \varphi + b(t) = 0$ has three roots $\varphi_1(t) < \varphi_2(t) < \varphi_3(t)$. In the interval $(0, t_0)$: $\varphi_1(t) < -\pi$, $\varphi_2(t) > 0$, $\varphi_3(t) < \pi$, and in the interval $(t_0, 1)$: $\vartheta_1(t) > -\pi$, $\varphi_2(t) < 0$, $\varphi_3(t) > \pi$ (Fig. 12 solid line).

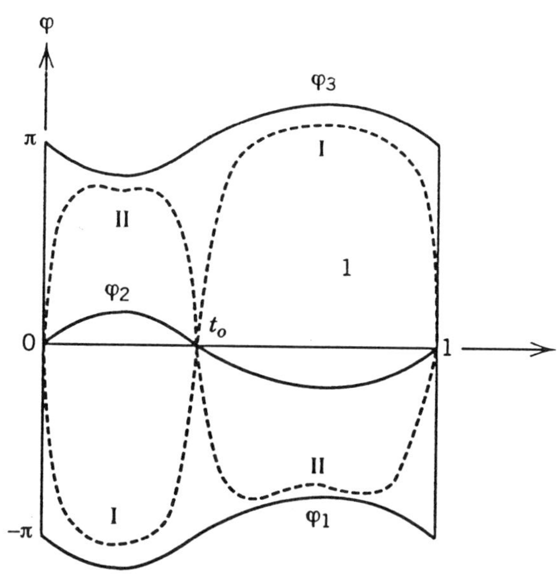

Figure 12. Two solutions I and II of step-type for the pendulum.

For $t = 0$, $t = t_0$, $t_0 = \pi$ these roots are $\varphi_1 = -\pi$, $\varphi_2 = 0$, $\varphi_3 = \pi$, and for these values of t there is a cell on the phase plane.

The conditions 1^0–4^0 are fulfilled, and there exist two solutions of step-type: I and II (Fig. 12, dashed line).

These solutions may be extended by periodicity for $t < 0$ and for $t > 0$ (in the presence of appropriate symmetry). Thus, we obtain periodic solutions having interior (transition) layers.

2. *Other Cases*

1. For some problems of type (5.4a) a phase plane cell occurs for any T. In such (critical) cases, under certain conditions, a solution of the form shown in Fig. 10 can also be constructed [4]. The equation defining T_0 will be more complicated than (5.17) ((5.17) will be satisfied identically in such cases):

$$\int_{\varphi_2(T)}^{\varphi_i(T)} \frac{\int_{\varphi_i(T)}^{z} F_t(s, T)\, ds}{\left(2 \int_{\varphi_i(T)}^{z} F(s, T)\, ds\right)^{1/2}}\, dz \Bigg|_{i=4}^{i=3} = 0$$

or

$$\int_{\varphi_2(T)}^{\varphi_i(T)} F_t(z, t) \int_{\varphi_2(T)}^{z} \frac{ds}{\left(2 \int_{\varphi_i(T)}^{s} F(p, T)\, dp\right)^{1/2}}\, dz \Bigg|_{i=1}^{i=3} = 0 \qquad (5.18)$$

To obtain this formula, we must equate the expressions (5.6) and (5.9) for z_1 at $t = T = T_0 + \mu T_1 + \cdots$ taking into account terms of first order with respect to μ. When Eq. (5.18) is also satisfied identically, we must write the new equation taking into account terms of second order, third order, and so on. The general rule is given in [10].

Notice that steps of this kind occur in some problems of galaxy theory [11, 12], where the right term of Eq. (5.4a) often has the form $F = \gamma(t)(g(t)z^2 - 1)z$; $\gamma(t) > 0$, $g(t) > 0$. Eq. (5.18) reduces to

$$\frac{\gamma'(T)}{\gamma(T)} = 2\frac{g'(T)}{g(T)}$$

The phase transition equation also belongs to this critical case. But this model is more interesting in its two-dimensional form (see Section VIII).

2. The results obtained in Section V.A can be generalized to the case of systems containing both "fast" and "slow" variables. Consider the

system [13]

$$
\mu^2 \frac{d^2 u}{dt^2} = f(u, v) \qquad \frac{d^2 v}{dt^2} = g(u, v)
$$
$$
u(0, \mu) = u^0 \quad u(1, \mu) = u^1 \quad v(0, \mu) = v^0 \quad v(1, \mu) = v^1 \tag{5.19}
$$

(one may also specify values of the first derivatives).

We suppose that the equation $f(u, v) = 0$ has three roots $\varphi_1(v) < \varphi_2(v) < \varphi_3(v)$ (cf. 1^0 from Section V.B). In this case, there may exist a solution having a step form with respect to the component u. The transition point T (the main term) can be defined according to the following rule:

a. Define the value v_0 (the main term of the asymptotic expansion of the v component at the transition point) from the equation

$$
\int_{\varphi_3(v_0)}^{\varphi_1(v_0)} f(u, v_0)\, du = 0 \tag{5.20}
$$

b. Solve the two problems (left problem (ℓ) and right problem (r)):

$$
\frac{d^2 \overset{(\ell)}{v}}{dt^2} = g(\varphi_3(\overset{(\ell)}{v}), \overset{(\ell)}{v}) \qquad \frac{d^2 \overset{(r)}{v}}{dt^2} = g(\varphi_1(\overset{(r)}{v}), \overset{(r)}{v})
$$
$$
\overset{(\ell)}{v}(0) = v_0 \quad \overset{(\ell)}{v}(T) = v_0 \quad \overset{(r)}{v}(T) = v_0 \quad \overset{(r)}{v}(1) = v^1
$$

where T is an unknown value.

c. Define T from the condition

$$
\frac{d\overset{(\ell)}{v}}{dt}(T) = \frac{d\overset{(r)}{v}}{dt}(T) \tag{5.21}
$$

(smooth matching).

Example 5.2. Consider the parabolic system with diffusion and non-linearity of the Van der Pol type [14]

$$
-D_1 \frac{\partial^2 u}{\partial x^2} + \frac{\partial u}{\partial t} = v + u - \frac{u^3}{3}
$$
$$
-D_2 \frac{\partial^2 u}{\partial x^2} + \frac{\partial v}{\partial t} = -u
$$

where $D_1 = \mu^2$ is small, $D_2 = 1$. The corresponding stationary problem is

$$\mu^2 \frac{d^2u}{dx^2} = -v - u + \frac{u^3}{3}$$

$$\frac{d^2v}{dx^2} = u$$

The independent variable is here denoted by x in contrast to (5.19). Let

$$u(-\ell) = u(\ell) = v(-\ell) = v(\ell) = 0$$

We cannot obtain $\varphi_1(v)$ and $\varphi_3(v)$ in simple explicit form. Assume that v is small. Then $\varphi_1(v) = -\sqrt{3} + v/2$, $\varphi_3(v) = \sqrt{3} + v/2$ (linear approximation). Equation (5.20) gives $v_0 = 0$. Then we can obtain $\overset{(\ell)}{v}$, $\overset{(r)}{v}$ and define $x_0 = 0$ from (5.21) (in (5.21) this value was denoted by T). Finally, we have

$$\overset{(\ell)}{v} = 2\sqrt{3}\left(1 - \operatorname{ch}\frac{x+\ell}{\sqrt{2}}\right) + C_2 \operatorname{sh}\frac{x+\ell}{\sqrt{2}}$$

$$\overset{(r)}{v} = -2\sqrt{3}\left(1 - \operatorname{ch}\frac{x-\ell}{\sqrt{2}}\right) + D_2 \operatorname{sh}\frac{x-\ell}{\sqrt{2}}$$

$$C_2 = D_2 = -\frac{2\sqrt{3}\left(1 - \operatorname{ch}\frac{\ell}{\sqrt{2}}\right)}{\operatorname{sh}\frac{\ell}{\sqrt{2}}}$$

When ℓ is small, $\overset{(\ell)}{v}$ and $\overset{(r)}{v}$ are small too, and our calculations are correct.

3. A contrast structure of step-type may arise in the system [15]

$$\mu \frac{dz}{dx} = A(y, x)z + b(y, x) \qquad \frac{dy}{dx} = z$$

$$y(0, \mu) = y^0 \qquad y(1, \mu) = y^1$$

For some $x = x_* = x_0 + \cdots$ the y component may have a jump and the corresponding z component is large as $\mu \to 0$. The leading term of y is

$$y = \begin{cases} \overset{(\ell)}{\bar{y}_0} + \overset{(\ell)}{Q}_0 y & x \le x_0 \\ \overset{(r)}{\bar{y}_0} + \overset{(r)}{Q}_0 y & x \ge x_0 \end{cases}$$

where $\overset{(\ell)}{\bar{y}}_0$, $\overset{(r)}{\bar{y}}_0$ are solutions of the problems

$$\frac{d}{dx}\overset{(\ell,r)}{\bar{y}_0} = -\frac{B(\overset{(\ell,r)}{\bar{y}_0}, x)}{A(\overset{(\ell,r)}{\bar{y}_0}, x)} \qquad \overset{(\ell)}{\bar{y}_0}(0) = y^0 \qquad \overset{(r)}{\bar{y}_0}(1) = y^1$$

(we assume that $A(\overset{(\ell)}{\bar{y}_0}, x) > 0$, $A(\overset{(r)}{\bar{y}_0}, x) < 0$)), and the leading term of z (the leading term of its interior layer series) is $(\overset{(\ell,r)}{Q}_{-1}z)/\mu$. The $\overset{(\ell)}{Q}_0 y$ and $\overset{(\ell)}{Q}_{-1}z$ satisfy the system

$$\frac{d}{d\tau}\overset{(\ell)}{Q}_{-1}z = A(\overset{(\ell)}{\bar{y}_0}(x_0) + \overset{(\ell)}{Q}_0 y, x_0)\overset{(\ell)}{Q}_{-1}z \qquad \frac{d}{d\tau}\overset{(\ell)}{Q}_0 y = \overset{(\ell)}{Q}_1 z$$

$$\overset{(\ell)}{Q}_0 y(0) = y^0 - \overset{(\ell)}{\bar{y}_0}(x_0) \qquad \overset{(\ell)}{Q}_{-1}z(0) = z_{-1}$$

$$\overset{(\ell)}{Q}_0 y(-\infty) = 0 \qquad \overset{(\ell)}{Q}_{-1}z(-\infty) = 0$$

where $\tau = (x - x_*)/\varepsilon$. For $\overset{(r)}{Q}_0 y$, $\overset{(\ell)}{Q}_{-1}z$ we can write an analogous system, but $-\infty$ must be changed to $+\infty$.

The unknown parameters x_0, y_0, z_{-1} are defined from the following equations

$$\int_{\overset{(\ell)}{\bar{y}_0}(x_0)}^{\overset{(r)}{\bar{y}_0}(x_0)} A(y, x_0)\, dy = 0 \qquad A(y_0, x_0) = 0$$

$$z_{-1} = -\int_{y_0}^{\overset{(\ell)}{\bar{y}_0}(x_0)} A(y, x_0)\, dy = -\int_{y_0}^{\overset{(r)}{\bar{y}_0}(x_0)} A(y, x_0)\, dy$$

4. Equations (5.17) and (5.18) become identities, if the system (5.4a) is autonomous, that is, if $F(z_2, t) = F(z_2)$. We briefly discussed a solution of step type for such cases in Section V.A. The solution with a transition from φ_1 to φ_3, to which a complete revolution along the cycle corresponds, is shown in Fig. 9 (solid line). The transition point T was not defined Section V.A. It turns out that, in this case the formula for T is

$$T = \frac{\lambda_3}{\lambda_3 + \lambda_1}$$

where $\lambda_i = \sqrt{F_z(\varphi_i)}$ $(i = 1, 3)$ [5].

5. There may exist more than three roots of the equation $F(z_2, t) = 0$.

In consequence of this may exist some solutions having transitions from φ_1 to φ_3, then, for example, from φ_3 to φ_5 and so on [16].

C. Contrast Structures of Spike-Type

1. *A Second-Order Equation*

This case is investigated in [17]. We consider again the boundary problem (5.4a).

Condition 1^0. *Let the equation* $F(z_2, t) = 0$ *have two roots* $z_2 = \varphi_1(t) = \varphi(t)$, $z_2 = \varphi_2(t) = \chi(t)$, $(\varphi(t) < \chi(t),\ F_{z_2}(\varphi(t), t) > 0,\ F_{z_2}(\chi(t), t) < 0)$. *Let the function* $\psi(t)$ *exist, defined by the equation*

$$\int_{\varphi(t)}^{\psi(t)} F(z, t)\, dz = 0 \tag{5.22}$$

Thus we have the situation shown in Fig. 13. We seek the solution having a spike at the point T. The z_2 component has an extremum at $t = T$, and hence the z_1 component is equal to zero. The value of T is an unknown priori as in the case of Section V.B.

By using the boundary function method we construct an asymptotic solution in the form

$$z_2(t, \mu) = \Pi_0 z_2(\tau_0) + Q_0 z_2(\tau) + R_0 z_2(\tau_1)$$

$$\tau_0 = \frac{t}{\mu} \qquad \tau_1 = \frac{t-1}{\mu} \qquad \tau = \frac{t-T}{\mu}$$

where the series $\Pi_0 z_2$ and $R_0 z_2$ are boundary series at $t = 0$ and $t = 1$, and its terms may be constructed as usual.

The series $Q_0 z_1$ must describe the spike. The $Q_0 z_2$ is defined as a solution of the problem [T_0 is the main term of the expansion $T(\mu) =$

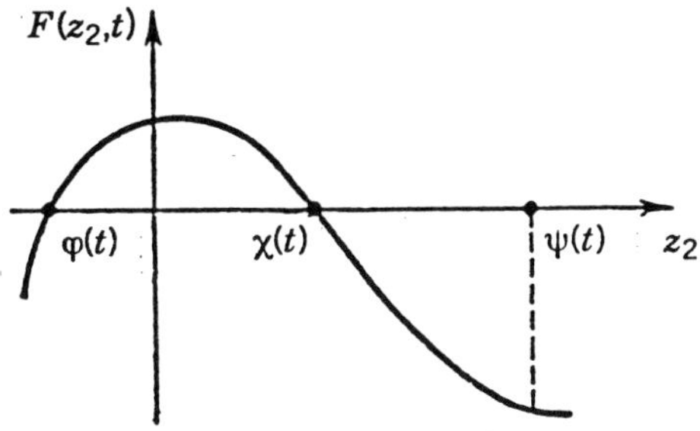

Figure 13. The graph of the function $F(z_2, t)$ with two zero points $\varphi(t)$ and $\chi(t)$ and the point $\psi(t)$ defined by Eq. (5.22).

$T_0 + \mu T_1 + \cdots]$.

$$\frac{d^2 Q_0 z_2}{d\tau^2} = f(\varphi(T_0) + Q_0 z_2, T_0) \tag{5.23}$$

$$\frac{dQ_0 z_2(0)}{d\tau} = 0 \qquad Q_0 z_2(-\infty) = Q_0 z_2(+\infty) = 0 \tag{5.24}$$

(we recall that z_2 has an extremum at $t = T$). The phase plane $Q_0 z_2$, $(d/d\tau)Q_0 z_2$ is shown in Fig. 14. There exists a unique solution of the problem (5.23), (5.24).

The value T_0 is still unknown. To define it we must consider the problem for $Q_1 z_2$. According to the general rule (see Section III.A) we have

$$\frac{d^2 Q_1 z_2}{d\tau^2} = F_{z_2}(\tau) Q_1 z_2 + F_1(\tau) \tag{5.25}$$

$$\frac{dQ_1 z_2(0)}{d\tau} = -\varphi'(T_0) \qquad Q_1 z_2(-\infty) = 0 \qquad Q_1 z_2(+\infty) = 0 \tag{5.26}$$

Here

$$F_1(\tau) = [F_{z_2}(\tau)\varphi'(T) + F_t(\tau)](T_1 + \tau) \qquad F_{z_2}(\tau) = F_{z_2}(\varphi(T_0) + Q_0 z_2, T_0)$$
$$F_t(\tau) = F_t(\varphi(T_0) + Q_0 z_2, T_0)$$

Multiplying Eq. (5.25) with $Z = Q_0' z_2(\tau)$ and integrating from $-\infty$ to 0 and from 0 to $+\infty$ we obtain

$$\begin{aligned} Q_1 z_2(-0) &= -\frac{1}{F(\psi_0(T_0), T_0)} \int_{-\infty}^{0} F_1(\tau) Z(\tau)\, d\tau \\ Q_1 z_2(+0) &= -\frac{1}{F(\psi_0(T_0), T_0)} \int_{+\infty}^{0} F_1(\tau) Z(\tau)\, d\tau \end{aligned} \tag{5.27}$$

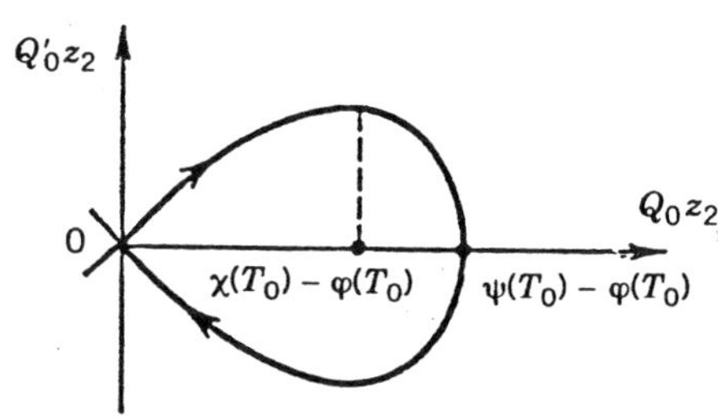

Figure 14. The phase plane $Q_0 z_2$, $Q_0' z_2$. The separatrix of the saddle point $(0, 0)$ makes the loop.

Hence,

$$\int_{-\infty}^{+\infty} F_1(\tau)Z(\tau)\,d\tau = 0 \tag{5.28}$$

Remark: The homogeneous equation, corresponding to (5.25), has obviously a nontrivial solution $Z(\tau) = (d/d\tau)/Q_0z_2$ satisfying the conditions $Z(-\infty) = Z(+\infty) = 0$. One can obtain it by differentiating Eq. (5.23). Notice that $Z(0) = 0$, $Z'(0) = (d^2/d\tau^2)Q_0z_2(0) = F(\psi(T_0), T_0) \neq 0$. Thus relation (5.28) may be considered to be an orthogonality condition: the necessary existence condition for the solution of the nonhomogeneous problem (5.25), (5.26).

The function Q_0z_2 is an even function, and $Z = (d/d\tau)Q_0z_2 = Q_0z_2$ is an odd function. Hence, the function $F_{11} = [F_{z_2}(\tau)\varphi'(T_0) + F_t(\tau)]\tau$ is odd, and $F_{12} = [F_{z_2}(\tau)\varphi'(T_0) + F_t(\tau)]T_1$ is even. Therefore,

$$\int_{-\infty}^{+\infty} F_{12}Z\,d\tau = 0 \qquad \int_{-\infty}^{+\infty} F_{11}Z\,d\tau = 2\int_{-\infty}^{0} [F_{z_2}\vartheta'(T_0)\tau F_t]\tau Z\,d\tau$$

We have in addition $Q_0'''z_2 = F_{z_2}Q_0'z_2$. Hence,

$$\int_{-\infty}^{0} F_{z_2}\varphi'(T_0)\tau Z\,d\tau = \varphi'(T_0)\int_0^{\infty} Q_0'''z_2(\tau)\tau\,d\tau$$

$$= \varphi'(T_0)\left(\tau Q_0''z_2|_0^{\infty} - \int_0^{\infty} Q_0''z_2\,d\tau\right) = \varphi'(T_0)(-Q_0'z_2)|_0^{\infty} = 0$$

Therefore,

$$\int_{-\infty}^{+\infty} F_1(\tau)Z(\tau)\,d\tau = 2\int_{-\infty}^{0} F_t(\tau)\tau Z\,d\tau$$

and Eq. (5.28) takes the form

$$\int_{-\infty}^{0} F_t(\tau)\tau Q_0'z_2\,d\tau = 0$$

which is just the equation giving the value T_0. This equation may be also written in the form

$$\Phi(T_0) = \int_{\varphi(T_0)}^{\psi(T_0)} F_t(z, T_0)\left(\int_{\psi(T_0)}^{z} \frac{ds}{\left(2\int_{\varphi(T_0)}^{s} F(u, T_0)\,du\right)^{1/2}}\right) dz = 0 \tag{5.29}$$

Condition 2^0. *Let Eq. (5.29) have the solution* $T_0 \in [0, 1]$, *and* $\Phi'(T_0) \neq 0$.

Condition 3^0. $\quad 0 < \psi(0) \quad 0 < \psi(1)$

(this condition is needed for constructing Πz_2 and Rz_2).

Under conditions 1^0–3^0, *for sufficiently small* μ, *there exists a solution of the boundary problem (5.4) for which*

$$\lim_{\mu \to 0} z_2(t, \mu) = \begin{cases} \varphi(t) & 0 < t < T_0 \quad T_0 < t < 1 \\ \psi(T_0) & t = T_0 \end{cases}$$

The proof of this statement, and the construction of the asymptotic expansion for $z_2(t, \mu)$, is done in detail in [17]. The corresponding graph is shown in Fig. 15.

2. *Other Cases*

1. Consider the system (5.19) with boundary conditions $u'(0, \mu) = v'(0, \mu) = u'(1, \mu) = v'(1, \mu) = 0$ having a solution with a spike in the component u. The independent variable is denoted here by x. Let the equation $f(u, v) = 0$ have two roots having the properties 1^0 in Section V.C.1 (t must be replaced by v).

The transition point can be found according to the following rule:

a. Consider the equation containing v_0 [the value of $v(x, \mu)$, its main term, at the spike point] and x_0 (the main term of the spike point; in Section V.C.1 it was denoted by T_0)

$$\Phi(v_0)\bar{v}_0'(x_0) = 0 \tag{5.30}$$

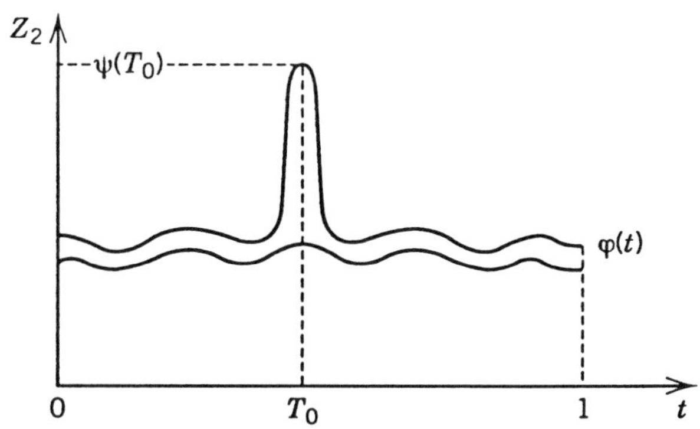

Figure 15. The graph of the spike-type solution. T_0: the spike point; $\psi(T_0)$: the magnitude of the spike.

where

$$\Phi(v_0) = \int_{\varphi(v_0)}^{\psi(v_0)} f_v(u, v_0) \left(\int_{\psi(v_0)}^{u} \frac{ds}{\left(2 \int_{\varphi(v_0)}^{s} f(\xi, v_0)\, d\xi\right)^{1/2}} \right) du \tag{5.31}$$

and $\bar{v}_0$ is the solution of the problem

$$\frac{d^2\bar{v}_0}{dx^2} = g(\varphi(\bar{v}_0), \bar{v}_0) \qquad \bar{v}_0'(0) = 0 \qquad \bar{v}_0'(1) = 0 \tag{5.32}$$

b. Let the equation $\Phi(v_0) = 0$ have a solution v_0 with $\Phi'(v_0) \neq 0$. For x_0 we then have the equation $\bar{v}_0(x) = v_0$. We suppose that $\bar{v}_0'(x) \neq$ const and $\bar{v}_0'(x_0) \neq 0$. We can then construct the main terms of the asymptotic expansion for $u(x, \mu)$ and $v(x, \mu)$. This solution is called the spike solution of the first kind.
c. Let the spike point x_0 be found from Eq. (5.30), that is, from the equation $\bar{v}_0'(\bar{x}_0) = 0$. We suppose that $\Phi(\bar{v}_0(x_0)) \neq 0$, $\bar{v}_0''(x_0) \neq 0$. The corresponding solution is called the spike solution of the second kind.
d. Let problem (5.32) have a solution $\bar{v}_0 = v_0 = \text{const}$. In this case, there may be a set of solutions having the spike points: (1) $x_0 = \frac{1}{2}$; (2) $x_0 = \frac{1}{3}, \frac{2}{3}$; (3) $x_0 = \frac{1}{4}, \frac{1}{2}, \frac{3}{4}$; and so on. These solutions are called the spike solutions of the third kind.

The contrast structure of spike type for the two equations (5.20) is considered in detail in [5, 18].

Example 5.3. Consider the model called the "brusselator," to which case (d) applies. The model describes a chemical reaction with two intermediate substances [19]. The nondimensional concentrations u and v of these substances satisfy the system

$$\frac{\partial u}{\partial t} = D_1\, \Delta u + A - (B+1)u + u^2 v \qquad \frac{\partial v}{\partial t} = D_2\, \Delta v + Bu - u^2 v$$

Here D_1, D_2, A, B are positive constants and Δ is the Laplace operator. If $D_1 \ll D_2$, then setting $D_1 = \mu^2$, $D_2 = 1$, we obtain the following system of equations for the stationary case (in the one-dimensional problem, where $\Delta = d^2/d^2$):

$$\mu^2 \frac{d^2u}{dx^2} = -A + (B+1)u - u^2 v \qquad \frac{d^2v}{dx^2} = -Bu + u^2 v \tag{5.33}$$

This system is of type (5.20), where $f(u, v) = -A + (B+1)u - u^2 v$, $g(u, v) = -Bu + u^2 v$, t is replaced by x. We investigate this system in the interval $[0, \ell]$ with the boundary conditions

$$u'(0, \mu) = u'(\ell, \mu) = v'(0, \mu) = v'(\ell, \mu) = 0 \tag{5.34}$$

The equation $f(u, v) = 0$, that is, $-A + (B+1)u - u^2 v = 0$, has two positive solutions (the conditions $u > 0$, $v > 0$ follow from the physical meaning of the variables u and v):

$$u = \frac{B + 1 - \sqrt{(B+1)^2 - 4Av}}{2v} \equiv \varphi(v)$$

$$u = \frac{B + 1 + \sqrt{(B+1)^2 - 4Av}}{2v} \equiv \chi(v)$$

Evidently, $\varphi(v) < \chi(v)$, and

$$f_u(\varphi(v), v) = \sqrt{(B+1)^2 - 4Av} > 0$$

$$f_u(\chi(v), v) = -\sqrt{(B+1)^2 - 4Av} < 0$$

The expression for the function $\psi(v)$ is defined as in the beginning of this section (see condition 1^0), and has the form

$$\psi(v) = \frac{B + 1 + 2\sqrt{(B+1)^2 - 4Av}}{2v}$$

The boundary value problem (5.32) can be written in the form

$$\bar{v}_0'' = \frac{-2A\bar{v}_0 + B + 1 - \sqrt{(B+1)^2 - 4A\bar{v}_0}}{2\bar{v}_0}$$
$$\bar{v}_0'(0) = \bar{v}_0'(\ell) = 0 \tag{5.35}$$

It is easy to see that if $B > 1$ the right-hand side of Eq. (5.35) is strictly negative, and therefore the boundary value problem (5.35) cannot have a solution. If $0 < B \leq 1$, then this problem has the unique solution

$$\bar{v}_0 = \frac{B}{A} = \text{const}$$

and therefore the problem (5.33), (5.34) does not have spike solutions of the first or second kind.

Thus, we obtain

$$\bar{v}_0 = \frac{B}{A} = v_0 \qquad \bar{v}_0 = \varphi(v_0) = A$$

The equation for $Q_0u(\tau)$ ($\tau = (x - x_*)/\mu$, $x_* = x_0 + \mu x_1 + \cdots$ is the spike-point) has the form (5.23) (we must change $z_2, \varphi(T_0), T_0$ to μ, $\varphi(v_0), v_0$, respectively). In our case, we get the equation

$$\frac{d^2}{d\tau^2} Q_0u = -A + (B+1)Q_0u - v_0(Q_0u)^2$$

From this we have

$$Q_0'u = \sqrt{2\int_A^{A+Q_0u} \left[-A + (B+1)\xi - \frac{B}{A}\xi^2\right] d\xi}$$

$$= Q_0u\sqrt{(1-B) - \frac{2B}{3A} Q_0u}$$

Solving this equation with the initial condition $Q_0u(0) = \psi(v_0) - \varphi(v_0) = 3(1-B)A/(2B)$, we obtain

$$Q_0u(\tau) = \frac{6\alpha^2 A \exp\{\alpha\tau\}}{B(1+\exp\{\alpha\tau\})^2} \qquad \tau \le 0$$

where $\alpha = \sqrt{1-B}$. The expression for $Q_0u(\tau)$, ($\tau \ge 0$) has the same form. Thus, the terms in the asymptotic expansion of the zeroth order are found explicitly.

For the terms in the asymptotic approximation of first order, we have

$$\bar{u}_1 = \frac{A^2}{\alpha^2}\bar{v}_1 \qquad \bar{v}_1'' = \frac{A^2}{\alpha^2}\bar{v}_1 \tag{5.36}$$

The boundary conditions at the ends of the interval are

$$\bar{v}_1'(0) = \bar{v}_1'(\ell) = 0 \tag{5.37}$$

Since v and v' must be continuous at the spike-point $x_* = x_0 + \mu x_1 + \cdots$ we obtain

$$\bar{v}_1(x_0-0)=\bar{v}_1(x+0)$$

$$\bar{v}_1'(x_0-0)+Q_2'v(-0)=\bar{v}_1'(x_0+0)+Q_2'v(+0)$$

Thus, to define $\bar{v}_1(x)$, we must first determine Q_2v. For Q_2v $(-\infty<\tau\leq 0)$ we have

$$\frac{d^2Q_2v}{d\tau^2}=Q_0g\equiv g(A+Q_0u(\tau),v_0)-g(A,v_0) \qquad Q_2v(-\infty)=0$$

This yields

$$Q_2v(\tau)=\int_{-\infty}^{\tau}\int_{-\infty}^{s}Q_0g(\sigma)\,d\sigma \qquad Q_2'v(\tau)=\int_{-\infty}^{\tau}Q_0g(\sigma)\,d\sigma$$

Analogous calculations for Q_2v when $\tau\geq 0$ provide the equalities

$$Q_2v(+0)=Q_2v(-0) \qquad Q_2'v(+0)=-Q_2'v(-0)$$

Hence, for $\bar{v}_1$ the following conditions must be satisfied at x_0

$$\begin{aligned}&\bar{v}_1(x_0+0)-\bar{v}_1(x_0-0)=0\\&\bar{v}_1'(x_0+0)-\bar{v}_1'(x_0-0)=2Q_2'v(-0)\end{aligned}\tag{5.38}$$

To find $\bar{v}_1(x)$, we have to solve the linear equation (5.36) with the additional conditions (5.37), (5.38). Computation of $Q_2'v(0)$ gives $Q_2'v(0)=3\alpha A/B$. After that we obtain $\bar{v}_1(x)$ explicitly from the formula $(k=A/\alpha,\ \alpha=\sqrt{1-B})$

$$\bar{v}_1(x)=-\frac{6\alpha^2}{B\,\mathrm{sh}\,k\ell}\begin{cases}\mathrm{ch}\,k(x_0-\ell)\mathrm{ch}\,kx & 0\leq x\leq x_0\\ \mathrm{ch}\,kx_0\mathrm{ch}\,k(x-\ell) & x_0\leq x\leq \ell\end{cases}\tag{5.39}$$

To calculate x_0 we consider the problem for Q_2u

$$\frac{d^2Q_2u}{d\tau^2}=f_u(\tau)Q_2u+f_2(\tau) \qquad \tau\leq 0$$

$$Q_2'u(0)=-\bar{u}_1'(x_0-0) \qquad Q_2u(-\infty)=0$$

where

$$f_2(\tau) = h_1(\tau)\tau\bar{v}_1'(x_0 - 0) + \tilde{f}_2(\tau) \qquad h_1(\tau) = f_v(\tau) + f_u(\tau)\varphi'(v_0)$$

$$\tilde{f}_2(\tau) = h_1(\tau)[\bar{v}_2(x_0 - 0) + \bar{v}_1'(x_0 - 0)x_1] + f_v(\tau)Q_2v(\tau)$$
$$+ \tfrac{1}{2} f_{uu}(\tau)[\bar{u}_1(x_0) + Q_1u(\tau)]^2 + f_{uv}(\tau)[\bar{u}_1(x_0) + Q_1u(\tau)]\bar{v}_1(x_0)$$
$$+ \tfrac{1}{2} f_{vv}(\tau)\bar{v}_1^2(x_0)$$

(here all the partial derivatives of the function f are evaluated at the point $(A + Q_0u(\tau), v_0)$). The problem for $\tau \geq 0$ has the same form with (-0) replaced by $(+0)$.

As in Section V.C.1, we obtain

$$Q_2u(\pm 0) = -\frac{1}{f(\psi(v_0), v_0)} \int_{\pm\infty}^{0} f_2(\tau)Q_0'u(\tau)\,d\tau$$

Making the change of variable of integration

$$\tau = \int_{\psi(v_0)-A}^{Q_0u} \frac{ds}{\left(2\int_A^s f(u, v_0)\,du\right)^{1/2}}$$

we obtain

$$Q_2u(\pm 0) = -\frac{1}{f(\psi(v_0), v_0)} [\mp\Phi(v_0)\bar{v}_1'(x_0 \pm 0) + \beta]$$

where $\Phi(v_0)$ is expressed by (5.31) and β is some constant (see [18]), its value being unimportant.

The condition that function u is continuous at the point x_* leads to the equality

$$\bar{u}_2(x_0 - 0) + \bar{u}_1'(x_0 - 0)x_1 + Q_2\bar{u}(-0)$$
$$= \bar{u}_2(x_0 + 0) + \bar{u}_1'(x_0 + 0)x_1 + Q_2u(+0)$$

From here, taking into account the relations

$$\bar{u}_1'(x) = \frac{A^2}{\alpha^2}\bar{v}_1'(x) \qquad \bar{u}_2(x) = \frac{A^2}{\alpha^2}\bar{v}_2(x) + \gamma\bar{v}_1^2(x)$$

(γ is some constant (see [18]), its value being unimportant), and the continuity of the function $\bar{v}_1(x)$ at x_* we obtain

$$\frac{A^2}{\alpha^2}[\bar{v}_2(x_0+0) - \bar{v}_2(x_0-0) + (\bar{v}_1' + 0) - \bar{v}_1'(x_0-0))x_1]$$
$$+ Q_2u(+0) - Q_2u(-0) = 0$$

But the expression in the square brackets is zero by virtue of the continuity of v at the point x_*. Thus, we arrive at the equation $Q_2u(+0) - Q_2u(-0) = 0$. Substituting the expressions for $Q_2u(\pm 0)$, we obtain

$$\frac{1}{f(\psi(v_0), v_0)}\Phi(v_0)[\bar{v}_1'(x_0+0) + \bar{v}_1'(x_0-0)] = 0$$

In our case $\Phi(v_0) \neq 0$, which is easily verified. Indeed, the function $\Phi(v_0)$ is defined by the formula (5.31). In our case $f_u(u, v_0) = -u^2 < 0$. Hence, the function under the integral in the expression (5.31) does not change sign, so $\Phi(v_0) \neq 0$. Then we obtain the equation

$$\bar{v}_1'(x_0+0) + \bar{v}_1'(x_0-0) = 0$$

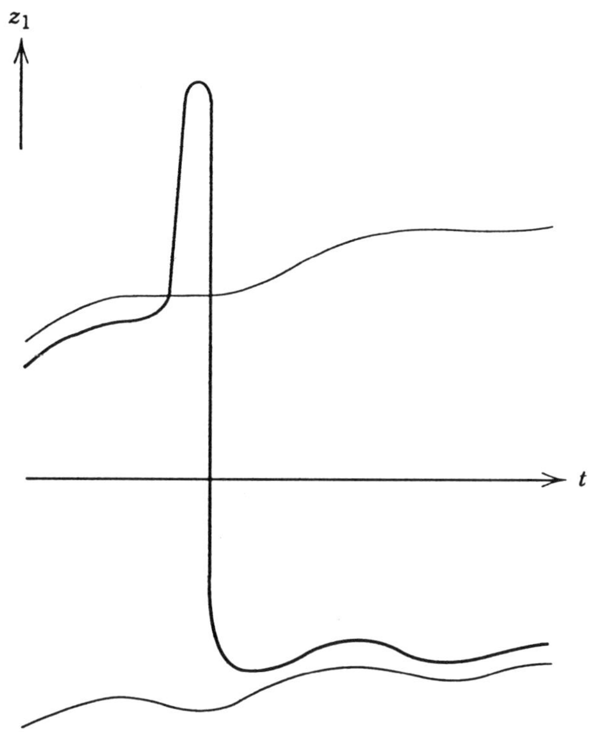

Figure 16. The graph of the solution having interior layer of "step with spike" type.

for x_0. Using the expression (5.39) for $\bar{v}_1(x)$, we arrive at $\operatorname{sh}\{k(2x_0 - \ell)\} = 0$ and, therefore, $x_0 = \ell/2$. There may also be the set of solutions having $2, 3, \ldots$, spike points (see Section V.C.2d).

All this is done in detail in [18].

2. We have considered the system (5.4b). For a more general system

$$\mu \frac{dz_1}{dt} = F(z_1, z_2, t)$$

$$\mu \frac{dz_2}{dt} = G(z_1, z_2, t)$$

there may be contrast structures of "step with spike" type (Fig. 16) [20].

D. Stability of Solutions Having Boundary and Interior Layers

The solutions investigated in the previous subsections may be considered as stationary solutions of the corresponding system of parabolic equations.

For one parabolic equation

$$D \frac{\partial^2 z}{\partial x^2} - \frac{\partial z}{\partial t} = F(z, x) \tag{5.40}$$

the stationary equation has the form

$$Dz'' = F(z, x) \tag{5.41}$$

This equation is of type (5.4) when the diffusion is small: $D = \mu^2$.

Depending on different properties of the function F and of the complementary conditions, the solution $z(x, \mu)$ of Eq. (5.41) may have boundary layers or interior layers (these layers can coexist).

The following question arises: Since the actual solution $z(x, t, \mu)$ of (5.40) can approach only *stable* stationary solutions as $t \to \infty$, which types of solutions are stable? We are interested in stability of the stationary solution $z(x, \mu)$ in the sense of Lyapunov, that is, whether for any $\varepsilon > 0$ there exists $\delta(\varepsilon)$ such that if

$$\sup_x |z(x, t, \mu) - z(x, \mu)| < \delta$$

the inequality

$$\sup_{x,t} |z(x, t, \mu) - z(x, \mu)| < \varepsilon$$

holds for $t \geq 0$. Some general principles for studying stability properties of stationary solutions of parabolic equations are discussed, for example, in [21]. By using the ideas of the first method of Lyapunov, the investigation of stability in the case of one parabolic equation can be reduced to studying the spectrum of the following Sturm–Liouville problem:

$$\mu\psi'' = [F_0(z(x, \mu), x) + \lambda]\psi \qquad \psi(0) = \psi(1) = 0$$

[we investigate solutions defined by boundary conditions of Dirichlet type as in (5.4)]. A stationary solution $z(x, \mu)$ will be stable if, for all the eigenvalues $\lambda_1 > \lambda_2 > \cdots > \lambda_k > \cdots$ of the Sturm–Liouville problem, the inequality $\lambda_k < 0$, $k = 1, 2, \ldots$, holds. If at least $\lambda_1 > 0$, then the solution $z(x, \mu)$ is unstable.

The following results are valid:

1. The solutions having only boundary layers (Section III) are *stable* since for them $\lambda_1 = O(1) < 0$ (except when the spike is at the boundary point).
2. The spike-type contrast structures (Fig. 15) for one equation (5.4) are *not stable* since for them $\lambda_1 = O(1) > 0$. The addition of the second "slow" equation, as in (5.20), cannot change the sign of λ.
3. The contrast structures of step type (Fig. 10) have $\lambda_1 = O(\mu)$, that is, $\lambda_1 = \mu\lambda_{11} + \mu^2\lambda_{12} + \cdots$, and for λ_{11} we have (for transition from φ_3 to φ_1)

$$\lambda_{11} = \left[\int_{\varphi_3(x_0)}^{\varphi_1(x_0)} F_x(z, x_0)\, dz\right] \Big/ \left[\int_{\varphi_1(x_0)}^{\varphi_3(x_0)} \left(2\int_{\varphi_1(x_0)}^{z} F_z(\xi, x_0)\, d\xi\right)^{1/2} dz\right] \tag{5.42}$$

   where x_0 is the leading term in the expansion for the transition point coordinate $x_* = x_0 + \mu x_1 + \cdots$ (x_* is the value of x for which $z(x, \mu) = \varphi_2(x)$). Under condition 4^0 in Section V.B.1 the value of λ_{11} differs from zero. Notice that the numerator in (5.42) is equal, apart from notation, to the expression in 4^0, which is nonzero. The sign of λ_{11} is defined by the sign of the numerator and may be either positive or negative.
4. In the case considered in Section V.B.2 (case 3), $\lambda_1 = O(1)$ and

$$\lambda_{10} = \left[\frac{d}{dx_0}\int_{\overset{(\ell)}{\bar{y}}_0(x_0)}^{\overset{(r)}{\bar{y}}_0(x_0)} A(y, x_0)\, dy\right] \Big/ [\overset{(r)}{\bar{y}}_0(x_0) - \overset{(\ell)}{\bar{y}}_0(x_0)]$$

These and some other results on contrast structures and their stability are discussed in [15, 16, 22–25, 26] and elsewhere.

The stability, without an estimate of λ (such an estimate can be said to determine the degree of stability) is easy to prove, under appropriate conditions, using the method of differential inequalities [26].

VI. THE METHOD OF VISHIK–LYUSTERNIK

A. Statement of the Problem

Beginning with this section, we will now consider partial differential equations. Let us consider a general approach for constructing asymptotic expansions of the solutions of singularly perturbed linear partial differential equations, which was proposed in the well-known fundamental work of Vishik and Lyusternik [27]. We will illustrate the idea of this approach (known in the literature as the *method of Vishik–Lyusternik*) on a simple example of an elliptic equation in a bounded domain with smooth boundary.

Consider the equation

$$L_\varepsilon u \equiv \varepsilon^2 \Delta u - k^2(x, y)u = f(x, y, \varepsilon) \qquad (x, y) \in \Omega \tag{6.1}$$

with the Dirichlet boundary condition

$$u|_{\partial\Omega} = 0 \tag{6.2}$$

Here $\varepsilon > 0$ is a small parameter, $\Delta = (\partial^2/\partial x^2) + (\partial^2/\partial y^2)$ is the Laplace operator, Ω is a bounded planar domain. The boundary of the domain is assumed to be a sufficiently smooth curve $\partial\Omega$. The functions $k(x, y)$ and $f(x, y, \varepsilon)$ are also sufficiently smooth and $k(x, y) > 0$ in $\bar{\Omega} = \Omega + \partial\Omega$. The smoothness of k, f, $\partial\Omega$ and the negative sign of the coefficient $-k^2(x, y)$ guarantee the existence of a unique classical solution $u(x, y, \varepsilon)$ of problem (6.1), (6.2). Our goal is to construct an asymptotic expansion (uniform in $\bar{\Omega}$) of this solution with respect to the parameter ε.

B. Local Coordinates

The asymptotic expansion of the solution of problem (6.1), (6.2) consists of regular and boundary layer parts

$$u = \bar{u} + \Pi$$

To construct the boundary layer functions we need to introduce new (local) coordinates in the vicinity of the boundary $\partial\Omega$. Let the parametric equations describing $\partial\Omega$ have the form

$$x = \varphi(\ell) \qquad y = \psi(\ell) \qquad 0 \le \ell \le \ell_0$$

where ℓ is a parameter. When the parameter ℓ increases from 0 to ℓ_0, the point $[\varphi(\ell), \psi(\ell)]$ moves along $\partial\Omega$ in such a way that the domain Ω remains to the left. In a δ-vicinity of the curve $\partial\Omega$, we introduce new (local) variables as follows. Let the point $M(x, y)$ belonging to $\bar{\Omega}$ lie on the normal to $\partial\Omega$ passing through the point $M'(\varphi(\ell), \psi(\ell))$ (see Fig. 17). We denote the distance MM' by r. The pair of numbers (r, ℓ) will be the new coordinates of the point M. If the δ-vicinity $\{(0 \le r \le \delta) \times (0 \le \ell \le \ell_0)\}$ is sufficiently narrow, that is, δ is small enough, the normals to $\partial\Omega$ drawn through various points of $\partial\Omega$ do not intersect within this neighborhood. Hence, in this vicinity there exists a one-to-one correspondence between the two coordinates (x, y) and (r, ℓ), expressed by the relations

$$x = \varphi(\ell) - r\frac{\psi'(\ell)}{\sqrt{\varphi'^2(\ell) + \psi'^2(\ell)}} \qquad y = \psi(\ell) + r\frac{\varphi'(\ell)}{\sqrt{\varphi'^2(\ell) + \psi'^2(\ell)}}$$

In the δ-vicinity of $\partial\Omega$, we have the following expression for the operator L_ε in terms of the new variables (r, ℓ)

$$L_\varepsilon = \varepsilon^2\left[\frac{\partial^2}{\partial r^2} + \alpha(r, \ell)\frac{\partial^2}{\partial \ell^2} + \beta(r, \ell)\frac{\partial}{\partial r} + \gamma(r, \ell)\frac{\partial^2}{\partial \ell}\right] - k^2(x(r, \ell), y(r, \ell))$$

where

$$\alpha(r, \ell) = \left(\frac{\partial \ell}{\partial x}\right)^2 + \left(\frac{\partial \ell}{\partial y}\right)^2 \qquad \beta(r, \ell) = \frac{\partial^2 r}{\partial x^2} + \frac{\partial^2 r}{\partial y^2}$$

$$\gamma(r, \ell) = \frac{\partial^2 \ell}{\partial x^2} + \frac{\partial^2 \ell}{\partial y^2}$$

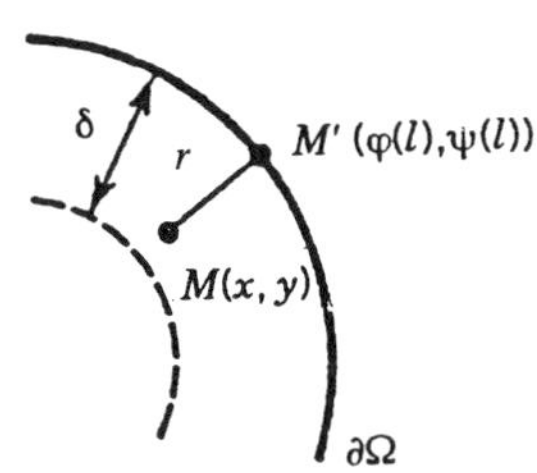

Figure 17. New variables (r, ℓ) in the vicinity of the boundary.

Let us now stretch the variable r: $r = \varepsilon\rho$. Expanding the coefficients in the expression for L_ε in power series in ε, we obtain

$$L_\varepsilon = \left[\frac{\partial^2}{\partial\rho^2} - k^2(\ell)\right] + \sum_{i=1}^{\infty} \varepsilon^i L_i$$

where $k^2(\ell) = k^2(\varphi(\ell), \psi(\ell))$, and the L_i are linear differential operators containing the derivatives $\partial/\partial\rho$, $\partial/\partial\ell$, $\partial^2/\partial\ell^2$.

C. The Regular Part of the Asymptotic Expansion

We seek an asymptotic expansion of the solution of (6.1), (6.2) in the form

$$u(x, y, \varepsilon) = \bar{u} + \Pi = \sum_{i=0}^{\infty} \varepsilon^i[\bar{u}_i(x, y) + \Pi_i(\rho, \ell)] \tag{6.3}$$

Substituting (6.3) into (6.1), (6.2), and expanding $f(x, y, \varepsilon)$ in a power series in ε, we arrive at the equalities

$$[\varepsilon^2 \Delta - k^2(x, y)] \sum_{i=0}^{\infty} \varepsilon^i \bar{u}_i(x, y)$$
$$+ \left\{\left[\frac{\partial^2}{\partial\rho^2} - k^2(\ell)\right] + \sum_{j=0}^{\infty} \varepsilon^j L_j\right\} \sum_{i=0}^{\infty} \varepsilon^i \Pi_i(\rho, \ell) = \sum_{i=0}^{\infty} \varepsilon^i f_i(x, y) \tag{6.4}$$

$$\sum_{i=0}^{\infty} \varepsilon^i[\bar{u}_i(\ell) + \Pi_i(0, \ell)] = 0 \tag{6.5}$$

Here $\bar{u}_i(\ell) = \bar{u}_i(\varphi(\ell), \psi(\ell))$. Equating coefficients of like powers of ε in (6.4) separately for terms depending on (x, y) and on (ρ, ℓ), we obtain equations for the coefficients of the asymptotic series (6.3).

For the regular terms of the asymptotic expansion, we have

$$-k^2(x, y)\bar{u}_0 = f_0(x, y) \qquad -k^2(x, y)\bar{u}_1 = f_1(x, y)$$
$$\Delta\bar{u}_{i-2} - k^2(x, y)\bar{u}_i = f_i(x, y) \qquad i = 2, 3, \ldots$$

Hence,

$$\bar{u}_0 = -f_0(x, y)/k^2(x, y) \qquad \bar{u}_1 = -f_1(x, y)/k^2(x, y)$$
$$\bar{u}_i = -[f_i(x, y) - \Delta\bar{u}_{i-2}]/k^2(x, y) \qquad i = 2, 3, \ldots$$

D. The Boundary Layer Part of the Asymptotic Expansion

The regular part of the asymptotic expansion does not generally satisfy the boundary condition (6.2). In the terminology of the paper of Vishik and Lyusternik [27], the regular terms of the asymptotics introduce a discrepancy into the boundary condition. The purpose of the boundary layer functions $\Pi_i(\rho, \ell)$ is to compensate for this discrepancy. The equality (6.5) shows that the boundary layer functions together with the regular terms must satisfy the boundary condition (6.2).

For the boundary layer functions $\Pi_i(\rho, \ell)$ we have the following ordinary differential equations (ℓ enters as a parameter)

$$\frac{\partial^2 \Pi_i}{\partial \rho^2} - k^2(\ell)\,\Pi_i = \pi_i(\rho, \ell) \qquad \rho \geq 0 \tag{6.6}$$

where $\pi_0 = 0$, $\pi_i(\rho, \ell) = -\Sigma_{j=1}^{i} L_j \Pi_{i-j}(\rho, \ell)$, $i \geq 1$. Therefore the right-hand sides of Eqs. (6.6) are expressed recursively through the Π_j with $j < i$. The boundary conditions at $\rho = 0$ follow from (6.5):

$$\Pi_i(0, \ell) = -\bar{u}_i(\ell) \tag{6.7}$$

In addition, we require that the Π-functions decay as $\rho \to \infty$

$$\Pi_i(\infty, \ell) = 0 \tag{6.8}$$

The explicit solutions of problems (6.6)–(6.8) can be found successively. For example,

$$\Pi_0(\rho, \ell) = -\bar{u}_0(\ell)\exp\{-k(\ell)\rho\}$$

and the other functions $\Pi_i(\rho, \ell)$ are products of polynomials in ρ and $\exp\{-k(\ell)\rho\}$. Consequently, all of the Π-functions satisfy the exponential estimate

$$|\Pi_i(\rho, \ell)| \leq C\exp\{-\alpha\rho\}$$

where, as before, we denote by C and α appropriate positive constants which, in general, are different for different Π-functions.

Notice that though formally Π-functions are defined for $\rho \geq 0$, they in fact make sense only for $0 \leq \rho \leq \delta/\varepsilon$, that is, in a δ vicinity of $\partial\Omega$ where we introduced the local coordinates (r, ℓ). The standard procedure of smooth continuation of these functions onto the entire domain $\bar{\Omega}$ can be described as follows. We multiply the constructed Π-functions by a smooth (infinitely differentiable) function $\Psi(r)$ such that $\Psi(r)=1$ when $0\leq r\leq\delta/3$, $0\leq\Psi(r)\leq 1$ when $\delta/3\leq r\leq 2\delta/3$, and $\Psi(r)=0$ when $r\geq 2\delta/3$. Then we consider $\Pi_k(\rho,\ell)\Psi(r)$ as the new Π-functions. These new boundary layer functions are uniquely defined in the entire domain $\bar{\Omega}$. They coincide with the original Π-functions for $0\leq r\leq\delta/3$ and they are asymptotically smaller than any power of ε for $r\geq\delta/3$:

$$|\Pi_i(\rho,\ell)|\leq C\exp\left\{-\frac{\alpha\delta}{3\varepsilon}\right\}=o(\varepsilon^N)$$

for any $N>0$, when $r\geq\delta/3$. Thus, the procedure described above changes each of the Π-functions by an amount that is less than any power of ε. In the problems considered below, without mentioning this fact explicitly, we assume (where necessary) that the boundary layer functions are defined in the entire domain with the help of the continuation procedure, similar to what has just been introduced.

Definition. *We say that the Π-functions are defined by the operator*

$$L_\Pi=\frac{\partial^2}{\partial\rho^2}-k^2(\ell)$$

and by the boundary conditions (6.7), (6.8). We call L_Π a boundary layer operator in the vicinity of $\partial\Omega$.

E. Asymptotic Approximations of the Solution

Let $U_n(x,y,\varepsilon)$ denote the n-th partial sum of the series (6.3):

$$U_n(x,y,\varepsilon)=\sum_{i=0}^{n}\varepsilon^i[\bar{u}_i(x,y)+\Pi_i(\rho,\ell)]$$

It is not difficult to prove that

> *$U_n(x,y,\varepsilon)$ is an asymptotic approximation of the solution $u(x,y,\varepsilon)$ of problem (6.1), (6.2) with accuracy of order ε^{n+1} uniformly in $\bar{\Omega}$, that is, the following estimate holds*
>
> $$\max_{\bar{\Omega}}|u(x,y,\varepsilon)-U_n(x,y,\varepsilon)|=O(\varepsilon^{n+1})$$

Thus, the constructed series (6.3) is in fact an asymptotic series for the solution $u(x,y,\varepsilon)$.

We illustrated the method of Vishik–Lyusternik with the example of a linear elliptic equation. Note that even for this simple problem it is impossible to find the solution explicitly for an arbitrary coefficient $k(x,y)$

and domain Ω. On the other hand, an asymptotic approximation of the solution can be easily found to any required accuracy in explicit form.

The method of Vishik–Lyusternik is well developed not only for elliptic but also for parabolic and hyperbolic partial differential equations. Some nonlinear problems can be solved using this method as well. The main advantage of this method is its simplicity. Usually the boundary layer functions are defined as solutions of ordinary differential equations in which the independent variable is the stretched distance along the normal to the boundary (the variable ρ in our case). However, there are some problems when the boundary layer functions are described by more complicated equations, for example, by parabolic equations. A survey of results obtained by the Vishik–Lyusternik method with many references can be found in [28].

VII. PROBLEMS WITH CORNER BOUNDARY LAYERS

A. Elliptic Equation in a Rectangle

The construction of an asymptotic solution in the previous section was carried out under an essential assumption: The boundary $\partial\Omega$ of the domain Ω is assumed to be a smooth curve. The normal to the curve exists at each point and the boundary layer functions were constructed from the solutions of ordinary differential equations with derivatives taken along these normals.

In the case when the boundary of the domain is no longer smooth, but contains corner points, the structure of the asymptotic solutions becomes more complicated in vicinities of these points. The boundary layer functions constructed in the previous section are not sufficient to describe the asymptotic behavior of the solution near the corners. We need to introduce a new type of boundary layer functions, *corner boundary functions*, in the vicinities of the corner points. We begin our discussion with a problem for the same elliptic equation as in Section VI, but now the domain Ω is a rectangle, whose boundary has four corner points.

Consider the problem

$$\varepsilon^2 \Delta u - k^2(x, y)u = f(x, y, \varepsilon) \tag{7.1}$$

$$(x, y) \in \Omega = \{(0 < x < a) \times (0 < y < b)\}$$

$$u|_{\partial\Omega} = 0 \tag{7.2}$$

The functions $k(x, y)$ and $f(x, y, \varepsilon)$ are assumed to satisfy the same conditions as in Section VI.

We seek an asymptotic expansion of the solution of (7.1), (7.2) in the form

$$u = \bar{u} + \Pi + P \tag{7.3}$$

Here $\bar{u} = \Sigma_{i=0}^{\infty} \varepsilon^i \bar{u}_i(x, y)$ is the regular part of the asymptotic solution that can be determined in exactly the same manner as in Section VI. The notation Π is used for boundary layer functions playing an important role near the sides of the rectangle, and the notation P is used for corner boundary functions important in the vicinities of the vertices of the rectangle. Corresponding to the sides of the rectangle, the Π part of the asymptotics can be represented as a sum of four terms:

$$\begin{aligned}\Pi &= \overset{(1)}{\Pi} + \overset{(2)}{\Pi} + \overset{(3)}{\Pi} + \overset{(4)}{\Pi} \\ &= \sum_{i=0}^{\infty} \varepsilon^i \Big(\overset{(1)}{\Pi}_i (x, \eta) + \overset{(2)}{\Pi}_i (\xi, y) + \overset{(3)}{\Pi}_i (x, \eta_*) + \overset{(4)}{\Pi}_i (\xi_*, y) \Big)\end{aligned}$$

where $\eta = y/\varepsilon$, $\xi = x/\varepsilon$, $\eta_* = (b - y)/\varepsilon$, $\xi_* = (a - x)/\varepsilon$ are the boundary layer variables. In the vicinity of the corresponding side of the rectangle, each of them plays the same role as the variable $\rho = r/\varepsilon$ near $\partial\Omega$ in Section VI. In order to avoid describing the standard procedure for defining Π-functions for each side (this procedure is quite similar to the one described in Section VI), we describe only the corresponding boundary layer operator and boundary conditions.

The functions $\overset{(1)}{\Pi}_i (x, \eta)$, describing the boundary layer near the side $y = 0$, are defined by the boundary layer operator

$$\frac{\partial^2}{\partial y^2} - k^2(x, 0) \qquad \eta > 0$$

and the boundary conditions

$$\overset{(1)}{\Pi}_i (x, 0) = -\bar{u}_i(x, 0) \qquad \overset{(1)}{\Pi}_i (x, \infty) = 0$$

Thus, the $\overset{(1)}{\Pi}$-functions compensate for the discrepancy introduced by the regular part of the asymptotic expansion to the boundary condition on the side $y = 0$.

For $\overset{(1)}{\Pi}_0(x, \eta)$, we obtain

$$\overset{(1)}{\Pi}_0 = -\bar{u}_0(x, 0)\exp\{-k(x, 0)\eta\} \tag{7.4}$$

and the explicit expressions for $\overset{(1)}{\Pi}_i(x, \eta)$ can be found successively for $i = 1, 2, \ldots$. All these functions satisfy the exponential estimate

$$\left|\overset{(1)}{\Pi}_i(x, \eta)\right| \leq C\exp\{-\alpha\eta\} \tag{7.5}$$

In a similar way, the functions $\overset{(2)}{\Pi}_i(\xi, y)$ are defined by the boundary layer operator

$$\frac{\partial^2}{\partial\xi^2} - k^2(0, y) \qquad \xi > 0$$

and the boundary conditions

$$\overset{(2)}{\Pi}_i(0, y) = -\bar{u}_i(0, y) \qquad \overset{(2)}{\Pi}_i(\infty, y) = 0$$

For $\overset{(2)}{\Pi}_0(\xi, y)$, we obtain

$$\overset{(2)}{\Pi}_0(\xi, y) = -\bar{u}_0(0, y)\exp\{-k(0, y)\xi\} \tag{7.6}$$

The functions $\overset{(2)}{\Pi}_i(\xi, y)$ $(i = 1, 2, \ldots)$ can also be found explicitly and satisfy estimates of the type (7.5)

$$\left|\overset{(2)}{\Pi}_i(\xi, y)\right| \leq C\exp\{-\alpha\xi\}$$

The boundary functions $\overset{(3)}{\Pi}_i(x, \eta_*)$ and $\overset{(4)}{\Pi}_i(\xi_*, y)$ are defined in a similar way.

B. Corner Boundary Functions

Note that the boundary functions $\overset{(1)}{\Pi}_i(x, \eta)$, compensating for the discrepancy in the boundary condition on the side $y = 0$, introduce additional discrepancies in the boundary conditions on the sides $x = 0$ and $x = a$.

These discrepancies are essential in the vicinities of the corner points (0, 0) and $(a, 0)$ and, according to (7.5), they decay exponentially as η increases. Similar discrepancies are introduced by the functions $\overset{(2)}{\Pi}_i(\xi, y)$ in the boundary conditions on the sides $y = 0$ and $y = b$, by the functions $\overset{(3)}{\Pi}_i(x, \eta_*)$ in the boundary conditions on $x = 0$ and $x = a$, and by the functions $\overset{(4)}{\Pi}_i(\xi_*, y)$ in the boundary conditions on $y = 0$ and $y = b$.

To compensate for these discrepancies we introduce corner boundary functions. One corner boundary function must correspond to each of the vertices of the rectangle, and therefore P contains terms of four types:

$$
\begin{aligned}
P &= \overset{(1)}{P} + \overset{(2)}{P} + \overset{(3)}{P} + \overset{(4)}{P} \\
&= \sum_{i=0}^{\infty} \left(\overset{(1)}{P}_i(\xi, \eta) + \overset{(2)}{P}_i(\xi, \eta_*) + \overset{(3)}{P}_i(\xi_*, \eta_*) + \overset{(4)}{P}_i(\xi_*, \eta) \right)
\end{aligned}
$$

The functions $\overset{(1)}{P}_i$, for example, are needed to compensate for the discrepancies introduced by the $\overset{(1)}{\Pi}$-functions in the boundary condition along $x = 0$ and by $\overset{(2)}{\Pi}$-functions in the boundary condition on $y = 0$. The equation for the functions $\overset{(1)}{P}_i(\xi, \eta)$ can be obtained from the original Eq. (7.1) [more precisely, from the homogeneous equation corresponding to (7.1)] in a standard way. We have to introduce the new variables $\xi = x/\varepsilon$, $\eta = y/\varepsilon$, expand the coefficient $k^2(\varepsilon\xi, \varepsilon\eta)$ into a power series in ε, and equate coefficients of like powers of ε on both sides of the homogeneous equation corresponding to (7.1). This leads to the following equations $(i = 0, 1, 2, \ldots)$:

$$
\frac{\partial^2 \overset{(1)}{P}_i}{\partial \xi^2} + \frac{\partial^2 \overset{(1)}{P}_i}{\partial \eta^2} - k^2(0, 0)\overset{(1)}{P}_i = p_i(\xi, \eta) \qquad \xi > 0 \qquad \eta > 0 \tag{7.7}
$$

Here the functions $p_i(\xi, \eta)$ are expressed recursively through $\overset{(1)}{P}_j(\xi, \eta)$ with subscripts $j < i$, in particular $p_0(\xi, \eta) = 0$. Since the purpose of $\overset{(1)}{P}_i(\xi, \eta)$ is to compensate for the discrepancies in the boundary conditions introduced by the boundary layer functions $\overset{(1)}{\Pi}_i$ and $\overset{(2)}{\Pi}_i$, we should prescribe additional conditions for (7.7) in the form

$$
\overset{(1)}{P}_i(0, \eta) = -\overset{(1)}{\Pi}_i(0, \eta) \qquad \overset{(1)}{P}_i(\xi, 0) = -\overset{(2)}{\Pi}_i(\xi, 0) \tag{7.8}
$$

Further, we demand that the $\overset{(1)}{P}$-functions tend to zero when either of the boundary layer variables tends to infinity.

$$\overset{(1)}{P}_i(\xi, \eta) \to 0 \qquad \text{as} \qquad (\xi + \eta) \to \infty \tag{7.9}$$

Solutions of problems (7.7)–(7.9) can be successively expressed in explicit form using a Green's function. Consider the problem for $\overset{(1)}{P}_0(\xi, \eta)$, that is, Eq. (7.7) and the boundary conditions (7.8), (7.9) with $i = 0$. By virtue of expressions (7.4) and (7.6), the boundary conditions (7.8) can be rewritten as

$$\overset{(1)}{P}_0(0, \eta) = q \exp\{-k\eta\} \qquad \overset{(1)}{P}_0(\xi, 0) = q \exp\{-k\xi\}$$

where $q = u_0(0, 0)$ and $k = k(0, 0)$. Introducing the new dependent variable v by the formula

$$\overset{(1)}{P}_0(\xi, \eta) = q \exp\{-k(\xi + \eta)\} + v(\xi, \eta) \qquad \xi > 0 \qquad \eta > 0 \tag{7.10}$$

we obtain the following problem for $v(\xi, \eta)$:

$$\frac{\partial^2 v}{\partial \xi^2} + \frac{\partial^2 v}{\partial \eta^2} - k^2 v = -qk^2 \exp\{-k(\xi + \eta)\} \qquad \xi > 0 \qquad \eta > 0$$

$$v(0, \eta) = 0 \qquad v(\xi, 0) = 0 \qquad v(\xi, \eta) \to 0 \qquad \text{as} \qquad (\xi + \eta) \to \infty$$

The solution of this problem has the form

$$v(\xi, \eta) = -\int_0^\infty \int_0^\infty G(\xi, \eta, \xi_0, \eta_0) qk^2 \exp\{-k(\xi_0 + \eta_0)\}\, d\xi_0\, d\eta_0$$

Here

$$G(\xi, \eta, \xi_0, \eta_0) = \frac{1}{2\pi} [K_0(kR_1) + K_0(kR_2) - K_0(kR_3) - K_0(kR_4)]$$

is the Green's function, with $K_0(z)$ being the Bessel function with imaginary argument,

$$R_1 = \sqrt{(\xi - \xi_0)^2 + (\eta - \eta_0)^2} \qquad R_2 = \sqrt{(\xi + \xi_0)^2 + (\eta + \eta_0)^2}$$

$$R_3 = \sqrt{(\xi - \xi_0)^2 + (\eta + \eta_0)^2} \qquad R_4 = \sqrt{(\xi + \xi_0)^2 + (\eta - \eta_0)^2}$$

To estimate $v(\xi,\eta)$, we will use a rough but adequate estimate of the function $K_0(z)$ for $z>0$

$$0<K_0(z)\le C(|\ln z|+1)\exp\{-z\}$$

where $C>0$ is a constant. This estimate follows from the asymptotics of $K_0(z)$ for small and large z. We also use the inequality $K_0(z_1)\le K_0(z_2)$ for $z_1\ge z_2$. Then we have

$$G(\xi,\eta,\xi_0,\eta_0)\le\frac{1}{\pi}K_0(kR_1)\le C(|\ln R_1|+1)\exp\{-kR_1\}$$

Taking into account the inequality

$$R_1=\sqrt{(\xi-\xi_0)^2+(\eta-\eta_0)^2}\ge\tfrac{1}{2}(|\xi-\xi_0|+|\eta-\eta_0|)$$

we obtain

$$|v(\xi,\eta)|\le\int_0^\infty\int_0^\infty C(|\ln R_1|+1)\exp\left\{-\frac{k}{2}R_1\right\}$$
$$\exp\left\{-\frac{k}{4}(|\xi-\xi_0|+|\eta-\eta_0|)-k(\xi_0+\eta_0)\right\}d\xi_0\,d\eta_0$$

It can be easily seen that

$$-\frac{k}{4}(|\xi-\xi_0|+|\eta-\eta_0)-k(\xi_0+\eta_0)\le-\alpha(\xi+\eta)$$

for any α such that $0<\alpha\le k/4$. Finally, taking into account the inequality

$$\int_0^\infty\int_0^\infty(|\ln R_1|+1)\exp\left\{-\frac{k}{2}R_1\right\}d\xi_0\,d\eta_0\le C$$

we obtain the estimate for $v(\xi,\eta)$:

$$|v(\xi,\eta)|\le C\exp\{-\alpha(\xi+\eta)\}$$

It follows from (7.10) that $\overset{(1)}{P}_0(\xi,\eta)$ satisfies the same estimate

$$|\overset{(1)}{P}_0(\xi,\eta)|\le C\exp\{\alpha(\xi+\eta)\}\tag{7.11}$$

For the other functions $\overset{(1)}{P}_i(\xi,\eta)$ $(i\ge1)$ this estimate can be shown easily by induction. For $\overset{(1)}{P}_i(\xi,\eta)$ we have Eq. (7.7), where the right-hand side depends linearly on the $\overset{(1)}{P}_j(\xi,\eta)$ with $j<i$. Consequently, if the $\overset{(1)}{P}_j(\xi,\eta)$ for $j<i$ satisfy estimates of type (7.11), the same bound will hold for $p_i(\xi,\eta)$. But then, following the steps taken above, an estimate, similar to that for $\overset{(1)}{P}_0(\xi,\eta)$, can be obtained for $\overset{(1)}{P}_i(\xi,\eta)$.

Note that, due to the estimate (7.11), the functions $\overset{(1)}{P}_i(\xi,\eta)$ are significant only in some small vicinity of the corner point $(0,0)$ of the rectangle Ω. These functions vanish exponentially as their arguments move away from the point $(0,0)$. This explains the name of these functions—*corner boundary functions*.

In what follows it is convenient to introduce the following:

Definition. *We will say that the* $\overset{(1)}{P}$*-functions are defined by the operator*

$$L_p = \frac{\partial^2}{\partial\xi^2} + \frac{\partial^2}{\partial\eta^2} - k^2(0,0)$$

and the boundary conditions (7.8), (7.9). We will call L_p *the boundary layer operator in the vicinity of the vertex (0, 0).*

The other corner boundary functions $\overset{(\ell)}{P}_i$ $(\ell=2,3,4)$, playing an important role in the vicinities of the other vertices of the rectangle Ω, are defined analogously and satisfy exponential estimates similar to (7.11).

The following statement holds:

Series (7.3) is an asymptotic series for the solution $u(x, y, \varepsilon)$ *of problem (7.1), (7.2) in the rectangle* $\bar{\Omega}$ *as* $\varepsilon\to 0$.

C. The Role of First-Order Derivatives

Equation (7.1) does not contain first-order derivatives. The boundary layer structure of the asymptotics may change substantially if terms with lower order derivatives (even with small factors multiplying them) are added to Eq. (7.1). As an example, consider the equation

$$\varepsilon^2\,\Delta u - \varepsilon^\alpha A(x, y)\frac{\partial u}{\partial y} - k^2(x, y)u = f(x, y, \varepsilon)$$

in the rectangle $\Omega=(0<x<a)\times(0<y<b)$, with the boundary conditions (7.2). Here, in comparison with (7.1), the term $\varepsilon^\alpha A(x, y)(\partial u/\partial y)$ is added. For $\alpha\geq 1$ the asymptotic structure of the solution is similar to that described in Sections VII.A and B. In the case $0<\alpha<1$, however, the boundary layer operators in the vicinities of the sides $y = 0$ and $y = b$ as well as near the vertices of Ω change substantially. Their form now depends on the sign of the coefficient $A(x, y)$ on these sides. Let

$$A(x, 0) > 0 \qquad A(x, b) > 0$$

and $\alpha = \frac{1}{2}$. In this case, the asymptotic expansion is a series in powers of $\sqrt{\varepsilon}$. The boundary layer functions $\overset{(1)}{\Pi}_i(x, \eta)$ in the vicinity of the side $y = 0$ depend on the new variable $\eta = y/\sqrt{\varepsilon}$ and are defined by the boundary layer operator $-A(x, 0)(\partial/\partial\eta) - k^2(x, 0)$ $(\eta > 0)$ and the boundary condition $\overset{(1)}{\Pi}_i(x, 0) = -\bar{u}_i(x, 0)$, where $\bar{u}_i(x, y)$ are the regular terms of the asymptotic solution. The boundary layer functions $\overset{(3)}{\Pi}_i(x, \eta_*)$, with $\eta_* = (b - y)/\varepsilon^{3/2}$, near the side $y = b$ are defined by the boundary layer operator

$$\frac{\partial^2}{\partial \eta_*^2} + A(x, b)\frac{\partial}{\partial \eta_*} \qquad (\eta_* > 0)$$

and the boundary conditions $\overset{(3)}{\Pi}_i(x, 0) = -\bar{u}_i(x, b)$, $\overset{(3)}{\Pi}_i(x, \infty) = 0$.

We see that the boundary layer operators in the neighborhoods of the sides $y = 0$ and $y = b$ are different from those of Section VII.A. Another distinguishing feature of this problem is the presence of boundary layer variables with different scales. Near the side $y = 0$ the variable y is stretched by $\sqrt{\varepsilon}$ and near the side $y = 0$ it is stretched by $\varepsilon^{3/2}$.

In the vicinities of the sides $x = 0$ and $x = a$ the boundary layer operators do not change. For example, the boundary functions $\overset{(2)}{\Pi}_i(\xi, y)$ $(\xi = x/\varepsilon)$ near the side $x = 0$ are defined as before, by the boundary layer operator

$$\frac{\partial^2}{\partial \xi^2} - k^2(0, y) \qquad (\xi > 0)$$

and the boundary conditions $\overset{(2)}{\Pi}_i(0, y) = -\bar{u}_i(0, y)$, $\overset{(2)}{\Pi}_i(\infty, y) = 0$.

Substantial changes take place in the corner boundary layer operators. Corner boundary functions $\overset{(1)}{P}_i(\xi, \eta)$ near the vertex $(0, 0)$ are now defined by the parabolic operator

$$\frac{\partial^2}{\partial \xi^2} - A(0, 0)\frac{\partial}{\partial \eta} - k^2(0, 0) \qquad (\xi > 0, \eta > 0)$$

and the boundary conditions

$$\overset{(1)}{P}_i(\xi, 0) = -\overset{(2)}{\Pi}_i(\xi, 0) \qquad \overset{(1)}{P}_i(0, \eta) = -\overset{(1)}{\Pi}_i(0, \eta)$$

A similar parabolic boundary layer operator appears in the vicinity of the vertex $(a, 0)$.

Two types of corner boundary functions appear near each of the vertices $(0, b)$ and (a, b). Let us consider them, for example, for the vertex $(0, b)$. Corner boundary functions of the first type [we will denote them by $\overset{(2)}{R}_i(\xi, \eta_*)$] are defined by the ordinary differential operator

$$\frac{\partial^2}{\partial \eta_*^2} + A(0, b)\frac{\partial}{\partial \eta_*} \qquad (\eta_* > 0)$$

and the boundary conditions $\overset{(2)}{R}_i(\xi, 0) = -\overset{(2)}{\Pi}_i(\xi, b)$, $\overset{(1)}{R}_i(\xi, \infty) = 0$. They introduce a discrepancy in the boundary condition at $x = 0$. Corner boundary functions of the second type $\overset{(2)}{P}_i(\zeta, \eta_*)$, $(\zeta = x/\varepsilon^{3/2})$ are defined with the help of the elliptic operator

$$\frac{\partial^2}{\partial \zeta^2} + \frac{\partial^2}{\partial \eta_*^2} + A(0, b)\frac{\partial}{\partial \eta_*} \qquad (\zeta > 0, \eta_* > 0)$$

and the boundary conditions

$$\overset{(2)}{P}_i(\zeta, 0) = 0 \qquad \overset{(2)}{P}_i(0, \eta_*) = -\left[\overset{(3)}{\Pi}_i(0, \eta_*) + \overset{(2)}{R}_i(0, \eta_*)\right]$$

$$\overset{(2)}{P}_i(\zeta, \eta_*) \to 0 \qquad \text{as} \qquad (\zeta + \eta_*) \to \infty$$

All the boundary functions satisfy exponential decay estimates of type (7.5) and (7.11) with respect to the boundary layer variables.

Note that the corner boundary functions $\overset{(1)}{P}_0(\xi, \eta)$ and $\overset{(1)}{P}_1(\xi, \eta)$ have unbounded second derivatives with respect to η in the vicinity of the point $(0, 0)$. [Likewise for the functions $\overset{(4)}{P}_i(\xi_*, \eta)$, $i = 0, 1$ in the vicinity of the vertex $(a, 0)$.] These second derivatives enter the right-hand sides of the equations for $\overset{(1)}{P}_2$ and $\overset{(1)}{P}_3$, respectively. In turn, the second-order derivatives with respect to η, of the functions $\overset{(1)}{P}_2$ and $\overset{(1)}{P}_3$ having higher order singularities, enter the right-hand sides of the equations for the subsequent terms of the asymptotics. Thus, the order of the singularities of the $\overset{(1)}{P}$-functions grow in proportion to the serial number of these functions. This does not permit the use of the iterative process for defining the functions $\overset{(1)}{P}_i$ (as well as $\overset{(4)}{P}_i$) beyond the first step. If we restrict the asymptotics to terms of only zeroth order, then their sum will give a uniform asymptotic approximation of the solution in $\bar{\Omega}$ with accuracy of order $\sqrt{\varepsilon}$.

If we exclude some small (but fixed as $\varepsilon \to 0$) vicinities of the vertices $(0, 0)$ and $(a, 0)$, then in the remaining part of the rectangle $\bar{\Omega}$, the n-th partial sum of the constructed expansion gives an approximation to the solution with an accuracy of order $\varepsilon^{(n+1)/2}$.

D. Parabolic Equations

Singularly perturbed parabolic systems of equations of the type

$$\begin{aligned} \varepsilon\left(\frac{\partial u}{\partial t} - a\,\Delta u\right) &= f(u, v, x, t, \varepsilon) \\ \frac{\partial v}{\partial t} - b\,\Delta v &= g(u, v, x, t, \varepsilon) \end{aligned} \tag{7.12}$$

appear naturally for describing chemical reactions with diffusion. Here $x = (x_1, x_2, x_3)$ is the three-dimensional spatial variable, t is time, the components of the vector functions u and v are the concentrations of the reacting substances. The small parameter ε is a quantity inversely proportional to the constants of the fast reactions. Systems of similar kind appear also in some other applied problems. We will consider the construction of an asymptotic expansion in ε for the solution of the boundary value problem for (7.12). To simplify the description of the algorithm, we will restrict our study to the case when the slow variables are absent (i.e., the second equation in (7.12) will be omitted), and x is the one-dimensional spatial variable. Further, for convenience, let us denote the small parameter in front of the derivatives by ε^2. Thus, we will consider a system of equations

$$\begin{gathered} \varepsilon^2\left(\frac{\partial u}{\partial t} - a(x, t)\frac{\partial^2 u}{\partial x^2}\right) = f(u, x, t, \varepsilon) \\ (x, y) \in \Omega = \{(0 < x < 1) \times (0 < t \le T)\} \end{gathered} \tag{7.13}$$

where u is an m-dimensional vector-function, subject to the initial condition

$$u(x, 0, \varepsilon) = \varphi(x) \tag{7.14}$$

and Neumann boundary conditions

$$\frac{\partial u}{\partial x}(0, t, \varepsilon) = \frac{\partial u}{\partial x}(1, t, \varepsilon) = 0 \tag{7.15}$$

(such boundary conditions are common in problems of chemical kinetics).

Let the following conditions be satisfied:

Condition 1^0. *The functions $a(x, t) > 0$, $f(u, x, t, \varepsilon)$, and φ are sufficiently smooth (they should be $n+2$ times continuously differentiable to enable us to construct the n-th order asymptotics), and conditions (7.14) and (7.15) are matched at the corner points (0, 0) and (1, 0) of the rectangle Ω: $\varphi'(0) = \varphi'(1) = 0$.*

Condition 2^0. *The reduced equation $f(\bar{u}, x, t, 0) = 0$ [obtained from (7.13) for $\varepsilon = 0$] has a solution $\bar{u}_0 = \bar{u}_0(x, t)$ in the rectangle $\Omega = \{(0 \le x \le 1) \times (0 \le t \le T)\}$.*

Notice that (because of the nonlinearity) this equation may also have other solutions.

Condition 3^0. *All the eigenvalues $\lambda_i(x, t)$ $(i = 1, \ldots, m)$ of the matrix $\bar{f}_u(x, t) = f_u(\bar{u}_0(x, t), x, t, 0)$ have negative real parts: Re $\lambda_i(x, t) < 0$, $(x, t) \in \bar{\Omega}$.*

Condition 4^0. *The solution $\Pi_0(x, \tau)$ of the initial value problem (x enters as a parameter, $0 \le x \le 1$).*

$$\begin{aligned} \frac{\partial \Pi_0}{\partial \tau} &= f(\bar{u}_0(x, 0) + \Pi_0, x, 0, 0) \qquad \tau \ge 0 \\ \Pi_0(x, 0) &= \varphi(x) - \bar{u}_0(x, 0) \end{aligned} \tag{7.16}$$

exists for $\tau \ge 0$ and satisfies the condition $\Pi_0(x, \infty) = 0$.

Condition 4^0 implies that the initial value $\varphi(x) - \bar{u}_0(x, 0)$ belongs to the domain of attraction of the rest point $\Pi_0 = 0$ for Eq. (7.16). This rest point is asymptotically stable by virtue 3^0.

E. Construction of an Asymptotic Expansion for the Parabolic Problem

Under conditions 1^0–4^0, we construct an asymptotic solution of (7.13)–(7.15) in the form

$$\begin{aligned} u &= \bar{u} + \Pi + Q + Q^* + P + P^* = \\ &= \sum_{i=0}^{\infty} \varepsilon^i [\bar{u}_i(x, t) + \Pi_i(x, \tau) + Q_i(\xi, t) + Q_i^*(\xi_*, t) + \\ &\quad + P_i(\xi, \tau) + P_i^*(\xi_*, \tau)] \end{aligned} \tag{7.17}$$

Here $\tau = t/\varepsilon^2$, $\xi = x/\varepsilon$, $\xi_* = (1-x)/\varepsilon$ are boundary layer variables; $\bar{u}_i$ are coefficients of the regular series, Π_i, Q_i, Q_i^* are the boundary layer functions describing the boundary layers in the vicinities of the sides $t=0$, $x=0$, and $x=1$, respectively. The corner boundary functions P_i and P_i^* are needed to describe the boundary layers near the vertices $(0,0)$ and $(1,0)$.

Let us substitute the series (7.17) into Eq. (7.13) and change the right-hand side $f(\bar{u} + \Pi + Q + Q^* + P + P^*, x, t, \varepsilon)$ into an expression of type (7.17)

$$\bar{f} + \Pi f + Qf + Q^*f + Pf + P^*f \tag{7.18}$$

where

$$\bar{f} = f(\bar{u}, x, t, \varepsilon) \qquad \Pi f = \left[f(\bar{u} + \Pi, x, t, \varepsilon) - \bar{f}\right]\Big|_{t=\varepsilon^2\tau}$$

$$Qf = \left[f(\bar{u} + Q, x, t, \varepsilon) - \bar{f}\right]\Big|_{x=\varepsilon\xi}$$

$$Pf = \left[f(\bar{u} + \Pi + Q + P, x, t, \varepsilon) - \Pi f - Qf - \bar{f}\right]\Big|_{\substack{t=\varepsilon^2\tau \\ x=\varepsilon\xi}}$$

Q^*f and P^*f are similar to Qf and Pf. Now, expanding the terms of (7.18) into power series in ε and equating coefficients of like powers of ε separately for the regular functions and for each type of boundary function, we obtain equations for the terms of the asymptotic expansion. Additional conditions are obtained by substituting (7.17) into (7.14) and (7.15). This gives

$$\bar{u}(x, 0, \varepsilon) + \Pi(x, 0, \varepsilon) = \varphi(x) \tag{7.19}$$

$$Q(\xi, 0, \varepsilon) + P(\xi, 0, \varepsilon) = 0 \qquad Q^*(\xi_*, 0, \varepsilon) + P^*(\xi_*, 0, \varepsilon) = 0 \tag{7.20}$$

$$\frac{\partial \bar{u}}{\partial x}(0, t, \varepsilon) + \frac{1}{\varepsilon}\frac{\partial Q}{\partial \xi}(0, t, \varepsilon) = 0 \qquad \frac{\partial \bar{u}}{\partial x}(1, t, \varepsilon) - \frac{1}{\varepsilon}\frac{\partial Q^*}{\partial \xi_*}(0, t, \varepsilon) = 0 \tag{7.21}$$

$$\frac{\partial \Pi}{\partial x}(0, \tau, \varepsilon) + \frac{1}{\varepsilon}\frac{\partial P}{\partial \xi}(0, \tau, \varepsilon) = 0 \qquad \frac{\partial \Pi}{\partial x}(1, \tau, \varepsilon) - \frac{1}{\varepsilon}\frac{\partial P^*}{\partial \xi_*}(1, \tau, \varepsilon) = 0 \tag{7.22}$$

For $\bar{u}_0(x, t)$ we obtain the equation

$$\bar{f} \equiv f(\bar{u}_0, x, t, 0) = 0$$

which coincides with the reduced equation. Hence, $\bar{u}_0 = \bar{u}_0(x, t)$. For $\bar{u}_i$

$(i \geq 1)$ we have linear equations of the type

$$\bar{f}_u(x, t)\bar{u}_i = f_i(x, t)$$

where the $f_i(x, t)$ are expressed recursively through the $\bar{u}_j(x, t)$ with $j < i$. Therefore, $\bar{u}_i = \bar{f}_u^{-1}(x, t) f_i(x, t)$. Note that $\bar{f}_u^{-1}(x, t)$ exists by virtue of condition 3^0.

For $\Pi_0(x, \tau)$ we obtain the equation (x enters as a parameter, $0 \leq x \leq 1$)

$$\frac{\partial \Pi_0}{\partial \tau} = \Pi_0 f \equiv f(\bar{u}_0(x, 0) + \Pi_0, x, 0, 0) \qquad \tau \geq 0$$

with the initial condition

$$\Pi_0(x, 0) = \varphi(x) - \bar{u}_0(x, 0)$$

This condition follows from (7.19). The problem for $\Pi_0(x, \tau)$ evidently coincides with problem (7.16) from condition 4^0. The following exponential estimate holds for $\Pi_0(x, \tau)$ by virtue of 3^0 and 4^0.

$$\|\Pi_0(x, \tau)\| \leq C \exp\{-\alpha\tau\} \tag{7.23}$$

Subsequent $\Pi_i(x, \tau)$ $(i \geq 1)$ are defined as solutions of the linear initial value problems

$$\frac{\partial \Pi_i}{\partial \tau} = f_u(x, \tau)\Pi_i + \pi_i(x, \tau) \qquad \Pi_i(x, 0) = -\bar{u}_i(x, 0) \tag{7.24}$$

where

$$f_u(x, \tau) = f_u(\bar{u}_0(x, 0) + \Pi_0(x, \tau), x, 0, 0) \tag{7.25}$$

and the $\pi_i(x, \tau)$ are expressed recursively through $\Pi_j(x, \tau)$, $j < i$. We will use the notation $\Phi(x, \tau)$ for the fundamental matrix of the corresponding homogeneous equation [$\Phi(x, 0) = E$, the identity $m \times m$ – matrix]. By virtue of condition 3^0, the following estimate holds:

$$\begin{gathered} \|\Phi(x, \tau)\Phi^{-1}(x, \tau_0)\| \leq C \exp\{-\alpha(\tau - \tau_0)\} \\ 0 \leq x \leq 1 \qquad 0 \leq \tau_0 \leq \tau \end{gathered} \tag{7.26}$$

By using this estimate and the explicit expression for $\Pi_i(x, \tau)$, which has the form

$$\Pi_i(x, \tau) = -\Phi(x, \tau)\bar{u}_i(x, 0) + \int_0^\tau \Phi(x, \tau)\,\Phi^{-1}(x, \tau_0)\pi_i(x, \tau_0)\,d\tau_0$$

it is easy to prove by induction that all of the Π-functions satisfy an exponential estimate of type (7.23).

For $Q_0(\xi, t)$ we obtain the equation (t enters as a parameter, $0 \le t \le T$)

$$-a(0, t)\frac{\partial^2 Q_0}{\partial \xi^2} = Q_0 f \equiv f(\bar{u}_0(0, t) + Q_0, 0, t, 0) \qquad \xi \ge 0$$

with the boundary condition

$$\frac{\partial Q_0}{\partial t}(0, t) = 0$$

This condition follows from the first relation (7.21). We will also impose the second boundary condition

$$Q_0(\infty, t) = 0$$

Then $Q_0(\xi, t) = 0$ and for $Q_i(\xi, t)$ $(i \ge 1)$ we obtain linear boundary value problems of the type

$$-a(0, t)\frac{\partial^2 Q_i}{\partial \xi^2} = \bar{f}_u(0, t)Q_i + q_i(\xi, t) \tag{7.27}$$

$$\frac{\partial Q_i}{\partial \xi}(0, t) = -\frac{\partial \bar{u}_{i-1}}{\partial x}(0, t) \qquad Q_i(\infty, t) = 0 \tag{7.28}$$

where the $q_i(\xi, t)$ are expressed recursively through $Q_j(\xi, t)$, $j < i$. The roots of the characteristic equation for (7.27) are

$$\pm\sqrt{-\lambda_k(0, t)/a(0, t)} \qquad k = 1, 2, \ldots, m$$

where $\lambda_k(x, t)$ are the eigenvalues of $\bar{f}_u(x, t)$. By virtue of condition 3^0 m roots have negative and m roots have positive real parts. Therefore the homogeneous equation corresponding to (7.27) has m exponentially decreasing and m exponentially increasing solutions as $\xi \to \infty$. This establishes the unique solvability of problems (7.27), (7.28). The solutions can be found successively in explicit form to arbitrary order by transforming the matrix $\bar{f}_u(0, t)$ to Jordan canonical form. All these functions satisfy the exponential estimate

$$\|Q_i(\xi, t)\| \le C \exp\{-\alpha\xi\}$$

The functions $Q_i^*(\xi_*, t)$ are defined analogously to the Q functions and satisfy a similar exponential estimate.

As in the case of elliptic equations, corner boundary functions $P_i(\xi, \tau)$ and $P_i^*(\xi_*, \tau)$ are needed to compensate for the discrepancies introduced by the Π-functions in the boundary conditions at $x = 0$ and $x = 1$, as well as by the Q- and Q^*-functions in the initial condition at $t = 0$. Equalities (7.20) and (7.22) reflect this role of the corner boundary functions. The initial and boundary conditions for P_i and P_i^* can be found from these equalities. Further, we demand that the P-functions approach zero as each of the boundary layer variables tends to infinity.

For $P_0(\xi, \tau)$ we obtain the following problem:

$$\frac{\partial P_0}{\partial \tau} - a\frac{\partial^2 P_0}{\partial \xi^2} = P_0 f \equiv f(\bar{u}_0(0,0) + \Pi_0(0,\tau) + P_0, 0, 0, 0)$$

$$- f(\bar{u}_0(0,0) + \Pi_0(0,\tau), 0, 0, 0) \qquad \xi > 0 \qquad \tau > 0$$

$$P_0(\xi, 0) = -Q_0(\xi, 0) = 0 \qquad \frac{\partial P_0}{\partial \xi}(0, \tau) = 0$$

where $a = a(0, 0)$. It can be easily seen that the unique solution of this problem is trivial: $P_0(\xi, \tau) = 0$. For $P_i(\xi, \tau)$ $(i \ge 1)$ we have the linear problems

$$\frac{\partial P_i}{\partial \tau} - a\frac{\partial^2 P_i}{\partial \xi^2} - f_u(0,\tau)P_i = p_i(\xi, \tau)$$

$$P_i(\xi, 0) = -Q_i(\xi, 0) \qquad \frac{\partial P_i}{\partial \xi}(0,\tau) = -\frac{\partial \Pi_{i-1}}{\partial x}(0, \tau) \tag{7.29}$$

where $f_u(0, \tau)$ is defined by (7.25), and the $p_i(\xi, \tau)$ are expressed recursively through Π_j, Q_j $(j \le i)$ and P_j $(j < i)$. Notice that the initial and boundary conditions for $P_i(\xi, t)$ are matched at the corner point $(0, 0)$, that is, $(\partial Q_i/\partial\xi)(0, 0) = [(\partial\Pi_{i-1}/(\partial x)](0, 0)$. This follows from the equalities

$$\frac{\partial Q_i}{\partial \xi}(0,0) = -\frac{\partial \bar{u}_{i-1}}{\partial x}(0,0)$$

[see (7.28)] and $\Pi_{i-1}(x, 0) = -\bar{u}_{i-1}(x, 0)$ [see (7.24)]; for $i = 1$ the condition $\varphi'(0) = 0$ should also be taken into account. The functions $P_i(\xi, \tau)$ can be found in the explicit form

$$P_i(\xi, \tau) = g_i(\xi, \tau) + \int_0^\tau \int_0^\infty G(\xi, \tau, \xi_0, \tau_0) h_i(\xi_0, \tau_0)\, d\xi_0\, d\tau_0$$

where $g_i(\xi, \tau)$ is an arbitrary sufficiently smooth function satisfying conditions (7.29), for example,

$$g_i(\xi, \tau) = -Q_i(\xi, 0) \exp\{-\alpha\tau\}$$
$$- \left[\frac{\partial \Pi_{i-1}}{\partial x}(0, \tau) - \frac{\partial \Pi_{i-1}}{\partial x}(0, 0) \exp\{-\alpha\tau\} \right] \frac{1}{\alpha} \exp\{-\alpha\xi\}$$

$$h_i(\xi, \tau) = p_i(\xi, \tau) - \frac{\partial g_i}{\partial \tau} + a \frac{\partial^2 g_i}{\partial \xi^2} + f_u(0, \tau) g_i$$

$G(\xi, \tau, \xi_0, \tau_0)$ is the Green's matrix which has the form

$$G(\xi, \tau, \xi_0, \tau_0) = \Phi(0, \tau)\Phi^{-1}(0, \tau_0) \frac{1}{2\sqrt{\pi a(\tau - \tau_0)}}$$
$$\cdot \left[\exp\left\{ -\frac{(\xi - \xi_0)^2}{4a(\tau - \tau_0)} \right\} + \exp\left\{ -\frac{(\xi + \xi_0)^2}{4a(\tau - \tau_0)} \right\} \right]$$

$\Phi(0, \tau)$ is the same fundamental matrix as in (7.26). By virtue of inequality (7.26) the following estimate for the Green's matrix holds:

$$\|G(\xi, \tau, \xi_0, \tau_0)\| \le \frac{C}{\sqrt{\tau - \tau_0}} \exp\{-\alpha(|\xi - \xi_0| + \tau - \tau_0)\}$$

By using this estimate it is easy to prove that all of the P-functions satisfy the exponential estimate

$$\|P_i(\xi, \tau)\| \le C \exp\{-\alpha(\xi + \tau)\}$$

The functions $P_i^*(\xi_*, \tau)$ are defined analogously to the P-functions and satisfy a similar estimate.

Thus, we have completed the construction of the series (7.17).

It can be proved that

Under Conditions 1^0–4^0 *and for sufficiently small* ε, *problem (7.13)–(7.15) has a unique solution* $u(x, t, \varepsilon)$, *and the series (7.17) is an asymptotic expansion of this solution in the rectangle* $\bar{\Omega}$ *as* $\varepsilon \to 0$.

In a similar way an asymptotic solution can be constructed for the case of system (7.12) containing both fast variables (function u) and slow variables (function v). However, the addition of the slow variables leads to a situation where the asymptotics can be constructed only to the zeroth- and first-order approximations. This is related to the fact that the initial and boundary conditions for the function $\bar{v}_2(x, t)$ [the coefficient of ε^2 in the regular series for $v(x, t, \varepsilon)$] are not matched at the corner points $(0, 0)$ and $(1, 0)$. As a result, the function $\bar{v}_2(x, t)$ is not smooth in $\bar{\Omega}$. The derivatives $\partial \bar{v}_2/\partial t$ and $\partial^2 \bar{v}_2/\partial x^2$ are unbounded in the vicinities of the corner points (they have a singularity of order $1/\sqrt{t}$). The order of the singularity grows with each step of the iteration process for defining the terms of the asymptotic expansion, and this does not allow us to use the algorithm described above beyond the first step.

F. Other Problems with Corner Boundary Layers

The method of corner boundary functions is well developed also for equations of hyperbolic type [29], for systems of elliptic equations [30], for systems of parabolic equations [31], for partial differential equations in the multidimensional case [32], as well as for difference equations [33]. This method works successfully for a variety of applied problems.

G. Nonisothermal Fast Chemical Reactions

Consider a nonisothermal chemical reaction where one molecule of Substance A transforms to one molecule of Substance B. In the case of one spatial dimension and in the presence of diffusion and thermal conduction, the process is described by the system of parabolic equations (see [34])

$$\begin{aligned} \frac{\partial u}{\partial t} - D_A \frac{\partial^2 u}{\partial x^2} &= -k_0 \exp\left\{-\frac{E}{RT}\right\}u \\ \frac{\partial v}{\partial t} - D_B \frac{\partial^2 v}{\partial x^2} &= k_0 \exp\left\{-\frac{E}{RT}\right\}u \\ \frac{\partial T}{\partial t} - \frac{\lambda}{c\rho}\frac{\partial^2 T}{\partial x^2} &= \frac{qk_0}{c\rho}\exp\left\{-\frac{E}{RT}\right\}u \end{aligned} \tag{7.30}$$

Here $u(0 \leq u \leq 1)$ and $v(0 \leq v \leq 1)$ are relative nondimensional concentrations of Substances A and B, T is the absolute temperature, D_A and D_B are the diffusion coefficients of Substances A and B, k_0 is the rate constant of the reaction, E is the activation energy, R is the universal gas constant, λ is the thermal conductivity coefficient, c is the specific heat capacity, ρ is the density of the medium in which the reaction occurs, and q is the thermal effect of reaction. We assume that all of these quantities, with the exception of u, v, and T, are constant.

Let the reaction be fast, that is, k_0 is large. Set $k_0 = k/\varepsilon^2$ where ε is a small positive parameter, and introduce the nondimensional temperature θ, time t', spatial variable x', and constants a, b, α, m, n by the formulas

$$\theta = \frac{E}{RT_*^2}(T - T_*) \qquad t' = kt \qquad x' = \frac{x}{\ell}$$

$$a = \frac{D_A}{k\ell^2} \qquad b = \frac{D_B}{k\ell^2} \qquad \alpha = \frac{\lambda}{c\rho k_0 \ell^2}$$

$$m = \exp\left\{-\frac{E}{RT_*}\right\} \qquad n = \frac{Eq}{RT_*^2 c\rho} m$$

where T_* is a characteristic temperature, for example, the temperature of the surrounding medium, and ℓ is the length of the domain in which the reaction takes place. In the new variables, system (7.30) has the form (we again use the notation t, x instead of t', x'):

$$\begin{aligned}
\varepsilon^2\left(\frac{\partial u}{\partial t} - a\frac{\partial^2 u}{\partial x^2}\right) &= -m\exp\left\{\frac{\theta}{1+\beta\theta}\right\}u \\
\varepsilon^2\left(\frac{\partial v}{\partial t} - b\frac{\partial^2 v}{\partial x^2}\right) &= m\exp\left\{\frac{\theta}{1+\beta\theta}\right\}u \\
\varepsilon^2\left(\frac{\partial \theta}{\partial t} - \alpha\frac{\partial^2 \theta}{\partial x^2}\right) &= n\exp\left\{\frac{\theta}{1+\beta\theta}\right\}u
\end{aligned} \tag{7.31}$$

We will investigate the system (7.31) for $0 \le t \le T$ under the initial conditions

$$u|_{t=0} = u^0(x) \qquad v|_{t=0} = v^0(x) \qquad \theta|_{t=0} = 0 \tag{7.32}$$

and the boundary conditions

$$\left.\frac{\partial u}{\partial x}\right|_{\substack{x=0\\x=1}} = \left.\frac{\partial v}{\partial x}\right|_{\substack{x=0\\x=1}} = \left.\frac{\partial \theta}{\partial x}\right|_{\substack{x=0\\x=1}} = 0 \tag{7.33}$$

These boundary conditions mean that there are no fluxes of mass and heat through the boundaries.

Since the first and third equations of the system (7.31) do not contain the function v, we can consider these equations separately. We seek asymptotics of the solution in the form (7.17). Let us construct the leading terms of these asymptotics.

For the regular terms $\bar{u}_0$ and $\bar{\theta}_0$ we obtain one equation [it follows from (7.31) for $\varepsilon = 0$]

$$\exp\left\{\frac{\bar{\theta}_0}{1+\beta\bar{\theta}_0}\right\}\bar{u}_0 = 0$$

Hence, $\bar{u}_0 = 0$, and $\theta_0(x, t)$ is still unknown. Thus, the reduced system has a family of solutions depending on an arbitrary function, that is, we have a *critical case*. This is an important difference between system (7.31) and system (7.13), for which the reduced equation has an isolated solution.

The algorithm for constructing the asymptotics for the solution of system (7.31) is in many ways similar to the algorithm for the system of ordinary differential equations in the critical case described in Section IV.

For the first order, we have $\bar{u}_1 = 0$, and θ_1 is an arbitrary function. At the next step of the asymptotic algorithm, we obtain (a typical situation for problems in the critical case) the equation for the as yet unknown function $\bar{\theta}_0(x, t)$

$$\frac{\partial\bar{\theta}_0}{\partial t} - \alpha\frac{\partial^2\bar{\theta}_0}{\partial x^2} = 0 \tag{7.34}$$

The initial and boundary conditions for $\bar{\theta}_0(x, t)$ are defined during the construction of the boundary layer functions.

For the boundary layer functions $\Pi_0 u(x, \tau)$ and $\Pi_0\theta(x, \tau)$ $(\tau = t/\varepsilon^2)$, describing the solution near the initial moment of time, we obtain the system

$$\frac{\partial\Pi_0 u}{\partial\tau} = -m\exp\left\{\frac{\bar{\theta}_0(x,0)+\Pi_0\theta}{1+\beta(\bar{\theta}_0(x,0)+\Pi_0\theta)}\right\}\Pi_0 u$$

$$\frac{\partial\Pi_0\theta}{\partial\tau} = n\exp\left\{\frac{\bar{\theta}_0(x,0)+\Pi_0\theta}{1+\beta(\bar{\theta}_0(x,0)+\Pi_0\theta)}\right\}\Pi_0 u$$

Comparing the right-hand sides of these equations and introducing the notation $n/m = \gamma$, we arrive at the equality

$$\frac{\partial\Pi_0\theta}{\partial\tau} = -\gamma\frac{\partial\Pi_0 u}{\partial\tau}$$

Hence,

$$\Pi_0\theta(x,\tau) = -\gamma\Pi_0 u(x,\tau) + C(x)$$

where $C(x)$ is arbitrary. Let us impose the usual condition on the boundary layer function at infinity

$$\Pi_0\theta(x,\infty) = \Pi_0 u(x,\infty) = 0$$

Then $C(x) = 0$, and

$$\Pi_0\theta(x,\tau) = -\gamma\Pi_0 u(x,\tau) \tag{7.35}$$

From the initial conditions (7.32) for the zeroth order, we have the following relations

$$\bar{u}_0(x,0) + \Pi_0 u(x,0) = u^0(x) \qquad \bar{\theta}_0(x,0) + \Pi_0\theta(x,0) = 0$$

Taking into account $\bar{u}_0 = 0$ and (7.35), we obtain initial conditions for $\Pi_0 u$ and $\bar{\theta}_0$

$$\Pi_0 u(x,0) = u^0(x) \tag{7.36}$$

$$\bar{\theta}_0(x,0) = \gamma u^0(x) \tag{7.37}$$

By virtue of (7.35) and (7.37), the equation for $\Pi_0 u(x,\tau)$ reduces to

$$\frac{\partial \Pi_0 u}{\partial \tau} = -m \exp\left\{\frac{\gamma\left(u^0(x) - \Pi_0 u\right)}{1 + \beta\gamma\left(u^0(x) - \Pi_0 u\right)}\right\}\Pi_0 u \equiv -k(x, \Pi_0 u)\Pi_0 u$$

Since $k(x, \Pi_0 u) > 0$ for any x and $\Pi_0 u$, this equation with the initial condition (7.36) has a unique solution for which the exponential estimate

$$|\Pi_0 u(x,\tau)| \le C \exp\{-\alpha\tau\} \; (0 \le x \le 1 \qquad \tau \ge 0)$$

holds. The equation for $\Pi_0 u(x,\tau)$ can be solved implicitly

$$\int_{u^0(x)}^{\Pi_0 u(x,\tau)} \frac{ds}{k(x,s)s} = -\tau$$

The function $\Pi_0\theta(x,\tau)$ is expressed through $\Pi_0 u(x,\tau)$ by (7.35) and also satisfies an exponential decay estimate. Thus, the Π-functions for the zeroth order are defined, and for the as yet unknown $\bar{\theta}_0(x,t)$ we have the initial condition (7.37).

The boundary Q, Q^*-functions and corner P, P^*-functions of the zeroth order are identically zero as in Section VII.E. During the construction of the first-order Q-functions, we obtain a relation similar to (7.35)

$$Q_1\theta(\xi, t) = -\gamma Q_1 u(\xi, t) \qquad \xi = \frac{x}{\varepsilon} \tag{7.38}$$

Therefore, from the boundary conditions (7.33) at $x = 0$, we have for the zeroth order

$$\frac{\partial \bar{u}_0}{\partial x}(0, t) + \frac{\partial Q_1 u}{\partial \xi}(0, t) = 0 \qquad \frac{\partial \bar{\theta}_0}{\partial x}(0, t) + \frac{\partial Q_1 \theta}{\partial \xi}(0, t) = 0$$

Taking into account $\bar{u}_0 = 0$ and (7.38), we obtain the boundary condition for $\bar{\theta}_0(x, t)$

$$\frac{\partial \bar{\theta}_0}{\partial x}(0, t) = 0 \tag{7.39}$$

The second boundary condition at $x = 1$ is obtained analogously

$$\frac{\partial \bar{\theta}_0}{\partial x}(1, t) = 0 \tag{7.40}$$

Thus, Eq. (7.34) and initial and boundary conditions (7.37), (7.39), (7.40) determine the function $\bar{\theta}_0(x, t)$. Solving this problem we obtain

$$\bar{\theta}_0(x, t) = \int_0^1 G(x, x_0, t)\gamma u^0(x_0)\, dx_0$$

where

$$G(x, x_0, t) = 1 + 2 \sum_{k=1}^{\infty} \exp\{-\alpha\pi^2 k^2 t\}\cos(\pi k x)\cos(\pi k x_0) \tag{7.41}$$

is the corresponding Green's function.

This completes the construction of the zeroth-order terms of the asymptotics. It can be shown that they approximate the exact solution with an accuracy of order ε

$$u(x, t, \varepsilon) = \Pi_0 u(x, t/\varepsilon^2) + O(\varepsilon)$$

$$\theta(x, t, \varepsilon) = \bar{\theta}_0(x, t) + \Pi_0\theta(x, t/\varepsilon^2) + O(\varepsilon)$$

Now it is easy to find the zeroth-order asymptotics for the function

$v(x, t, \varepsilon)$. From (7.31) we have for $\Pi_0 v$ the relation

$$\frac{\partial \Pi_0 v}{\partial \tau} = -\frac{\partial \Pi_0 u}{\partial \tau}$$

Therefore, taking into account the condition at infinity, we obtain

$$\Pi_0 v(x, \tau) = -\Pi_0 u(x, \tau)$$

The functions $Q_0 v$, $Q_0^* v$, $P_0 v$, $P_0^* v$ are identically zero. For $\bar{v}_0(x, t)$ we obtain the problem

$$\frac{\partial \bar{v}_0}{\partial t} - b\frac{\partial^2 \bar{v}_0}{\partial x^2} = 0$$

$$\bar{v}_0(x, 0) = v^0(x) - \Pi_0 v(x, 0) = v^0(x) + u^0(x) \qquad \left.\frac{\partial \bar{v}_0}{\partial x}\right|_{\substack{x=0\\x=1}} = 0$$

Its solution is

$$\bar{v}_0(x, t) = \int_0^1 G_1(x, x_0, t)[v^0(x_0) + u^0(x_0)]\, dx_0$$

where $G_1(x, x_0, t)$ has the form (7.41) with α replaced by b.

The above zeroth-order approximation allows us to find such characteristics of the reaction as the temperature distribution and the typical time of the process. In particular, the maximal temperature of the reaction mixture is expressed by the formula

$$T_{\max} = T_* + \frac{RT_*^2}{E}\max \bar{\theta}_0(x, t) = T_* + \frac{RT_*^2}{T}\gamma \max u^0(x)$$

For the singular perturbation analysis of some other chemical kinetics problems in the critical case see, for example, [35].

VIII. CONTRAST STRUCTURES IN PARTIAL DIFFERENTIAL EQUATIONS

A. Step-Type Solutions in the Noncritical Case

Contrast structures of step type and spike type were considered in Section V for ordinary differential equations. In this section, we construct asymptotics for step-type and spike-type solutions of partial differential equations.

Consider the equation

$$\varepsilon^2 \Delta u = f(u, x, y) \qquad (x, y) \in \Omega \tag{8.1}$$

with the Dirichlet boundary condition

$$u|_{\partial\Omega} = g(x, y) \tag{8.2}$$

Here

$$\Delta = \frac{\partial^2}{\partial x^2} + \frac{\partial^2}{\partial y^2}$$

is the Laplace operator, Ω is a bounded planar domain. The functions f, g and the boundary $\partial\Omega$ are assumed to be sufficiently smooth. We construct the asymptotics of contrast step-type structure for this problem.

A contrast step-type structure is defined (analogously to Section V) as a solution of the (8.1), (8.2), which is close to the distinct solutions $\varphi_1(x, y)$ and $\varphi_2(x, y)$ of the reduced equation

$$f(u, x, y) = 0 \tag{8.3}$$

on different sides of some closed curve Γ located inside Ω, and thus, the interior transition layer occurs in a small vicinity of the curve Γ. The location of the curve Γ is not known a priori, it is defined during the construction of the asymptotics.

Condition I. *Let Eq. (8.3) have three solutions* $u = \varphi_i(x, y)$ $(i = 1, 2, 3)$ *such that*

a. $\varphi_1(x, y) < \varphi_2(x, y) < \varphi_3(x, y)$ *for* $(x, y) \in \bar{\Omega} = \Omega + \partial\Omega$ *and there are no other solutions of (8.3) in the intervals* (φ_1, φ_2) *and* (φ_2, φ_3).
b. $f_u(\varphi_i(x, y), x, y) > 0$, $i = 1, 3$; $f_u(\varphi_2(x, y), x, y) < 0$ *for* $(x, y) \in \bar{\Omega}$.

It follows from this condition that for each fixed point $(x, y) \in \bar{\Omega}$ the graph of the function $f(u, x, y)$ has the same form as in Fig. 6.

We introduce the function $I(x, y)$:

$$I(x, y) = \int_{\varphi_1(x,y)}^{\varphi_3(x,y)} f(u, x, y)\, du$$

Condition II. *Let the equation* $I(x, y) = 0$ *define some closed smooth simple curve* Γ_0 *located inside* Ω *and let*

$$\frac{\partial}{\partial n} I(x, y) \neq 0 \qquad \text{for} \qquad (x, y) \in \Gamma_0 \tag{8.4}$$

where $\partial/\partial n$ *is the derivative along the normal to* Γ_0.

A case where condition (8.4) holds is said to be *noncritical*. In a critical case $I(x, y) \equiv 0$ for $(x, y) \in \bar{\Omega}$. Such cases are discussed in Section VIII.B.

The curve Γ_0 is the main "term" in the asymptotic representation of the above curve Γ. We define the location of Γ by the condition

$$u(x, y, \varepsilon) = \varphi_2(x, y) \qquad \text{for} \qquad (x, y) \in \Gamma \tag{8.5}$$

The curve Γ (as yet unknown) divides the domain Ω into two subdomains: Ω_i (interior to Γ) and Ω_e (exterior to Γ).

The following condition is a standard one that is necessary for the construction of boundary functions.

Condition III. *Let*

$$\int_{\varphi_1(x,y)}^{s} f(u, x, y)\, du > 0 \qquad \text{for} \qquad (x, y) \in \partial\Omega \qquad s \in (\varphi_1(x, y), g(x, y)]$$

Under conditions I–III, we seek an asymptotic expansion of the solution in the form

$$u(x, y, \varepsilon) = \begin{cases} \bar{u}^i + Q^i = \sum_{k=0}^{\infty} \varepsilon^k [\bar{u}_k^i(x, y) + Q_k^i(\tau, \theta)] \\ \text{for} \qquad (x, y) \in \bar{\Omega}_i \\ \bar{u}^e + Q^e + \Pi = \sum_{k=0}^{\infty} \varepsilon^k [\bar{u}_k^e(x, y) + Q_k^e(\tau, \theta) + \Pi_k(\rho, \ell)] \\ \text{for} \qquad (x, y) \in \bar{\Omega}_e \end{cases} \tag{8.6}$$

Here $\bar{u}_k^i, \bar{u}_k^e$ are the regular terms of the asymptotics, Q_k^i and Q_k^e are transition layer functions, and Π_k are boundary functions. The variables ρ, ℓ are defined in an analogous way to that in Section VI.B. The variables τ, θ are introduced as follows. First we introduce new (local) coordinates (r, θ) in the vicinity of the curve Γ_0 similarly to the introduction of local coordinates (r, ℓ) in the vicinity of $\partial\Omega$ (see Section VI.B): θ plays the same role as ℓ (we assume that $0 \le \theta \le 2\pi$), r is the distance from given point M to the curve Γ_0 along the normal to Γ_0 with a plus sign if $M \in \Omega_i$ and with a minus sign if $M \in \Omega_e$.

We seek an equation for Γ in the form

$$r = \lambda(\theta, \varepsilon) = \varepsilon\lambda_1(\theta) + \varepsilon^2\lambda_2(\theta) + \cdots$$

where the functions $\lambda_i(\theta)$ must be found during construction of the asymptotics. Then we introduce the transition layer variable

$$\tau = \frac{r - \lambda(\theta, \varepsilon)}{\varepsilon}$$

The operator $\varepsilon^2 \Delta$ in terms of the new variables τ, θ has the form

$$\varepsilon^2 \Delta - (1 + \varepsilon^2 \alpha) \frac{\partial^2}{\partial \tau^2} + \sum_{j=1}^{\infty} \varepsilon^j L_j$$

where $\alpha = \alpha(\varepsilon\tau, \theta, \varepsilon)$ is a known function, L_j are linear differential operators containing the differentiations $\partial/\partial\tau$, $\partial/\partial\theta$, $\partial^2/\partial\theta^2$; for example, $L_1 = k(\theta)(\partial/\partial\tau)$ and $k(\theta)$ is a known function.

Let us substitute series (8.6) into Eq. (8.1) and change the right-hand side f into an expression of type (8.6): $\bar{f}^i + Q^i f$ for $\bar{\Omega}_i$ and $\bar{f}^e + Q^e f + \Pi f$ for $\bar{\Omega}_e$ (see Section VII.E). Now, expanding $\bar{f}^i, \ldots, \Pi f$ into power series in ε and equating coefficients of like powers of ε separately for the regular terms, the transition layer functions and the boundary functions, we obtain equations for the terms of the asymptotic expansion of the solution.

For $\bar{u}_0(x, y)$, we have the degenerate equation:

$$f(\bar{u}_0, x, y) = 0$$

We choose

$$\bar{u}_0 = \bar{u}_0^i = \varphi_3(x, y) \quad \text{for} \quad (x, y) \in \bar{\Omega}_i$$

$$\bar{u}_0 = \bar{u}_0^e = \varphi_1(x, y) \quad \text{for} \quad (x, y) \in \bar{\Omega}_e$$

Then it is very simple to find subsequent terms of the regular part of the asymptotics, for example

$$\bar{u}_1^i = \bar{u}_1^e = 0 \qquad \bar{u}_2^i = \bar{f}_u^{-1}(\varphi_3, x, y)\, \Delta\varphi_3 \qquad \bar{u}_2^e = \bar{f}_u^{-1}(\varphi_1, x, y)\, \Delta\varphi_1$$

For $Q_0^i(\tau, \theta)$, $Q_0^e(\tau, \theta)$ we obtain the following problems [the functions $\varphi_i(x, y)$ and $f(u, x, y)$, after changing the variables (x, y) to the variables

(r, θ), are denoted by $\varphi_i(r, \theta)$ and $f(u, r, \theta)$]

$$\frac{\partial^2 Q_0^i}{\partial \tau^2} = f(\varphi_3(0, \theta) + Q_0^i, 0, \theta) \qquad \tau > 0 \tag{8.7}$$

$$Q_0^i(0, \theta) = \varphi_2(0, \theta) - \varphi_3(0, \theta) \qquad Q_0^i(+\infty, \theta) = 0 \tag{8.8}$$

$$\frac{\partial^2 Q_0^e}{\partial \tau^2} = f(\varphi_1(0, \theta) + Q_0^e, 0, \theta) \qquad \tau < 0 \tag{8.9}$$

$$Q_0^e(0, \theta) = \varphi_2(0, \theta) - \varphi_1(0, \theta) \qquad Q_0^e(-\infty, 0) = 0 \tag{8.10}$$

Here θ enters as a parameter. The first conditions in (8.8) and (8.10) follow from (8.5). The systems (8.7), (8.8), and (8.9), (8.10) are similar to (5.7), (5.8) and (5.10), (5.11).

Let us introduce the function

$$\tilde{Q}_0(\tau, \theta) = \begin{cases} \varphi_1(0, \theta) + Q_0^e(\tau, \theta) & \tau \leq 0 \\ \varphi_3(0, \theta) + Q_0^i(\tau, \theta) & \tau \geq 0 \end{cases}$$

For this function we obtain the following problem from (8.7)–(8.10)

$$\frac{\partial^2 \tilde{Q}_0}{\partial \tau^2} = f(\tilde{Q}_0, 0, \theta) \qquad -\infty < \tau < +\infty$$

$$\tilde{Q}_0(0, \theta) = \varphi_2(0, \theta) \qquad \tilde{Q}_0(-\infty, \theta) = \varphi_1(0, \theta) \qquad \tilde{Q}_0(\infty, \theta) = \varphi_3(0, \theta) \tag{8.11}$$

It follows from the conditions I–II that for each fixed θ in the phase plane $(\tilde{Q}_0, \partial \tilde{Q}_0 / \partial \tau)$ of Eq. (8.11) there is a cell: the saddle points $(0, \varphi_1)$ and $(0, \varphi_3)$ are connected by separatrices [see Fig. 7(*c*)]. Thus, problem (8.11) and, hence, problems (8.7)–(8.10) are solvable. The functions Q_0^i and Q_0^e satisfy the exponential estimates

$$|Q_0^i(\tau, \theta)| \leq C \exp\{-\alpha \tau\} \qquad |Q_0^e(\tau, \theta)| \leq C \exp\{\alpha \tau\} \tag{8.12}$$

As before, C and α denote some suitable positive constants independent of ε. Note that

$$\frac{\partial Q_0^i}{\partial \tau}(0, \theta) = \frac{\partial Q_0^e}{\partial \tau}(0, \theta) = \frac{\partial \tilde{Q}_0}{\partial \tau}(0, \theta)$$

that is, the derivatives of the zeroth approximation are matched on the curve Γ.

For $Q_1^i(\tau, \theta)$, $Q_1^e(\tau, \theta)$ we have the problems

$$\begin{aligned}
&\frac{\partial^2 Q_1^i}{\partial \tau^2} = f_u(\tau, \theta) Q_1^i + f_1^i(\tau, \theta) \qquad \tau > 0 \\
&Q_1^i(0, \theta) = -\lambda_1(\theta) \frac{\partial \varphi_3}{\partial r}(0, \theta) \qquad Q_1^i(\infty, \theta) = 0 \\
&\frac{\partial^2 Q_1^e}{\partial \tau^2} = f_u(\tau, \theta) Q_1^e + f_1^e(\tau, \theta) \qquad \tau < 0 \\
&Q_1^e(0, \theta) = -\lambda_1(\theta) \frac{\partial \varphi_1}{\partial r}(0, \theta) \qquad Q_1^e(-\infty, \theta) = 0
\end{aligned} \tag{8.13}$$

where

$$f_1^i = \left[f_u(\tau, \theta) \frac{\partial \varphi_3}{\partial r}(0, \theta) + f_r(\tau, \theta) \right](\tau + \lambda_1(\theta)) - k(\theta) \frac{\partial \tilde{Q}_0}{\partial \tau}$$

f_1^e has a similar expression with φ_3 changed to φ_1, $f_u(\tau, \theta)$ and $f_r(\tau, \theta)$ are calculated at the point $(\tilde{Q}_0, 0, \theta)$, $k(\theta)$ is a known function entering the expression for the operator L_1 (see above). The solutions of these problems can be represented in explicit form using the function $\Phi(\tau, \theta) = (\partial \tilde{Q}_0 / \partial \tau)(\tau, \theta)$. Write, for example, the expression for Q_1^i

$$\begin{aligned}
Q_1^i(\tau, \theta) = &-\lambda_1(\theta) \frac{\partial \varphi_3}{\partial r}(0, \theta) \Phi^{-1}(0, \theta) \Phi(\tau, \theta) \\
&- \Phi(\tau, \theta) \int_0^\tau \Phi^{-2}(\xi, 0) \int_\xi^\infty \Phi(\sigma, \theta) f_1^i(\sigma, \theta) \, d\sigma \, d\xi
\end{aligned} \tag{8.14}$$

The matching condition for the derivatives of the first approximation gives the equality

$$\frac{\partial \varphi_1}{\partial r}(0, \theta) + \frac{\partial Q_1^e}{\partial \tau}(0, \theta) = \frac{\partial \varphi_3}{\partial r}(0, \theta) + \frac{\partial Q_1^i}{\partial \tau}(0, \theta)$$

By using (8.14), and an analogous expression for $Q_1^e(\tau, \theta)$, we can reduce this equality to the form

$$\frac{\partial \varphi_3}{\partial r}(0, \theta) - \frac{\partial \varphi_1}{\partial r}(0, \theta) - \Phi^{-1}(0, \theta) \int_{-\infty}^\infty \Phi(\tau, \theta) f_1(\tau, \theta) \, d\tau = 0$$

Next, using the expression for $f_1(\tau, \theta)$, which is equal to $f_1^i(\tau, \theta)$ for $\tau \geq 0$

and is equal to $f_1^e(\tau, \theta)$ for $\tau \leq 0$, after some transformations we obtain the equation

$$\left\{\frac{\partial}{\partial r}\int_{\varphi_1(r,\theta)}^{\varphi_2(r,\theta)} f(u, r, \theta)\, du\right\}\Bigg|_{r=0} \lambda_1(\theta) =$$

$$= \int_{-\infty}^{\infty} [k(\theta)\Phi(\tau, \theta) - f_r(\tau, \theta)\tau]\Phi(\tau, \theta)\, d\tau \qquad (8.15)$$

The coefficient in front of $\lambda_1(\theta)$ is the derivative $(\partial/\partial n)I(x, y)$ for $(x, y) \in \Gamma_0$ from condition (8.4). Since this coefficient is not equal to zero, $\lambda_1(\theta)$ is defined uniquely from Eq. (8.15). All the known functions entering (8.15) are 2π periodic with respect to θ, therefore $\lambda_1(\theta)$ is also a 2π-periodic function.

In a similar way, the transition layer functions Q_k^i, Q_k^e can be found successively for $k = 2, 3, \ldots$. While finding them, the $\lambda_k(\theta)$ $(k = 2, 3, \ldots)$ are determined. All of the functions Q_k^i, Q_k^e satisfy estimates of type (8.12).

For the boundary function $\Pi_0(\rho, \ell)$ we have the following problem, where ℓ enters as a parameter (the notation for the functions f, φ, g in terms of the variables ρ, ℓ is analogous to the notation in terms of the variables τ, θ)

$$\frac{\partial^2 \Pi_0}{\partial \rho^2} = f(\varphi_1(0, \ell) + \Pi_0, 0, \ell) \qquad \rho > 0$$

$$\Pi_0(0, \ell) = g(0, \ell) - \varphi_1(0, \ell) \qquad \Pi_0(\infty, \ell) = 0$$

By virtue of condition III, this problem has a unique solution and $\Pi_0(\rho, \ell)$ satisfies the exponential estimate

$$|\Pi_0(\rho, \ell)| \leq C \exp\{-\alpha\rho\} \qquad (8.16)$$

The following $\Pi_k(\rho, \ell)$ $(k = 1, 2, \ldots)$ are defined similarly to $Q_k(\tau, \theta)$. Formulas analogous to (8.14) may be obtained for $\Pi_k(\rho, \ell)$. All the $\Pi_k(\rho, \ell)$ satisfy estimates of type (8.16).

As was described in Section VI.D all of the functions Q_k^i, Q_k^e, Π_k must be multiplied by smoothing functions.

Set $\tau = [r - (\varepsilon\lambda_1(\theta) + \cdots + \varepsilon^{n+1}\lambda_{n+1}(\theta))]/\varepsilon$ and denote the n-th partial sum of the series (8.6) by $U_n(x, y, \varepsilon)$. It can be proved that

Under conditions I–III for sufficiently small ε, problem (8.1), (8.2) has a step-type solution $u(x, y, \varepsilon)$ for which the series (8.6) gives an asymptotic expansion in the domains $\bar{\Omega}_i$ and $\bar{\Omega}_e$ as $\varepsilon \to +0$, that is, *the following estimate holds:*

$$\max_{\bar{\Omega}} |u - U_n| = O(\varepsilon^{n+1}) \tag{8.17}$$

It follows from this statement that

$$\lim_{\varepsilon \to 0} u(x, y, \varepsilon) = \begin{cases} \varphi_3(x, y) & \text{for} \quad (x, y) \in \Omega_i \\ \varphi_1(x, y) & \text{for} \quad (x, y) \in \Omega_e \end{cases}$$

This result was first obtained in [36]. Its proof is very complicated. The estimatc (8.17) was obtained in [37] by another, simpler method.

Remark: If instead of the Dirichlet boundary condition (8.2) we have the Neumann boundary condition

$$\left. \frac{\partial u}{\partial n} \right|_{\partial\Omega} = 0$$

the construction of an asymptotic expansion for the step-type solution is obtained in the same way. One difference is that $\Pi_0(\rho, \ell) = 0$ and therefore it is not necessary to impose condition III.

B. Step-Type Solutions in the Critical Case

Consider again problem (8.1), (8.2). We retain conditions I and III but replace condition II by condition II′.

Condition II′.

$$I(x, y) = \int_{\varphi_1(x,y)}^{\varphi_3(x,y)} f(u, x, y)\, du \equiv 0 \qquad \textit{for} \qquad (x, y) \in \bar{\Omega}$$

This case is said to be *critical*. Now the equation $I(x, y) = 0$ does not define a curve Γ_0 in contrast to the noncritical case.

For finding Γ_0 (the main term of curve Γ), we introduce a polar system of coordinates (R, θ) with the pole at some point inside Ω and we seek the equation of Γ_0 in the form

$$R = R(\theta) \qquad 0 \le \theta \le 2\pi$$

For convenience, we also use the function $\sigma(\theta) = R^{-1}(\theta)$.

As before, we introduce new variables r, θ instead of x, y in a vicinity of Γ_0 and we seek the equation of Γ in terms of the new variables:

$$r = \lambda(\theta, \varepsilon) = \varepsilon\lambda_1(\theta) + \varepsilon^2\lambda_2(\theta) + \cdots$$

The location of Γ is determined by condition (8.5).

As in the noncritical case, the asymptotics of a step-type solution are sought in the form (8.6). The regular terms of the asymptotics are the same as in Section VIII.A. The transition layer functions $Q_0^i(\tau, \theta)$ and $Q_0^e(\tau, \theta)$ are defined as before and satisfy estimates (8.12). For the functions $Q_1^i(\tau, \theta)$ and $Q_1^e(\tau, \theta)$ we again have problems (8.13) but now the function $k(\theta)$ in the expressions for f_1^i and f_1^e is not known. It is given by the following expression:

$$k(\theta) = -\frac{\sigma'' + \sigma}{[1 + (\sigma'/\sigma)^2]^{3/2}} \tag{8.18}$$

where $\sigma = \sigma(\theta)$ is the as yet unknown function mentioned above. The matching condition for the derivatives of the first approximation on Γ again gives Eq. (8.15). By virtue of condition II′ the left-hand side of (8.15) is equal to zero. Taking into account expression (8.18) for $k(\theta)$ we obtain from (8.15) the following differential equation for the function $\sigma(\theta)$

$$\frac{\sigma'' + \sigma}{[1 + (\sigma'/\sigma)^2]^{3/2}} = -\left[\int_{-\infty}^{+\infty} \Phi^2(\tau, \theta)\, d\tau\right]^{-1} \cdot \int_{-\infty}^{+\infty} f_r(\tau, \theta)\tau\Phi(\tau, \theta)\, d\tau \tag{8.19}$$

Here $\Phi(\tau, \theta) = \partial\tilde{Q}_0/\partial\tau$, $\tilde{Q}_0(\tau, \theta)$ is the solution of problem (8.11). Note that the left-hand side of (8.19) is the curvature of Γ_0 at the point $(R(\theta), \theta)$ and the right-hand side of (8.19) is a function of $\sigma(\theta), \theta$. Therefore, from the geometric point of view the problem of finding the curve Γ_0 is reduced to one of determining a curve when its curvature at each point is a given function of the point.

Remark: If the function f depends on ε, $f = f(u, x, y, \varepsilon)$, then the function under the integral sign in Eq. (8.19) will be $(f_r(\tau, \theta)\tau + f_\varepsilon(\tau, \theta))$ instead of $f_r(\tau, \theta)\tau$.

Condition IV. *Let Eq. (8.19), with the periodicity condition* $\sigma(\theta) = \sigma(\theta + 2\pi)$, *have a solution* $\sigma = \sigma(\theta)$.

Thus, the curve Γ_0 is found.

It can be shown that the matching condition for the derivatives of the $(k+1)$-th approximation on Γ gives a linear differential equation of second order for $\lambda_k(\theta)$ $(k = 1, 2, \ldots)$:

$$a(\theta)\lambda_k'' + b(\theta)\lambda_k' + c(\theta)\lambda_k = F_k(\theta) \tag{8.20}$$

where $a(\theta) > 0$, $b(\theta)$, $c(\theta)$, $F_k(\theta)$ are known 2π-periodic functions.

Condition V. *Let* $c(\theta) < 0$.

Then Eq. (8.20) has a unique 2π-periodic solution.

All the functions $Q_k^i(\tau, \theta)$, $Q_k^e(\tau, \theta)$ possess exponential estimates of type (8.12).

The boundary functions $\Pi_k(\rho, \ell)$ are the same as in Section VIII.A.

Under conditions I, II′, III–V for sufficiently small ε, problem (8.1), (8.2) has a step-type solution, for which the constructed series (8.6) are asymptotic expansions in the domains $\bar{\Omega}_i$ and $\bar{\Omega}_e$ as $\varepsilon \to 0$.

C. Spike-Type Solutions

We consider again problem (8.1), (8.2) but now under different conditions than in Sections VIII.A and B.

Condition I. *Let the equation $f(u, x, y) = 0$ have two solutions $u = \varphi(x, y)$ and $u = \chi(x, y)$. Let*

a. *$\varphi(x, y) < \chi(x, y)$ for $(x, y) \in \bar{\Omega}$ and assume that there are no other solutions of this equation in the interval (φ, χ).*

b. *$f_u(\varphi(x, y), x, y) > 0$, $f_u(\chi(x, y), x, y) < 0$ for $(x, y) \in \bar{\Omega}$.*

Condition II. *Assume that a function $\psi(x, y)$ exists such that*

$$\int_{\varphi(x,y)}^{\psi(x,y)} f(u, x, y)\, du = 0$$

$$\int_{\varphi(x,y)}^{s} f(u, x, y)\, du > 0 \qquad \text{for} \qquad (x, y) \in \bar{\Omega},\ s \in (\varphi(x, y), \psi(x, y))$$

It follows from conditions I and II that for each fixed point $(x, y) \in \bar{\Omega}$ the graph of the function $f(u, x, y)$ has the same form as in Fig. 13.

We retain condition III from Section VIII.A, which is required for the construction of the boundary functions. Under conditions I–III we construct an asymptotic expansion for a spike-type solution or, in other words, for a contrast spike-type structure. This is a solution of problem (8.1), (8.2), which is close to the solution $u = \varphi(x, y)$ of the equation $f(u, x, y) = 0$ everywhere in the domain Ω except in a small vicinity of some closed curve Γ where the solution has a spike. The location of the curve Γ is not known a priori just as in Section VIII.B. This curve is defined during the construction of the asymptotics.

An asymptotic expansion of the spike-type solution is sought in the form

$$u = \sum_{k=0}^{\infty} \varepsilon^k [\bar{u}_k(x, y) + Q_k(\tau, \theta) + \Pi_k(\rho, \ell)] \tag{8.21}$$

where $\bar{u}_k(x, y)$ are the regular terms of the asymptotics, $Q_k(\tau, \theta)$ are the functions describing the spike of the solution in a small vicinity of the curve Γ (the spike functions), $\Pi_k(\rho, \ell)$ are the boundary functions. The variables τ, θ are introduced in the same way as in Section VIII.B: the equation of Γ_0 (the main term of the curve Γ) is sought in the form $R = R(\theta)$, where (R, θ) are polar coordinates with the pole at some point inside Ω, the variables (r, θ) are local coordinates in a vicinity of Γ_0. The equation for Γ is sought in the form

$$r = \lambda(\theta, \varepsilon) = \varepsilon\lambda_1(\theta) + \varepsilon^2\lambda_2(\theta) + \cdots$$

and finally $\tau = (r - \lambda(\theta, \varepsilon))/\varepsilon$. The location of Γ is determined now by the condition

$$\left.\frac{\partial u}{\partial r}\right|_{\Gamma} = 0$$

which shows that the solution u has an extremum on Γ along each straight line $\theta = \text{const}$. The variables (ρ, ℓ) are the same as in Sections VIII.A and B.

For $\bar{u}_0(x, y)$ we have the reduced equation $f(\bar{u}_0, x, y) = 0$. We set

$$\bar{u}_0 = \varphi(x, y)$$

The following $\bar{u}_k(x, y)$ $(k \geq 1)$ are defined explicitly.

The problem for $Q_0(\tau, \theta)$ has the form

$$\frac{\partial^2 Q_0}{\partial \tau^2} = f(\varphi(0, \theta) + Q_0, 0, \theta) \qquad -\infty < \tau < \infty \tag{8.22}$$

$$\frac{\partial Q_0}{\partial \tau}(0, \theta) = 0 \tag{8.23}$$

$$Q_0(-\infty, \theta) = Q_0(+\infty, 0) = 0 \tag{8.24}$$

Condition (8.23) is obtained from the condition $(\partial u/\partial r)|_\Gamma = 0$. As before, the functions $\varphi(x, y)$, $f(u, x, y)$ are denoted by $\varphi(r, \theta)$, $f(u, r, \theta)$ after changing the variables (x, y) to the variables (r, θ).

By virtue of conditions I and II on the phase plane $[Q_0, (\partial Q_0/\partial \tau)]$ of Eq. (8.22) there is a loop on which the extreme right point is $(0, \psi(0, \theta) - \varphi(0, \theta))$ (Fig. 18). The upper and lower parts of the loop

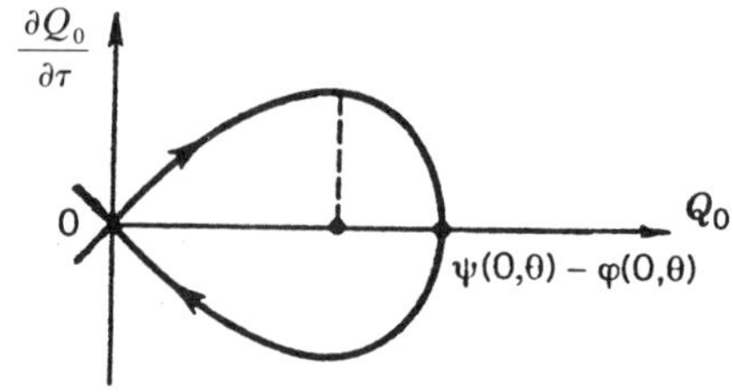

Figure 18. Phase plane of Eq. (8.22).

have the equations

$$\frac{\partial Q_0}{\partial \tau} = \pm\left[2\int_{\varphi(0,\theta)}^{\varphi(0,\theta)+Q_0} f(u,0,\theta)\,du\right]^{1/2} \tag{8.25}$$

The solution of (8.25) with the + sign for $\tau \le 0$ and the initial condition $Q_0(0,\theta) = \psi(0,\theta) - \varphi(0,\theta)$ together with the solution of (8.25) with the $-$ sign for $\tau \ge 0$ with the same initial condition form the solution of problem (8.22)–(8.24). It satisfies the estimate

$$|Q_0(\tau,\theta)| \le C\exp\{-\alpha|\tau|\} \qquad -\infty < \tau < \infty \tag{8.26}$$

Note that the function $R(\theta)$ describing the curve Γ_0 is not yet known. This function is found in the next step by considering the problem for $Q_1(\tau,\theta)$ which has the form

$$\frac{\partial^2 Q_1}{\partial \tau^2} = f_u(\tau,\theta)Q_1 + f_1(\tau,\theta) \qquad -\infty < \tau < +\infty \tag{8.27}$$

$$\frac{\partial Q_1}{\partial \tau}(0,\theta) = -\frac{\partial \varphi}{\partial r}(0,\theta) \tag{8.28}$$

$$Q_1(-\infty,\theta) = Q_1(+\infty,\theta) = 0 \tag{8.29}$$

Here

$$f_1(\tau,\theta) = \left[f_u(\tau,\theta)\frac{\partial\varphi}{\partial r}(0,\theta) + f_r(\tau,\theta)\right][\tau + \lambda_1(\theta)] - k(\theta)\frac{\partial Q_0}{\partial \tau}(\tau,\theta)$$

$f_u(\tau,\theta)$ and $f_r(\tau,\theta)$ are calculated at the point $(\varphi(0,\theta) + Q_0(\tau,\theta), 0, \theta)$, $k(\theta)$ is defined by formula (8.18), where $\sigma(\theta) = R^{-1}(\theta)$, that is, $k(\theta)$ is the curvature of Γ_0 at the point $(R(\theta),\theta)$.

The homogeneous equation corresponding to (8.27) with conditions (8.29) has a nontrivial solution $\Phi(\tau,\theta) = (\partial Q_0/\partial\tau)(\tau,\theta)$. This fact can be obtained by differentiating Eq. (8.22). Hence, for solvability of the inhomogeneous equation (8.27) with conditions (8.29), it is necessary that the function $f_1(\tau, \theta)$ is orthogonal to the function $\Phi(\tau, \theta)$

$$\int_{-\infty}^{+\infty} f_1(\tau, \theta)\Phi(\tau, \theta)\, d\tau = 0 \tag{8.30}$$

This equation can be reduced to the form (8.19) with $\Phi(\tau, \theta)$ replaced by $(\partial Q_0/\partial\tau)(\tau, \theta)$.

We retain condition IV from Section VIII.B, which allows us to obtain the function $\sigma(\theta)$ and, therefore, to find the curve Γ_0. Since condition (8.30) holds, Eq. (8.27) with the conditions (8.29) has a set of solutions

$$Q_1 = \tilde{Q}_1(\tau, \theta) + \alpha(\theta)\Phi(\tau, \theta)$$

where $\tilde{Q}_1(\tau, \theta)$ is a particular solution and $\alpha(\theta)$ is an arbitrary function of θ. For $\tau \le 0$ the solution $\tilde{Q}_1(\tau, \theta)$ can be chosen in the form

$$\tilde{Q}_1(\tau, \theta) = -Y_1(\tau, \theta)\int_0^{\tau} Y_2(s, \theta)f_1(s, \theta)\, ds + Y_2(\tau, \theta) \times$$

$$\int_{-\infty}^{\tau} Y_1(s, \theta)f_1(s, \theta)\, ds$$

where $Y_1 = \Phi(\tau, \theta)$ and $Y_2 = Y_1(\tau, \theta)\int_{-1}^{\tau} Y_1^{-2}(s, \theta)\, ds$ is a fundamental system of solutions for the homogeneous equation corresponding to (8.27). It is not difficult to show that $\tilde{Q}_1(\tau, \theta)$ satisfies the estimate (8.26) for $\tau \le 0$.

The function $\alpha(\theta)$ is chosen in order that $Q_1(\tau, \theta)$ satisfies condition (8.28). Taking into account the fact that

$$\frac{\partial\Phi}{\partial\tau}(0, \theta) = \frac{\partial^2 Q_0}{\partial\tau^2}(0, \theta) = f(\psi(0, \theta), 0, \theta) \neq 0$$

we obtain

$$\alpha(\theta) = -\frac{\varphi_r(0, \theta) + \dfrac{\partial\tilde{Q}_1}{\partial\tau}(0, \theta)}{f(\psi(0, \theta), 0, \theta)}$$

Thus, $Q_1(\tau, \theta)$ is found for $\tau \le 0$. In the same way $Q_1(\tau, \theta)$ is defined for $\tau \ge 0$.

Next, it can be shown that the solvability condition of the problem for $Q_{k+1}(\tau, \theta)$ gives a second-order differential equation of type (8.20) for $\lambda_k(\theta)$ $(k = 1, 2, \ldots)$.

Condition V from Section B guarantees the existence of a unique 2π-periodic solution $\lambda_k(\theta)$.

All of the functions $Q_k(\tau, \theta)$ satisfy estimates of type (8.26).

The boundary functions $\Pi_k(\rho, \ell)$ are the same as in Sections VIII.A and B.

Under conditions I–V and some other condition (see [38]), it can be proved that

Problem (8.1), (8.2) for sufficiently small ε has a spike-type solution for which the constructed series (8.21) is an asymptotic expansion in the domain $\bar{\Omega}$ as $\varepsilon \to 0$.

D. Applications

1. *Phase Transition Models*

Such models are often considered in the form

$$\frac{\partial T}{\partial t} + \lambda \frac{\partial u}{\partial t} = \Delta T + g(x) \tag{8.31}$$

$$\varepsilon^2 \Delta u - \varepsilon^2 \alpha \frac{\partial u}{\partial t} = f(u, T, \varepsilon) \tag{8.32}$$

Here T is the temperature of a substance, u is the so-called structural parameter, which shows that the substance is in the solid phase if u is close to -1 and in the liquid phase if u is close to $+1$. The parameters ε, λ and α are a dimensionless interaction length, a latent heat, and a relaxation time, respectively. A physical discussion of such models is given in [39].

System (8.31), (8.32) is usually considered in some bounded domain Ω with impermeability conditions on the boundary of Ω:

$$\left.\frac{\partial T}{\partial n}\right|_{\partial\Omega} = 0 \qquad \left.\frac{\partial u}{\partial n}\right|_{\partial\Omega} = 0$$

Here $\partial/\partial n$ is the derivative along the normal to $\partial\Omega$.

Under some standard conditions ε is a small parameter and $f(u, T, \varepsilon)$ is represented in the form

$$f(u, T, \varepsilon) = u^3 - u + \varepsilon T$$

If we consider a stationary process, then $\partial u/\partial t = 0$ and the problem for T is separated. Finding T we obtain the following problem for u from (8.32) (we assume that Ω is a two-dimensional domain):

$$\begin{aligned} \varepsilon^2 \Delta u &= u^3 - u + \varepsilon T(x, y) \\ \left.\frac{\partial u}{\partial n}\right|_{\partial\Omega} &= 0 \end{aligned} \tag{8.33}$$

For physical applications it is of interest to find the location of the interphase transition zone.

The degenerate equation for (8.33) has three solutions: $\varphi_1 = -1$, $\varphi_2 = 0$, $\varphi_3 = 1$, and conditions I and II′ of Section VIII.B hold.

The leading term describing the interphase transition [the function $\tilde{Q}_0(\tau, \theta)$] is the solution of the problem [see (8.11)].

$$\frac{\partial^2 \tilde{Q}_0}{\partial \tau^2} = \tilde{Q}_0^3 - \tilde{Q}_0 \qquad -\infty < \tau < \infty$$

$$\tilde{Q}_0(0, \theta) = 0 \qquad \tilde{Q}_0(-\infty, \theta) = -1 \qquad \tilde{Q}_0(+\infty, \theta) = 1$$

The solution can be found in the explicit form:

$$\tilde{Q}_0(\tau, \theta) = \frac{\exp\{\sqrt{2}\tau\} - 1}{\exp\{\sqrt{2}\tau\} + 1}$$

The equation for the curve Γ_0 has the form [see (8.19) and the remark before condition IV in Section VIII.B]:

$$\frac{\sigma'' + \sigma}{\left[1 + \left(\dfrac{\sigma'}{\sigma}\right)^2\right]^{3/2}} = -\left[\int_{-\infty}^{\infty} \left(\frac{\partial \tilde{Q}_0}{\partial \tau}\right)^2 d\tau\right]^{-1} \left[\int_{-\infty}^{\infty} \frac{\partial \tilde{Q}_0}{\partial \tau}\, d\tau\right] \cdot T(0, \theta)$$

Denoting

$$\left[\int_{-\infty}^{\infty} \left(\frac{\partial \tilde{Q}_0}{\partial \tau}\right)^2 d\tau\right]^{-1}$$

by γ and taking into account the fact that $\int_{-\infty}^{\infty} (\partial \tilde{Q}_0 / \partial \tau)\, d\tau = 2$ we obtain the equation

$$\frac{\sigma'' + \sigma}{\left[1 + \left(\dfrac{\sigma'}{\sigma}\right)^2\right]^{3/2}} = -2\gamma T(0, \theta) \tag{8.34}$$

Note that $T(0, \theta)$ is the temperature on the curve Γ_0 [its equation is

$R = R(\theta) = \sigma^{-1}(\theta)]$. Solving Eq. (8.34) we find the curve Γ_0. The phase transition takes place in a vicinity of Γ_0.

In the particular case when T depends on R and does not depend on θ [radially symmetric case: $T = T(R)$], Eq. (8.34) may have solutions independent of θ, which are defined by the equation

$$\frac{1}{R} = -2\gamma T(R)$$

In such a case the curve Γ_0 is a circle.

2. *Fisher's Equation*

Consider the equation with a quadratic nonlinearity

$$\varepsilon^2 \Delta u = u[a(x, y) - u] \qquad (x, y) \in \Omega \tag{8.35}$$

under the boundary condition

$$u|_{\partial\Omega} = 0$$

where $a(x, y) > 0$. Equation (8.35) is often used in mathematical models of biology. It is called *Fisher's equation*. This problem under some condition on $a(x, y)$ has a spike-type solution for which $\bar{u}_0(x, y) = 0$ and $Q_0(\tau, \theta)$ is defined as the nontrivial solution of the problem [see (8.22)–(8.24)]:

$$\frac{\partial^2 Q_0}{\partial \tau^2} = Q_0[a(0, \theta) - Q_0] \qquad -\infty < \tau < \infty$$

$$\frac{\partial Q_0}{\partial \tau}(0, \theta) = 0 \qquad Q_0(-\infty, \theta) = Q_0(+\infty, \theta) = 0$$

The solution can be found in the explicit form

$$Q_0(\tau, \theta) = 6a_0 \frac{\exp\{\sqrt{a_0}\tau\}}{(1 + \exp\{\sqrt{a_0}\tau\})^2} \qquad a_0 = a(0, \theta)$$

Using this expression we obtain

$$\int_{-\infty}^{+\infty} \left(\frac{\partial Q_0}{\partial \tau}\right)^2 d\tau = \frac{1}{30} a^{5/2}(0,\theta)$$

$$\int_{-\infty}^{+\infty} f_r(\tau,\theta)\tau \frac{\partial Q_0}{\partial \tau} d\tau = \int_{-\infty}^{+\infty} a_r(0,\theta) Q_0 \tau \frac{\partial Q_0}{\partial \tau} d\tau = -\tfrac{1}{12} a_r(0,\theta) a^{3/2}(0,\theta)$$

Therefore, Eq. (8.19) for the curve Γ_0 has the form

$$\frac{\sigma'' + \sigma}{\left[1 + \left(\frac{\sigma'}{\sigma}\right)^2\right]^{3/2}} = \frac{5}{2} \frac{a_r(0,\theta)}{a(0,\theta)}$$

IX. MATHEMATICAL MODEL OF COMBUSTION IN THE CASE OF AUTOCATALYTIC REACTION

A. Statement of the Problem

The system of equations describing the combustion process in the case of a first-order autocatalytic reaction has the form

$$\begin{aligned} c\rho \frac{\partial T}{\partial t} - \lambda \, \Delta T &= q(v_0 + v)(1-v) k_0 \exp\left\{-\frac{E}{RT}\right\} \\ \frac{\partial v}{\partial t} - D \, \Delta v &= (v_0 + v)(1-v) k_0 \exp\left\{-\frac{E}{RT}\right\} \end{aligned} \tag{9.1}$$

(see, e.g., [40]). Here T is the absolute temperature, v is the depth of conversion of the combustible component that characterizes the relative amount of substance burnt ($v = 1 - u$, where u is the relative concentration of the combustible component, $0 \le u \le 1$, $0 \le v \le 1$), c, ρ, λ are physical characteristics of the medium: the specific heat capacity, density of the medium in which reaction occurs and thermal conductivity, respectively, q is the thermal effect of the reaction, v_0 is the criterion of autocatalytic behavior (the ratio of the initial rate of reaction to the autocatalytic constant, $0 < v_0 < 1$), k_0 is the reaction constant, E is the activation energy, R is the universal gas constant, and D is the diffusion coefficient of the combustible component. We assume that all of these quantities, with the exception of T and v, are constants.

The functions v and $u = 1 - v$ are already dimensionless. Let us introduce nondimensional temperature θ, time t', coordinates x', y', z'

and constants $\beta, \varepsilon, a_0, b_0$ by the formulas

$$\theta = \frac{E}{RT_*^2}(T - T_*) \qquad \beta = \frac{RT_*}{E} \qquad t' = tk_0 \exp\left\{-\frac{1}{\beta}\right\}$$

$$x' = \frac{x}{r} \qquad y' = \frac{y}{r} \qquad z' = \frac{z}{r} \qquad \varepsilon = \frac{c\rho}{q}\frac{RT_*^2}{E}$$

$$a_0 = \frac{\lambda}{c\rho r^2 k_0 \exp\left\{-\frac{1}{\beta}\right\}} \qquad b_0 = \frac{D}{r^2 k_0 \exp\left\{-\frac{1}{\beta}\right\}}$$

Here T_* is a characteristic temperature, for example, the temperature of the surrounding medium, and r is the characteristic size of the domain in which the reaction takes place. In the new variables, system (9.1) has the form (we again use the notations t, x, y, z instead of t', x', y', z'):

$$\varepsilon\left(\frac{\partial\theta}{\partial t} - a_0\,\Delta\theta\right) = (v_0 + v)(1 - v)\exp\left\{\frac{\theta}{1+\beta\theta}\right\}$$

$$\frac{\partial v}{\partial t} - b_0\,\Delta v = (v_0 + v)(1 - v)\exp\left\{\frac{\theta}{1+\beta\theta}\right\}$$

For combustion processes, the parameters ε and β are usually small. We investigate the case of a fast reaction (large k_0) such that $k_0 \times \exp\{1/\beta\} = O(1)$.

Since the dependence on β on the right-hand side of these equations is regular, we can simplify the system by setting $\beta = 0$ (this will not have much impact on the characteristics of the combustion process). Further, we consider only the one-dimensional case, when θ and v do not depend on the variables y and z (and therefore $\Delta = \partial^2/\partial x^2$). We also assume that the nondimensional coefficients of heat transfer a_0 and diffusion b_0 are large: $a_0 \sim b_0 \sim 1/\varepsilon$. From the physical point of view, this means that either the dimensional coefficients of heat transfer and diffusion are large, or the combustion process occurs in a thin layer (the characteristic length r is small). Introducing the notations $a_0 = a/\varepsilon$ and $b_0 = b/\varepsilon$, we arrive at a system of singularly perturbed equations

$$\begin{aligned} \varepsilon\frac{\partial\theta}{\partial t} - a\frac{\partial^2\theta}{\partial x^2} &= (v_0 + v)(1 - v)\exp\{\theta\} \\ \varepsilon\frac{\partial v}{\partial t} - b\frac{\partial^2 v}{\partial x^2} &= \varepsilon(v_0 + v)(1 - v)\exp\{\theta\} \end{aligned} \qquad (9.2)$$

Let us consider this system in the rectangle

$$\Omega = \{(0<x<1) \times (0<t \leq T)\}$$

We impose the natural initial and boundary conditions

$$\theta|_{t=0} = v|_{t=0} = 0 \qquad \theta|_{x=0} = \theta|_{x=1} = 0 \qquad \left.\frac{\partial v}{\partial x}\right|_{x=0} = \left.\frac{\partial v}{\partial x}\right|_{x=1} = 0 \quad (9.3)$$

The conditions for θ mean that at the initial time $t=0$, and on the boundary of the domain ($x=0$ and $x=1$), the temperature is that of the surrounding medium: $\theta=0$, that is, $T=T_*$. The initial condition for v shows that at $t=0$ the amount of burnt substance is zero (the reaction has not yet started). The boundary conditions for v describe the absence of a flow of the combustible component through the boundary (the impermeability condition).

B. Construction of the Leading Terms of the Asymptotic Expansion

We will construct an asymptotic expansion of the solution of the problem (9.2), (9.3) in the form a sum of regular and boundary layer series:

$$\begin{aligned} \theta &= \bar{\theta} + \Pi\theta = \sum_{i=0}^{\infty} \varepsilon^i[\bar{\theta}_i(x,t) + \Pi_i\theta(x,\tau)] \\ v &= \bar{v} + \Pi v = \sum_{i=0}^{\infty} \varepsilon^i[\bar{v}_i(x,t) + \Pi_i v(x,\tau)] \end{aligned} \qquad (9.4)$$

where $\tau = t/\varepsilon$. Substituting (9.4) into (9.2) and (9.3), and representing the right-hand sides of (9.2) in the form $f = \bar{f} + \Pi f$, in the standard way, we obtain equations for the terms of the series (9.4).

For $\bar{v}_0$ and $\bar{\theta}_0$, we have the system

$$\frac{\partial^2 \bar{\theta}_0}{\partial x^2} = -\frac{1}{a}(v_0 + \bar{v}_0)(1-\bar{v}_0)\exp\{\bar{\theta}_0\} \qquad \frac{\partial^2 \bar{v}_0}{\partial x^2} = 0$$

with the boundary conditions

$$\bar{\theta}_0(0,t) = \bar{\theta}_0(1,t) = 0 \qquad \frac{\partial \bar{v}_0}{\partial x}(0,t) = \frac{\partial \bar{v}_0}{\partial x}(1,t) = 0 \qquad (9.5)$$

This implies that

$$\bar{v}_0 = \alpha_0(t)$$

where $\alpha_0(t)$ is an arbitrary function. We can now rewrite the equation for

$\bar{\theta}_0$ as

$$\frac{\partial^2 \bar{\theta}_0}{\partial x^2} = -\delta(\alpha_0(t), a)\exp\{\bar{\theta}_0\} \tag{9.6}$$

where $\delta(\alpha_0, a) = (v_0 + \alpha_0)(1 - \alpha_0)/a$. Such an equation has been considered in [41] and [42]. Let us use the results of [41], where the boundary value problem

$$\frac{\partial^2 \theta}{\partial x^2} = \delta \exp\{\theta\} \qquad \theta(-1) = \theta(+1) = 0 \tag{9.7}$$

was discussed. It is shown in [41] that for $\delta < \delta_{cr} \cong 0.878$, there exist two positive solutions of this problem, the smaller of which is stable. Applying this result to the problem for $\bar{\theta}_0$ [Eq. (9.6) with boundary conditions (9.5)], whose solution depends on $\alpha_0(t)$ and a as parameters, we find that when

$$a > a_1 = \frac{\max\limits_{\alpha_0} (v_0 + \alpha_0)(1 - \alpha_0)}{4\delta_{cr}} = \frac{(1 + v_0)^2}{16\delta_{cr}} \tag{9.8}$$

there exist two positive solutions of this problem. Let us take the smaller solution and denote it by $\bar{\theta}_0 = \tilde{\theta}_0(x, \alpha_0(t), a)$. It is known (see [41]) that the solution of (9.7), which we have chosen is an increasing function of δ for $\delta < \delta_{cr}$. Therefore, $\tilde{\theta}_0$ decreases with growing a, and $\tilde{\theta}_0 \to 0$ for $a \to \infty$. It should also be noted that

$$\max_{0 \le x \le 1} \tilde{\theta}_0(x, \alpha_0(t), a) = \tilde{\theta}_0(\tfrac{1}{2}, \alpha_0(t), a)$$

Thus, the leading terms $\bar{\theta}_0$ and $\bar{v}_0$ of the regular part of the asymptotic expansion depend on the as yet unknown function $\alpha_0(t)$. This means that (9.2), (9.3) is a problem in the critical case: the corresponding reduced system has a family of solutions.

Let us now define the boundary layer functions of zeroth order. For $\Pi_0 v$, we have the problem

$$\frac{\partial \Pi_0 v}{\partial \tau} - b\frac{\partial^2 \Pi_0 v}{\partial x^2} = 0$$

$$\Pi_0 v(x, 0) = -\bar{v}_0(x, 0) = -\alpha_0(0) \qquad \frac{\partial \Pi_0 v}{\partial x}(0, \tau) = \frac{\partial \Pi_0 v}{\partial x}(1, \tau) = 0$$

This implies that $\Pi_0 v(x, \tau) = -\alpha_0(0)$. It is natural to demand that the

Π-functions approach zero as $\tau \to \infty$, so $\Pi_0 v(x, \tau) = 0$, and the initial condition $\alpha_0(0) = 0$ for the as yet unknown function $\alpha_0(t)$ has been found. This function is completely determined at the next step of the asymptotic algorithm [during the study of the equation for $\bar{v}_1(x, t)$].

For $\Pi_0 \theta$, we obtain the problem

$$\frac{\partial \Pi_0 \theta}{\partial \tau} - a \frac{\partial^2 \Pi_0 \theta}{\partial x^2} = v_0[\exp\{\bar{\theta}_0(x, 0) + \Pi_0 \theta\} - \exp\{\bar{\theta}_0(x, 0)\}]$$
$$\Pi_0 \theta(x, 0) = -\bar{\theta}_0(x, 0) \qquad \Pi_0 \theta(0, \tau) = \Pi_0 \theta(1, \tau) = 0 \tag{9.9}$$

Notice that initial and boundary conditions are matched so as to be continuous, since $\bar{\theta}_0(0, 0) = \bar{\theta}_0(1, 0) = 0$. We should also mention that, though the function $\bar{\theta}_0(x, t) = \tilde{\theta}_0(x, \alpha_0(t), a)$ is still not defined [it depends on the unknown $\alpha_0(t)$], its value at $t = 0$ is known: $\bar{\theta}_0(x, 0) = \tilde{\theta}_0(x, \alpha_0(0), a) = \tilde{\theta}_0(x, 0, a)$. Let us make the change of variable $w(x, \tau) = \bar{\theta}_0(x, 0) + \Pi_0 \theta(x, \tau)$. For $w(x, \tau)$, we obtain

$$\frac{\partial w}{\partial \tau} - a \frac{\partial^2 w}{\partial x^2} = v_0 \exp\{w\} \qquad w(x, 0) = 0 \qquad w(0, \tau) = w(1, \tau) = 0$$

It has been shown in [40] that under condition (9.8), a unique solution of this problem exists and $w(x, \tau) \to \bar{\theta}_0(x, 0)$ monotonically as $\tau \to \infty$. Consequently, $\Pi_0 \theta(x, \tau) \to 0$ when $\tau \to \infty$, $\Pi_0 \theta(x, \tau) \leq 0$.

Let us prove that for sufficiently large a the function $\Pi_0 \theta(x, \tau)$ exponentially approaches zero as $\tau \to \infty$. We introduce $\tilde{\Pi} = -\Pi_0 \theta$. From (9.9) we obtain the following problem for $\tilde{\Pi}$

$$\frac{\partial \tilde{\Pi}}{\partial \tau} - a \frac{\partial^2 \tilde{\Pi}}{\partial x^2} = v_0 \exp\{\bar{\theta}_0(x, 0)\}[1 - \exp\{-\tilde{\Pi}\}]$$
$$\tilde{\Pi}(x, 0) = \bar{\theta}_0(x, 0) \qquad \tilde{\Pi}(0, \tau) = \tilde{\Pi}(1, \tau) = 0$$

Let us introduce the notation

$$k = v_0 \max_{0 \leq x \leq 1} \exp\{\bar{\theta}(x, 0)\} = v_0 \exp\{\tilde{\theta}_0(\tfrac{1}{2}, 0, a)\}$$

and consider the problem

$$\frac{\partial Q}{\partial \tau} - a \frac{\partial^2 Q}{\partial x^2} = kQ \qquad Q(x, 0) = \bar{\theta}_0(x, 0) \qquad Q(0, \tau) = Q(1, \tau) = 0$$

If we use a comparison theorem for parabolic equations [43] and taking into account the inequality $\tilde{\Pi} \geq 0$, we obtain

$$0 \leq \tilde{\Pi}(x, \tau) \leq Q(x, \tau)$$

The function $Q(x, \tau)$ can be represented by the Fourier series

$$Q(x, \tau) = \sum_{n=1}^{\infty} q_n \exp\{-(a\lambda_n - k)\tau\} \sin(\pi n x)$$

$$\text{where} \qquad \lambda_n = \pi^2 n^2,\ q_n = 2 \int_0^1 \bar{\theta}_0(x, 0) \sin(\pi n x)\, dx$$

Hence,

$$Q(x, \tau) \leq C \exp\{-(a\lambda_1 - k)\tau\} \leq C \exp\{-\alpha\tau\}$$

if

$$a > \frac{k}{\lambda_1} = \frac{v_0 \exp\{\tilde{\theta}_0(1/2, 0, a)\}}{\pi^2} \tag{9.10}$$

Since $\tilde{\theta}_0(\frac{1}{2}, 0, a) \to 0$ as $a \to \infty$, there exists an a_2 such that for $a > a_2$ inequality (9.10) holds. Taking into account (9.8), we can write the condition on a as

$$a > \max(a_1, a_2) \tag{9.11}$$

Then, for a satisfying (9.11), the function $\tilde{\Pi}$ (and therefore $\Pi_0\theta$) satisfies the exponential estimate

$$|\Pi_0\theta(x, \tau)| \leq C \exp\{-\alpha\tau\} \qquad 0 \leq x \leq 1 \qquad \tau \geq 0 \tag{9.12}$$

For $\bar{v}_1$ and $\bar{\theta}_1$ we have the system

$$\begin{aligned}
\frac{\partial^2 \bar{v}_1}{\partial x^2} &= \frac{1}{b}[\alpha_0'(t) - (v_0 + \alpha_0(t))(1 - \alpha_0(t)) \exp\{\tilde{\theta}_0(x, \alpha_0(t), a)]\\
&\equiv \psi_0(x, \alpha_0)\\
\frac{\partial^2 \bar{\theta}_1}{\partial x^2} &= \frac{1}{a}\left\{\frac{\partial \bar{\theta}_0}{\partial t} - [(v_0 + \alpha_0(t))(1 - \alpha_0(t))\bar{\theta}_1\right.\\
&\qquad \left. + (1 - v_0 - 2\alpha_0(t))\bar{v}_1] \exp\{\tilde{\theta}_0(x, \alpha_0(t), a)\}\right\}
\end{aligned} \tag{9.13}$$

with boundary conditions

$$\frac{\partial \bar{v}_1}{\partial x}(0,t)=0 \qquad \frac{\partial \bar{v}_1}{\partial x}(1,t)=0 \qquad \bar{\theta}_1(0,t)=\bar{\theta}_1(1,t)=0 \tag{9.14}$$

Integrating (9.13) with the first condition of (9.14), we obtain

$$\frac{\partial \bar{v}_1}{\partial x}=\int_0^x \psi_0(x_0,\alpha_0)\,dx_0 \tag{9.15}$$

The second condition of (9.14) is satisfied only if

$$\int_0^1 \psi_0(x_0,\alpha_0)\,dx_0=0 \tag{9.16}$$

This provides an equation for the unknown function $\alpha_0(t)$. Taking into account the explicit form for $\psi_0(x,\alpha_0)$, we can write this equation as

$$\alpha_0'=(v_0+\alpha_0)(1-\alpha_0)g(\alpha_0) \tag{9.17}$$

where

$$g(\alpha_0)=\int_0^1 \exp\{\tilde{\theta}_0(x_0,\alpha_0,a)\,dx_0$$

Since $g(\alpha_0)>0$, the solution of (9.17) with the initial condition $\alpha_0(0)=0$ monotonically approaches the rest point $\alpha_0=1$ as $t\to\infty$. Hence,

$$0\le\alpha(t)<1 \qquad \text{for} \qquad 0\le t\le T$$

Notice that Eq. (9.17) can be integrated by quadratures.

Thus, $\alpha_0(t)$ is constructed, and hence all of the terms of the zeroth-order approximation are defined.

C. Construction of Subsequent Terms of the Asymptotic Expansion

Integrating (9.15), we obtain $\bar{v}_1(x,t)=\tilde{v}_1(x,t)+\alpha_1(t)$ where

$$\tilde{v}_1(x,t)=\int_0^x\left\{\int_0^{x_1}\psi_0(x_0,\alpha(t))\,dx_0\right\}dx_1$$

is known and $\alpha_1(t)$ is an arbitrary function.

Let us rewrite the equation for $\bar{\theta}_1$ in the form

$$\frac{\partial^2\bar{\theta}_1}{\partial x^2}+\beta(x,t)\bar{\theta}_1=\gamma(x,t)\alpha_1(t)+g(x,t)$$

where

$$\beta(x, t) = \frac{1}{a}(v_0 + \alpha_0(t))(1 - \alpha_0(t))\exp\{\tilde{\theta}_0(x, \alpha_0(t), a)\}$$

$$\gamma(x, t) = -\frac{1}{a}(1 - v_0 - 2\alpha_0(t))\exp\{\tilde{\theta}_0(x, \alpha_0(t), a)\}$$

$$g(x, t) = \frac{1}{a}\left\{\frac{\partial\tilde{\theta}_0}{\partial t} - [1 - v_0 - 2\alpha_0(t)]\tilde{v}_1\exp\{\tilde{\theta}_0(x, \alpha_0(t), a\}\right\}$$

are given. We seek $\bar{\theta}_1$ in the form

$$\bar{\theta}_1 = A(x, t) + B(x, t)\alpha_1(t) \tag{9.18}$$

For A and B we obtain linear equations (t enters as a parameter)

$$\frac{\partial^2 A}{\partial x^2} + \beta(x, t)A = g(x, t) \qquad \frac{\partial^2 B}{\partial x^2} + \beta(x, t)B = \gamma(x, t) \tag{9.19}$$

with boundary conditions

$$A(0, t) = A(1, t) = 0 \qquad B(0, t) = B(1, t) = 0$$

The question of solvability of these boundary problems arises. It is well known that for the existence of a unique solution of the boundary value problem

$$u'' + d(x)u = f(x) \qquad u(0) = u(1) = 0$$

it is necessary and sufficient that the corresponding homogeneous problem has only the trivial solution. Likewise, the homogeneous problem has only the trivial solution if and only if there exists a nonnegative function $w(x)$ such that

$$w'' + d(x)w \leq 0 \qquad 0 < x < 1$$

$$w(0) \geq 0 \qquad w(1) \geq 0$$

and at least one of these inequalities is strict (see [44]). For Eqs. (9.19) such a function w can be easily found for sufficiently large a. Let us take $w(x) = \sin \pi x$. Then $w(x) \geq 0$ for $0 \leq x \leq 1$ and, furthermore,

$$w'' + \beta(x, t)w = [-\pi^2 + \beta(x, t)]\sin \pi x < 0 \qquad 0 < x < 1$$

if $\beta(x, t) < \pi^2$. It follows from the expression for $\beta(x, t)$ that there exists an a_3 such that for $a > a_3$, $\beta(x, t) < \pi^2$ holds. Thus, condition (9.11) should be changed to

$$a > \max(a_1, a_2, a_3) \tag{9.20}$$

Then the boundary value problems for A and B are uniquely solvable, and we obtain $\bar{\theta}_1$ in the form (9.18). Therefore, $\bar{v}_1$ and $\bar{\theta}_1$ depend linearly on the yet unknown function $\alpha_1(t)$.

Let us now construct boundary functions of the first order. For Π_1 we have

$$\frac{\partial \Pi_1 v}{\partial \tau} - b\frac{\partial^2 \Pi_1 v}{\partial x^2} = v_0[\exp\{\bar{\theta}_0(x, 0) + \Pi_0\theta(x, \tau)\} - \exp\{\bar{\theta}_0(x, 0\}]$$
$$\equiv h(x, \tau)$$
$$\Pi_1 v(x, 0) = -\bar{v}_1(x, 0) = -\tilde{v}_1(x, 0) - \alpha_1(0)$$
$$\frac{\partial \Pi_1 v}{\partial x}(0, \tau) = \frac{\partial \Pi_1 v}{\partial x}(1, \tau) = 0$$

Notice that the initial and boundary conditions are matched since

$$\frac{\partial \bar{v}_1}{\partial x}(0, 0) = \frac{\partial \bar{v}_1}{\partial x}(1, 0) = 0$$

The function $h(x, \tau)$ satisfies an estimate of type (9.12). The solution of this problem can be found as a Fourier series

$$\Pi_1 v(x, \tau) = \sum_{n=0}^{\infty} c_n(\tau)\cos(\pi n x) \tag{9.21}$$

For the coefficients $c_n(\tau)$, we obtain equations

$$c_n' = -b\pi^2 n^2 c_n + h_n(\tau) \qquad n = 0, 1, 2, \dots \tag{9.22}$$

with initial conditions

$$c_0(0) = -\int_0^1 \tilde{v}_1(x, 0)\, dx - \alpha_1(0) = \varphi_0 - \alpha_1(0)$$
$$c_n(0) = \varphi_n \qquad n = 1, 2, \dots$$

Here $h_n(\tau)$ and φ_n are the coefficients of the Fourier series expansions of $h(x, \tau)$ and $[-\tilde{v}_1(x, 0)]$ in $\cos(\pi n x)$, respectively; the $h_n(\tau)$ evidently satisfy an estimate of type (9.12).

Solving (9.22) for $n=0$, we obtain

$$c_0(\tau)=\varphi_0-\alpha_1(0)+\int_0^\tau h_0(\tau_0)\,d\tau_0$$

For $\Pi_1 v(x,\tau)$ to approach zero as $\tau\to\infty$, it is necessary that $c_0(\infty)=0$. This condition allows us to find the initial value $\alpha_1(0)$ for the as yet unknown function $\alpha_1(t)$

$$\alpha_1(0)=\varphi_0+\int_0^\infty h_0(\tau)\,d\tau \tag{9.23}$$

The function $\alpha_1(t)$ is completely defined during the study of the problem for $\bar{v}_2(x,t)$.

The solution of (9.22) for $n=1,2,\ldots$ can be written as

$$c_n(\tau)=\varphi_n\exp\{-b\pi^2n^2\tau\}+\int_0^\tau \exp\{-b\pi^2n^2(\tau-\tau_0)\}h_n(\tau_0)\,d\tau_0$$

The exponential estimate for $c_n(\tau)$ follows in an elementary fashion from this formula. It can be easily verified that the $c_n(\tau)$ decay like $1/n^4$ when $n\to\infty$. Therefore, series (9.21) converges. It can be differentiated termwise, once with respect to τ and twice with respect to x, and it satisfies an estimate of type (9.12).

For the $\Pi_1\theta$ we have the problem

$$\frac{\partial\Pi_1\theta}{\partial\tau}-a\frac{\partial^2\Pi_1\theta}{\partial x^2}=v_0\exp\{\bar{\theta}_0(x,0)+\Pi_0\theta(x,\tau)\}\Pi_1\theta+\sigma(x,\tau)$$

$$\Pi_1\theta(x,0)=-\bar{\theta}_1(x,0)\qquad \Pi_1\theta(0,\tau)=\Pi_1\theta(1,\tau)=0$$

Here $\sigma(x,\tau)$ is known and satisfies an estimate of type (9.12). As for $\Pi_1\theta$, an estimate such as (9.12) for $\Pi_1\theta$ can be obtained using a comparison theorem.

Consider the problem for $\bar{v}_2(x,t)$

$$\begin{aligned}\frac{\partial^2\bar{v}_2}{\partial x^2}&=\frac{1}{b}\{\alpha_1'(t)-[(v_0+\alpha_0(t))(1-\alpha_0(t))\bar{\theta}_1\\&\quad+(1-v_0-2\alpha_0(t))\alpha_1(t)]\exp\{\bar{\theta}_0(x,t)\}+F(x,t)\}\\&\equiv\psi_1(x,\alpha_1(t),t)\end{aligned}$$

$$\frac{\partial\bar{v}_2}{\partial x}(0,t)=\frac{\partial\bar{v}_2}{\partial x}(1,t)=0$$

where $F(x, t)$ is known, and $\bar{\theta}_1$ has the form (9.18). The solvability condition for this problem [analogous to (9.16)]

$$\int_0^1 \psi_1(x, \alpha_1(t), t)\, dx = 0$$

provides a linear differential equation for the as yet unknown $\alpha_1(t)$

$$\alpha_1' = k_1(t)\alpha_1 + k_2(t)$$

where k_1 and k_2 are known. Solving this equation with the initial condition (9.23), we obtain $\alpha_1(t)$. Thus, all of the terms of the first-order approximation have been completely determined.

Higher order terms of the series (9.4) can be obtained analogously.

Let θ_n and V_n denote the n-th partial sums of the series (9.4).

For sufficiently small ε, and sufficiently large a, there exists a unique solution θ, v of problem (9.2), (9.3), and the series (9.4) is an asymptotic series for this solution in the rectangle $\bar{\Omega}$ as $\varepsilon \to 0$, that is, the following estimates hold:

$$\max_{\bar{\Omega}} |\theta - \theta_n| = O(\varepsilon^{n+1}) \qquad \max_{\bar{\Omega}} |v - V_n| = O(\varepsilon^{n+1})$$

The detailed proof of this statement is presented in [45]. It is important to mention that during the proof of the statement one more condition on a appears:

$$a > \frac{(1 + v_0)^2}{32} \max_{0 \le t \le T} \exp\{\tilde{\theta}_0(\tfrac{1}{2}, \alpha_0(t), a)\} \tag{9.24}$$

By virtue of the behavior of $\tilde{\theta}_0$ for $a \to \infty$, it is clear that for sufficiently large a this inequality holds. Thus, we add (9.24) to the condition (9.20) on a. It can be shown that (9.20) is a consequence of (9.24). It should also be mentioned that of all the conditions on a, only (9.8) is a necessary one for the solvability of problem for $\bar{\theta}(x, t)$. The other conditions are sufficient. They are associated with obtaining estimates and with the method of proof. Condition (9.24) on a could possibly be somewhat relaxed.

D. Physical Interpretation of the Asymptotics of the Solution

Notice that the quantity a has quite an important practical meaning. If the conditions on a are satisfied, a nonexplosive reaction occurs.

The constructed asymptotics allows one to analyze the behavior of the temperature θ and the degree of conversion of the combustible component v. Since Π-functions satisfy an exponential estimate of type (9.12), the temperature θ during a short period of time changes rapidly from zero to values close to $\bar{\theta}_0(x, t)$. This corresponds to the fast stage of reaction. During this stage, the value of v changes very little since $\Pi_0 v = 0$. Then the slow stage of reaction begins: $\theta = \bar{\theta}_0(x, t) + O(\varepsilon)$ and $v = \alpha_0(t) + O(\varepsilon)$. As we mentioned earlier, the function $\alpha_0(t)$ grows monotonically from zero at $t = 0$ to one as $t \to \infty$. Thus, during a sufficiently long time, the combustible component burns out almost completely. The process of combustion is uniform in the domain $0 \le x \le 1$ since the leading term $\alpha_0(t)$ of the asymptotic expansion for v does not depend on x. The nonuniformity appears in the process starting with the terms of order ε.

As $\alpha_0(t)$ grows from 0 to 1, the function $\delta(\alpha_0(t), a) = (v_0 + \alpha_0(t)) \cdot (1 - \alpha_0(t))/a$ first increases (for $0 \le \alpha_0(t) \le (1 + v_0)/2$), and then decreases (for $(1 + v_0)/2 \le \alpha_0(t) \le 1$). The fact that $\bar{\theta}_0(x, t)$ is an increasing function of δ implies that the temperature θ during the slow stage of reaction first grows, and then, after $\alpha_0(t)$ reaches $(1 + v_0)/2$, starts to decrease.

We considered a spatially one-dimensional case. One can construct analogously the asymptotics of the solution for the combustion problem (9.1) in a cylindrical domain. See also [46] where a problem for the nonautocatalytic reaction is discussed.

X. HEAT CONDUCTION IN THIN BODIES

A. Statement of the Problem

A body is called *thin* if one (or more) of its characteristic dimensions is much smaller than the others. A thin rod and a thin plate are examples of such bodies. We consider a boundary value problem describing a heat conduction process in a thin rod, where the ratio ε of the thickness of the rod to its length is a small parameter. To simplify the presentation, we consider the problem for a planar rod, that is, in the thin rectangle $\{(0 \le x \le 1) \times (0 \le y \le \varepsilon)\}$

$$\frac{\partial u}{\partial t} - a\left(\frac{\partial^2 u}{\partial x^2} + \frac{\partial^2 u}{\partial y^2}\right) = f(u, x, t) \tag{10.1}$$

$$u|_{t=0} = \varphi(x, y) \qquad u|_{x=0} = \psi_0(y, t) \qquad u|_{x=1} = \psi_1(y, t) \tag{10.2}$$

$$\left.\frac{\partial u}{\partial y} - \varepsilon A u\right|_{y=0} = 0 \qquad \left.\frac{\partial u}{\partial y} + \varepsilon A u\right|_{y=\varepsilon} = 0 \tag{10.3}$$

The function $f(u, x, t)$ describes the heat sources (in general, nonlinear) within the rod. Conditions (10.2) give the temperature at the initial time, as well as at the ends $x = 0$ and $x = 1$. Conditions (10.3) mean that on the lateral surfaces $y = 0$ and $y = \varepsilon$, a weak heat exchange with the surrounding medium occurs.

It seems possible to reason in the following way: Since the rod is thin ($0 \le y \le \varepsilon$) and the heat exchange through its lateral surfaces is small (the corresponding heat exchange coefficient is εA) the temperature in each cross-section $x = \text{const}$ at every time t does not depend on y, at least in the zeroth approximation. So, in order to obtain an approximation for the solution of the problem, we should omit the second derivative with respect to y in (10.1), as well as conditions (10.3), and solve the resulting *truncated* (one-dimensional) equation

$$\frac{\partial u}{\partial t} - a\frac{\partial^2 u}{\partial x^2} = f(u, x, t) \tag{10.4}$$

with the additional conditions

$$u|_{t=0} = \varphi(x, 0) \qquad u|_{x=0} = \psi_0(0, t) \qquad u|_{x=1} = \psi_1(0, t)$$

However, as the asymptotic analysis will show, the correct one-dimensional equation describing the temperature in the zeroth-order approximation is different from the *truncated* Eq. (10.4). This correct equation is obtained during the construction of the asymptotics of the solution.

By the change of variable $y = \varepsilon y'$ problem (10.1)–(10.3) is reduced to the singularly perturbed problem (which we consider on the time interval $0 \le t \le T$)

$$\varepsilon^2\frac{\partial u}{\partial t} - a\left(\varepsilon^2\frac{\partial^2 u}{\partial x^2} + \frac{\partial^2 u}{\partial y^2}\right) = \varepsilon^2 f(u, x, t) \tag{10.5}$$

$$(x, y, t) \in \Omega = \{(0 < x < 1) \times (0 < y < 1) \times (0 < t \le T)\}$$

$$u|_{t=0} = \varphi(x, y, \varepsilon) \qquad u|_{x=0} = \psi_0(y, t, \varepsilon) \qquad u|_{x=1} = \psi_1(y, t, \varepsilon) \tag{10.6}$$

$$\left.\frac{\partial u}{\partial y} - \varepsilon^2 Au\right|_{y=0} = 0 \qquad \left.\frac{\partial u}{\partial y} + \varepsilon^2 Au\right|_{y=1} = 0 \tag{10.7}$$

Here we use the old notation y for the new variable y'. We assume that all the given functions are sufficiently smooth, that $a > 0$, and that the initial and boundary values (10.6) are matched at the corner points, that is,

$$\varphi(0, y, \varepsilon) = \psi_0(y, 0, \varepsilon) \qquad \varphi(1, y, \varepsilon) = \psi_1(y, 0, \varepsilon)$$

B. Construction of the Asymptotics of the Solution

We construct the asymptotics of the solution of (10.5)–(10.7) in the form

$$u(x, y, t, \varepsilon) = \sum_{i=0}^{\infty} \varepsilon^i [\bar{u}_i(x, y, t) + \Pi_i(x, y, \tau) + Q_i(\xi, y, t) + Q_i^*(\xi_*, y, t) + P_i(\xi, y, \tau) + P_i^*(\xi_*, y, \tau)] \quad (10.8)$$

where $\bar{u}_i$ are terms of the regular part of the asymptotics; Π_i, Q_i and Q_i^* are boundary layer functions describing the boundary layers near $t = 0$, $x = 0$, and $x = 1$, respectively; P_i and P_i^* are corner boundary functions; $\tau = t/\varepsilon^2$, $\xi = x/\varepsilon$, and $\xi_* = (1 - x)/\varepsilon$ are boundary layer variables. In the standard way, substituting the series (10.8) into (10.5)–(10.7) and representing f in the form $\bar{f} + \Pi f + Qf + Q^*f + Pf + P^*f$ (see Section VII.E), we obtain equations for the terms of the asymptotics. We consider in detail only the construction of the zeroth approximation.

We have for $\bar{u}_0$ the reduced equation that is obtained from (10.5) for $\varepsilon = 0$:

$$a \frac{\partial^2 \bar{u}_0}{\partial y^2} = 0$$

with boundary conditions that follow from (10.7).

$$\left.\frac{\partial \bar{u}_0}{\partial y}\right|_{y=0} = \left.\frac{\partial \bar{u}_0}{\partial y}\right|_{y=1} = 0$$

The solution of this problem is an arbitrary function of x and t: $\bar{u}_0 = \alpha_0(x, t)$. Therefore, (10.5)–(10.7) is a problem in the critical case. Analogously, we obtain $\bar{u}_1 = \alpha_1(x, t)$, an arbitrary function. For $\bar{u}_2(x, t)$ we have the problem

$$a \frac{\partial^2 \bar{u}_2}{\partial y^2} = \frac{\partial \alpha_0}{\partial t} - a \frac{\partial^2 \alpha_0}{\partial x^2} - f(\alpha_0, x, t) \equiv h(\alpha_0, x, t)$$

$$\left.\frac{\partial \bar{u}_2}{\partial y}\right|_{y=0} = A\alpha_0(x, t) \qquad \left.\frac{\partial \bar{u}_2}{\partial y}\right|_{y=1} = -A\alpha_0(x, t)$$

Integration of this equation with the first boundary condition implies

$$\frac{\partial \bar{u}_2}{\partial y} = A\alpha_0(x, t) + \frac{h(\alpha_0, x, t)y}{a}$$

Substituting this expression into the second boundary condition, we obtain

$$h(\alpha_0, x, t) = -2Aa\alpha_0$$

Thus, the problem for $\bar{u}_2$ is solvable only if the above equality holds. This equality provides an equation for the unknown function $\alpha_0(x, t)$, that is, for the leading term of the regular part of the asymptotics. Taking into account the form of $h(\alpha_0, x, t)$, this relation can be rewritten as

$$\frac{\partial \alpha_0}{\partial t} - a\frac{\partial^2 \alpha_0}{\partial x^2} = f(\alpha_0, x, t) - 2Aa\alpha_0 \tag{10.9}$$

This equation is exactly the correct one-dimensional equation describing the leading term of the temperature distribution in the rod. As we can see, Eq. (10.9) differs substantially from the truncated one-dimensional Eq. (10.4). We discuss in detail this difference in Section X.C.

The initial and boundary conditions for α_0 are obtained during the construction of the boundary layer functions Π_0, Q_0, and Q_0^*. Then we will be able to define $\alpha_0(x, t)$ completely.

For $\Pi_0(x, y, \tau)$, we have the problem (x enters as a parameter, $0 \le x \le 1$)

$$\frac{\partial \Pi_0}{\partial \tau} - a\frac{\partial^2 \Pi_0}{\partial y^2} = 0 \qquad 0 < y < 1 \qquad \tau > 0$$

$$\Pi_0(x, y, 0) = \varphi_0(x, y) - \alpha_0(x, 0) \qquad \frac{\partial \Pi_0}{\partial y}(x, 0, \tau) = \frac{\partial \Pi_0}{\partial y}(x, 1, \tau) = 0$$

Here $\varphi_0(x, y)$ is the leading term of the expansion for $\varphi(x, y, \varepsilon)$ into a power series in ε. Notice that the initial condition contains the as yet unknown function $\alpha_0(x, 0)$. Solving this problem by Fourier series, we obtain

$$\Pi_0(x, y, \tau) = \sum_{n=0}^{\infty} b_n \exp\{-a\pi^2 n^2 \tau\} \cos(\pi n y)$$

where

$$b_0(x) = \int_0^1 \varphi_0(x, y)\, dy - \alpha_0(x, 0)$$

$$b_n(x) = 2\int_0^1 \varphi_0(x, y) \cos(\pi n y)\, dy \qquad n \ge 1$$

The standard condition, that Π-functions must tend to zero as $\tau \to \infty$, leads to $b_0 = 0$, which allows us to determine $\alpha_0(x, 0)$:

$$\alpha_0(x, 0) = \int_0^1 \varphi_0(x, y)\, dy \equiv \hat{\varphi}_0(x) \tag{10.10}$$

Note that $\hat{\varphi}_0(x)$ is the leading term of the initial temperature averaged over the cross-section $x = \text{const}$. Thus, Π_0 is completely determined and it satisfies the estimate

$$|\Pi_0(x, y, \tau)| \leq C \exp\{-\alpha\tau\}$$

For the boundary layer function $Q_0(\xi, y, t)$, we obtain the problem (t enters as a parameter, $0 \leq t \leq T$):

$$\frac{\partial^2 Q_0}{\partial \xi^2} + \frac{\partial^2 Q_0}{\partial y^2} = 0 \qquad \xi > 0 \qquad 0 < y < 1$$

$$Q_0(0, y, t) = \psi_{00}(y, t) - \alpha_0(0, t) \qquad \frac{\partial Q_0}{\partial y}(\xi, 0, t) = \frac{\partial Q_0}{\partial y}(\xi, 1, t) = 0$$

where $\psi_{00}(y, t)$ is the leading term of the expansion for $\psi_{00}(y, t, \varepsilon)$ into a power series in ε. Notice that the boundary condition at $\xi = 0$ contains the unknown function $\alpha_0(0, t)$. Using the Fourier method, we find

$$Q_0(\xi, y, t) = \sum_{n=0}^{\infty} d_n(t) \exp\{-\pi n \xi\} \cos(\pi n y)$$

where

$$d_0(t) = \int_0^1 \psi_{00}(y, t)\, dy - \alpha_0(0, t)$$

$$d_n(t) = 2 \int_0^1 \psi_{00}(y, t) \cos(\pi n y)\, dy \qquad n \geq 1$$

The standard assumption $Q_0(\infty, y, t) = 0$ leads to the condition $d_0 = 0$, which gives the value of the function $\alpha_0(0, t)$:

$$\alpha_0(0, t) = \int_0^1 \psi_{00}(y, t)\, dy \equiv \hat{\psi}_{00}(t) \tag{10.11}$$

Thus, Q_0 is completely determined and it satisfies the estimate

$$|Q_0(\xi, y, t)| \leq C \exp\{-\alpha\xi\}$$

The boundary layer function $Q_0^*(\xi_*, y, t)$ is defined analogously to the $Q_0(\xi, y, t)$ and has an exponential estimate with respect to the variable ξ_*. During the construction of Q_0^*, we find the boundary condition at $x = 1$ for $\alpha_0(x, t)$:

$$\alpha_0(1, t) = \int_0^1 \psi_{10}(y, t)\, dy \equiv \hat{\psi}_{10}(t) \tag{10.12}$$

where $\psi_{10}(y, t)$ is the leading term of the expansion of $\psi_1(y, t, \varepsilon)$ into a power series in ε.

So, for the unknown function $\alpha_0(x, t)$, we have obtained Eq. (10.9) with the initial condition (10.10) and the boundary conditions (10.11) and (10.12). It can be easily verified that these conditions are matched to be continuous at the corner points, that is,

$$\hat{\varphi}_0(0) = \hat{\psi}_{00}(0) \qquad \hat{\varphi}_0(1) = \hat{\psi}_{10}(0)$$

These equalities follow directly from (10.10)–(10.12), and conditions $\varphi_0(0, y) = \psi_{00}(y, 0)$, $\psi_0(1, y) = \psi_{10}(y, 0)$.

Suppose that Eq. (10.9) (notice, that it is in general nonlinear) with the conditions (10.10)–(10.12) has a solution $\alpha_0(x, t)$ for $0 \le t \le T$.

The corner boundary functions $P_0(\xi, y, \tau)$ and $P_0^*(\xi_*, y, \tau)$ are needed to compensate for the discrepancies introduced by the boundary layer function $\Pi_0(x, y, \tau)$ in the boundary conditions at $x = 0$ and $x = 1$, and by the boundary layer functions $Q_0(\xi, y, t)$ and $Q_0^*(\xi_*, y, t)$ in the initial condition at $t = 0$.

For $P_0(\xi, y, \tau)$, we obtain the problem

$$\frac{\partial P_0}{\partial \tau} - a\left[\frac{\partial^2 P_0}{\partial \xi^2} + \frac{\partial^2 P_0}{\partial y^2}\right] = 0 \qquad \xi > 0 \qquad 0 < y < 1 \qquad \tau > 0$$

$$P_0(\xi, y, 0) = -Q_0(\xi, y, 0) = -\sum_{n=1}^{\infty} d_n(0)\exp\{-\pi n\xi\}\cos(\pi n y)$$

$$P_0(0, y, \tau) = -\Pi_0(0, y, \tau) = -\sum_{n=1}^{\infty} b_n(0)\exp\{-a\pi^2 n^2 \tau\}\cos(\pi n y)$$

$$\frac{\partial P_0}{\partial y}(\xi, 0, \tau) = \frac{\partial P_0}{\partial y}(\xi, 1, \tau) = 0$$

From the expressions for $d_n(t)$ and $b_n(x)$ together with the equality $\varphi_0(0, y) = \psi_{00}(y, 0)$, it follows that $d_n(0) = b_n(0)$. Thus, the initial and boundary (at $\xi = 0$) conditions for P_0 are matched so as to be continuous at the points $(0, y, 0)$. We seek the solution of this problem in the form

$$P_0(\xi, y, \tau) = \sum_{n=1}^{\infty} [v_n(\xi, \tau) - d_n(0)\exp\{-\pi n\xi - a\pi^2 n^2 \tau\}]\cos(\pi n y)$$

Then for $v_n(\xi, \tau)$, we have the problem

$$\frac{\partial v_n}{\partial \tau} - a\frac{\partial^2 v_n}{\partial \xi^2} + a\pi^2 n^2 v_n = h_n(\xi, \tau) \qquad \xi > 0 \qquad \tau > 0$$

$$v_n(\xi, 0) = 0 \qquad v_n(0, \tau) = 0$$

where

$$h_n(\xi, \tau) = -ad_n(0)\pi^2 n^2 \exp\{-\pi n\xi - a\pi^2 n^2 \tau\}$$

The solution of this problem can be represented as

$$v_n = \int_0^{\tau}\int_0^{\infty} G_n(\xi, \xi_0, \tau - \tau_0)h_n(\xi_0, \tau_0)\, d\tau_0\, d\xi_0$$

where G_n is the Green's function:

$$G_n(\xi, \xi_0, \tau - \tau_0) = \frac{1}{2\sqrt{a\pi(\tau - \tau_0)}} \exp\{-a\pi^2 n^2(\tau - \tau_0)\}$$
$$\times \left[\exp\left\{-\frac{(\xi - \xi_0)^2}{4a(\tau - \tau_0)}\right\} - \exp\left\{-\frac{(\xi + \xi_0)^2}{4a(\tau - \tau_0)}\right\}\right]$$

Using this representation for $v_n(\xi, \tau)$, it is not difficult to prove that $|v_n| \le C|d_n(0)|\exp\{-a(\xi + \tau)\}$. Hence, $P_0(\xi, y, \tau)$ satisfies the estimate

$$|P_0(\xi, y, \tau)| \le C\exp\{-\alpha(\xi + \tau)\}$$

The corner boundary function $P_0^*(\xi_*, y, \tau)$ is constructed in a similar way, and it satisfies an analogous exponential estimate.

C. The Main Result

Let $U_0(x, y, t, \varepsilon)$ denote the sum of the leading terms of the series (10.8):

$$U_0(x, y, t, \varepsilon) = \alpha_0(x, t) + \Pi_0(x, y, t/\varepsilon^2) + Q_0(x/\varepsilon, y, t)$$
$$+ Q_0^*((1 - x)/\varepsilon, y, t) + P_0(x/\varepsilon, y, t/\varepsilon^2)$$
$$+ P_0^*((1 - x)/\varepsilon, y, t/\varepsilon^2)$$

For sufficiently small ε, problem 10.5)–(10.7) has a unique solution $u(x, y, t, \varepsilon)$, and the function $U_0(x, y, t, \varepsilon)$ is an asymptotic approximation for this solution in the parallelepiped $\bar{\Omega}$ as $\varepsilon \to 0$ with an accuracy of order ε, that is, the following estimate holds:

$$\max_{\bar{\Omega}} |u - U_0| = O(\varepsilon)$$

This statement can be proved by applying the method of successive approximations to the equation for the remainder $w = u - U_0$ and using the maximum principle to estimate these successive approximations. A detailed proof is presented in [47]. Under more severe constraints of matching of the initial and boundary conditions (10.6) in the corner points $(0, y, 0)$ and $(1, y, 0)$, it is possible to construct asymptotic approximations for the solution with greater accuracy.

By virtue of the corresponding exponential estimates, the boundary layer function Π_0 and the corner boundary functions P_0 and P_0^* become arbitrarily small for $t \geq \delta > 0$, and an approximation of the solution with an accuracy of order ε (such an approximation is often sufficient for practical purposes) is given by the sum

$$\alpha_0(x, t) + Q_0(x/\varepsilon, y, t) + Q_0^*((1 - x)/\varepsilon, y, t)$$

The boundary layer functions Q_0 and Q_0^* describe the fast change of temperature near the ends $x = 0$ and $x = 1$ of the rod. However, inside the rod (for $\delta \leq x \leq 1 - \delta$) they are arbitrarily small, and the distribution of temperature there is approximated by the function $\alpha_0(x, t)$, which is the solution of the one-dimensional heat conduction Eq. (10.9) with the additional conditions (10.10)–(10.12). As we can see, this equation differs substantially from the truncated one-dimensional Eq. (10.4), obtained from the original Eq. (10.1) by omitting the second derivative with respect to y [such a truncated version of (10.1) is often inappropriately used in practical calculations]. In comparison with the truncated Eq. (10.4), there is an additional term $-2Aa\alpha_0$ on the right-hand side of (10.9). This term has obvious physical meaning: Since the rod is thin the heat exchange with the surrounding medium on the lateral surface of the rod influences the temperature not only near the surface but also inside the rod. The influence is large for a large thermal diffusion coefficient a, and for a large coefficient A in the heat exchange term. Notice also that the leading term of the asymptotics $\alpha_0(x, t)$ does not depend on y. This is again connected with the fact that the rod is thin, therefore the temperature is the same along the cross section of the rod in the zeroth approximation.

Thus, the asymptotic method allows us to construct a correct one-dimensional model which provides a good approximation to the two-dimensional problem.

D. The Problem for a Thin Three-Dimensional Rod

Above we considered the heat conduction problem in a thin rectangle. The asymptotics of the solution can be constructed similarly for the problem in a thin three-dimensional rod of constant cross-section S. In this case, by stretching the variables y and z with the coefficient ε, we obtain the problem

$$\varepsilon^2 \frac{\partial u}{\partial t} - a\left(\varepsilon^2 \frac{\partial^2 u}{\partial x^2} + \frac{\partial^2 u}{\partial y^2} + \frac{\partial^2 u}{\partial z^2}\right) = \varepsilon^2 f(u, x, t)$$

$$0<x<1 \qquad (y, z) \in S \qquad 0<t \le T$$

$$u|_{t=0} = \varphi(x, y, z, \varepsilon) \qquad u|_{x=0} = \psi_0(y, z, t, \varepsilon) \qquad u|_{x=1} = \psi_1(y, z, t, \varepsilon)$$

$$\frac{\partial u}{\partial n} + \varepsilon^2 A u|_\Gamma = 0$$

Here Γ is the lateral surface of the rod and $\partial/\partial n$ is the derivative along the outer normal to Γ.

The asymptotics of the solution can be constructed in just the same way as in the case of a rectangle. In particular, the equation for the function $\alpha_0(x, t)$ has the same form (10.9), the only difference being that the coefficient 2 in the term $2Aa\alpha_0$ should be replaced by ℓ/s, where ℓ is the length of the boundary of the cross-section S and s is its area.

The initial and boundary conditions for the function $\alpha_0(x,t)$ are defined, as before, during the construction of the boundary layer functions. For example, for $Q_0(\xi, y, z, t)$ we obtain the problem

$$\frac{\partial^2 Q_0}{\partial \xi^2} + \frac{\partial^2 Q_0}{\partial y^2} + \frac{\partial^2 Q_0}{\partial z^2} = 0 \qquad \xi > 0 \qquad (y, z) \in S$$

$$Q_0(0, y, z, t) = \psi_{00}(y, z, t) - \alpha_0(0, t) \qquad \left.\frac{\partial Q_0}{\partial n}\right|_\Gamma = 0$$

The solution can be found by the Fourier method in the form

$$Q_0(\xi, y, z, t) = \sum_{n=0}^{\infty} d_n(t) \exp\{-\lambda_n \xi\} F_n(y, z)$$

where λ_n $(0 = \lambda_0 < \lambda_1 \le \lambda_2 \le \cdots \le \lambda_n \le \cdots)$ and $F_n(y, z)$ are the eigenvalues and eigenfunctions of the problem

$$\frac{\partial^2 F}{\partial y^2} + \frac{\partial^2 F}{\partial z^2} + \lambda^2 F = 0 \qquad (y, z) \in S \qquad \left.\frac{\partial F}{\partial n}\right|_{\Gamma} = 0$$

In particular, $F_0 = 1$, and

$$d_n(t) = \frac{1}{\|F_n\|} \int_S \int [\psi_{00}(y, z, t) - \alpha_0(0, t)] F_n(y, z)\, dy\, dz$$

The condition $Q_0(\infty, y, z, t) = 0$ leads to $d_0(t) = 0$, which, in turn, allows us to find the boundary condition for the as yet unknown function $\alpha_0(x, t)$ at $x = 0$

$$\alpha_0(0, t) = \frac{1}{s} \int_S \int \psi_{00}(y, z, t)\, dy\, dz \equiv \hat{\psi}_{00}(t)$$

The initial condition at $t = 0$ and the second boundary condition at $x = 1$ are defined analogously. Thus, the process of constructing the boundary layer functions is similar to that described in Section X.B. A distinction is that we use now expansions in a Fourier series in terms of the eigenfunctions $F_n(y, z)$ instead of $\cos(\pi n y)$.

E. The Case of a Small Thermal Diffusion Coefficient

Consider problem (10.1)–(10.3) for the case when the coefficient a is small. More precisely, we replace this coefficient by ε. In addition, assume that the heat exchange on the lateral surface of the rod is not weak, that is, the coefficient εA in conditions (10.3) is replaced by A.

In this case, the change of variable $y = \varepsilon y'$ leads to the following problem.

$$\begin{gathered} \varepsilon \frac{\partial u}{\partial t} - \varepsilon^2 \frac{\partial^2 u}{\partial x^2} - \frac{\partial^2 u}{\partial y^2} = \varepsilon f(u, x, t, \varepsilon) \\ (x, y, t) \in \Omega = \{(0 < x < 1) \times (0 < y < 1) \times (0 < t \leq T)\} \\ u|_{t=0} = \varphi(x, y, \varepsilon) \quad u|_{x=0} = \psi_0(y, t, \varepsilon) \quad u|_{x=1} = \psi_1(y, t, \varepsilon) \\ \left.\frac{\partial u}{\partial y} - \varepsilon A u\right|_{y=0} = 0 \qquad \left.\frac{\partial u}{\partial y} + \varepsilon A u\right|_{y=1} = 0 \end{gathered} \tag{10.13}$$

Here we once more use the old notation y for the new variable y'. The asymptotics of the solution of this problem can be constructed by the above method. However, it has substantial differences from the asymptotics considered in Sections X.B and X.C.

The first difference is that the equation for $\alpha_0(x, t)$ now has the form

$$\frac{\partial \alpha_0}{\partial t} = f(\alpha_0, x, t, 0) - 2A\alpha_0$$

that is, it is an ordinary differential equation in which x enters as a parameter ($0 \le x \le 1$). As in Section X.B, the initial condition for $\alpha_0(x, t)$ is obtained during the construction of the boundary layer function $\Pi_0(x, y, \tau)$. The function $\Pi_0(x, y, \tau)$ and the initial value $\alpha_0(x, 0)$ have the same forms as before.

The second difference is that the boundary layer near $x = 0$ (and also the boundary layer near $x = 1$) has two scales. In other words, the boundary layer near $x = 0$ is described by boundary layer functions of two types. The function $Q_0(\xi, y, t)$ ($\xi = x/\varepsilon$) is determined, like that from the Section X.B, as the solution of the problem

$$\frac{\partial^2 Q_0}{\partial \xi^2} + \frac{\partial^2 Q_0}{\partial y^2} = 0 \qquad \xi > 0 \qquad 0 < y < 1$$

$$Q_0(0, y, t) = \psi_{00}(y, t) - \hat{\psi}_{00}(t) \qquad \frac{\partial Q_0}{\partial y}(\xi, 0, t) = \frac{\partial Q_0}{\partial y}(\xi, 1, t) = 0$$

where

$$\hat{\psi}_{00}(t) = \int_0^1 \psi_{00}(y, t)\, dy$$

The function Q_0 can be found by the Fourier method and it satisfies an exponential estimate with respect to the variable ξ.

The second function, describing the boundary layer near $x = 0$ (denote it by R_0), depends on a new boundary layer variable $\zeta = x/\sqrt{\varepsilon}$ and on t. For this function, we obtain the problem

$$\frac{\partial R_0}{\partial t} - \frac{\partial^2 R_0}{\partial \zeta^2} = F(R_0, t) \qquad 0 < t \le T \qquad \zeta > 0$$

$$R_0(\zeta, 0) = 0 \qquad R_0(0, t) = \hat{\psi}_{00}(t) - \alpha_0(0, t)$$

where $F(R_0, t) = f(\alpha_0(0, t) + R_0, 0, t, 0) - f(\alpha_0(0, t), 0, t, 0) - 2AR_0$. As we can see, the equation for R_0 is a semilinear one-dimensional parabolic equation. It can be proved that the function $R_0(\zeta, t)$ satisfies an exponential estimate with respect to variable ζ.

The next term in the series of R-functions is of order $\sqrt{\varepsilon}$. For the function $R_{1/2}(\zeta, t)$, we obtain the linear parabolic equation

$$\frac{\partial R_{1/2}}{\partial t} - \frac{\partial^2 R_{1/2}}{\partial \zeta^2} = \beta(\zeta, t) R_{1/2} + \sigma(\zeta, t)$$

with the initial and boundary conditions

$$R_{1/2}(\zeta, 0) = 0 \qquad R_{1/2}(0, t) = 0$$

where $\beta(\zeta, t) = f_u(\alpha_0(0, t) + R_0(\zeta, t), 0, t, 0) - 2A$, $\sigma(\zeta, t)$ is a known function having an exponential estimate with respect to the variable ζ. The solution of this problem satisfies the same estimate as $R_0(\zeta, t)$. The boundary layer functions $Q_0^*(\xi_*, y, t)$ $[\xi_* = (1 - x)/\varepsilon]$, $R_0^*(\zeta_*, t)$ and $R_{1/2}^*(\zeta_*, t)$ $[\zeta_* = (1 - x)/\sqrt{\varepsilon}]$ are constructed in a similar way, and they satisfy analogous exponential estimates. The corner boundary functions $P_0(\xi, y, \tau)$ and $P_0^*(\xi_*, y, \tau)$ are determined in just the same way as in Section X.B.

The function

$$\begin{aligned} U = {} & \alpha_0(x, t) + \Pi_0(x, y, t/\varepsilon^2) + Q_0(x/\varepsilon, y, t) + R_0(x/\sqrt{\varepsilon}, t) \\ & + \sqrt{\varepsilon} R_{1/2}(x/\sqrt{\varepsilon}, t) + Q_0^*((1 - x)/\varepsilon, y, t) \\ & + R_0^*((1 - x)/\sqrt{\varepsilon}, t) + \sqrt{\varepsilon} R_{1/2}^*((1 - x)/\sqrt{\varepsilon}, t) \\ & + P_0(x/\varepsilon, y, t/\varepsilon^2) + P_0^*((1 - x)/\varepsilon, y, t/\varepsilon^2) \end{aligned}$$

is an asymptotic approximation for the solution of problem (10.13) in the domain $\bar{\Omega}$ with an accuracy of order ε.

F. The Problem of Thermoelasticity in Thin Bodies

The above algorithm for the construction of the asymptotics of the solutions of heat conduction problems in this bodies can also be applied to some more complicated problems, for example, to problems of thermoelasticity in thin bodies. The corresponding system of equations for the displacement vector $\mathbf{u}(x, y, z, t)$ and temperature $\Theta(x, y, z, t)$ in a domain G in the linear approximation has the form (see [48])

$$\mu\, \Delta \mathbf{u} + (\lambda + \mu) \text{grad div}\, \mathbf{u} + \mathbf{F} = \gamma \text{ grad}\, \theta + \rho_0 \frac{\partial^2 \mathbf{u}}{\partial t^2}$$

$$\Delta \theta - \frac{1}{k} \frac{\partial \theta}{\partial t} - \eta \text{ div} \frac{\partial \mathbf{u}}{\partial t} = -\frac{f}{k}$$

where μ and λ are the elastic moduli, **F** is the vector of mass forces, $\gamma = (3\lambda + 2\mu)/\alpha$, α is the coefficient of thermal extension, $\rho_0(x, y, z)$ is the volume density, k is the coefficient of thermal diffusion, $\eta = \gamma\Theta_0/\lambda_0$, Θ_0 is the average temperature of the body, λ_0 is the coefficient of heat conduction, and f represents thermal sources. In the case of a thin body (for a thin rod or plate), this system can be transformed to a singularly perturbed one by an appropriate stretching of the variables. This allows us to apply the boundary function method to construct the asymptotics of the solution. In [49] the asymptotics of the solution for a thin rod is constructed. Asymptotic analysis allows one to choose the correct simplified model of lower dimension, which provides a good approximation to the solution of the original problem.

ACKNOWLEDGMENT

The research described here was made possible in part by Grant No. JI0100 from the International Science Foundation and Russian Government.

REFERENCES

1. A. Tikhonov, *Mat. Sbornik*, **31**, 575 (1952) (in Russian).
2. A. Tikhonov, A. Vasil'eva, and A. Svcshnikov, *Differential Equations*, Springer-Verlag, Berlin, Heidelberg, New York, Tokyo, 1985.
3. A. Vasil'eva, *Vestnik MGU*, **6**, 39 (1954) (in Russian).
4. A. Vasil'eva and V. Butuzov, *Asymptotic Expansions of the Solutions of Singularly Perturbed Equations*, Nauka, Moscow, Russia, 1973 (in Russian).
5. A. Vasil'eva, V. Butuzov, and L. Kalachev, *The Boundary Function Method for Singular Perturbation Problems*, SIAM, Philadelphia, PA, 1995.
6. V. Esipova, *Differential Equations*, **11**, 1457 (1975).
7. L. Pontriagin, *Ordinary Differential Equations*, Addison-Wesley, Reading, MA, 1962.
8. A. Vasil'eva and V. Butuzov, *Singularly Perturbed Equations in Critical Case*, Technical Report MRC-TSR 2039, University of Wisconsin, Madison, WI, 1980.
9. V. Vasil'ev, A. Vol'pert, and S. Khudyaev, *Comput. Math. Math. Phys.*, **13**, 187 (1973).
10. A. Vasil'eva, A. Nikitin, and A. Petrov, *Mat. Model.*, **7**, 61 (1995), (in Russian).
11. M. Belyanin, D. Sokoloff, A. Shukurov, *Geophys. Astrophys. Fluid Dyn.*, **68**, 237 (1993).
12. A. Vasil'eva, A. Nikitin, and A. Petrov, *Geophys. Astrophys. Fluid Dyn.*, **78**, 261 (1994).
13. P. Fife, *Math. Anal. Appl.*, **54**, 497 (1976).
14. V. Vasil'ev, Yu. Romanovskii, D. Chernavskii, and V. Yakhno, *Autowave Processes in Kinetic Systems*, VEB Deutscher Verlag der Wissenschaften, Berlin, 1987.
15. A. Vasil'eva, *Comput. Math. Math. Phys.*, **35**, 520 (1995).
16. A. Vasil'eva, *Comput. Math. Math. Phys.*, **32**, 1582 (1992).
17. V. Butuzov and A. Vasil'eva, *Math. Notes*, **42**, 761 (1987).
18. V. Butuzov and A. Vasil'eva, *Comput. Math. Math. Phys.*, **28**, 346 (1988).

19. G. Nicolis and I. Prigogine, *Self-Organization in Nonequilibrium Systems: from Dissipative Structures to Order through Fluctuation*, Wiley, New York, 1977.
20. A. Vasil'eva, *Comput. Math. Math. Phys.*, **34**, 1401 (1994).
21. D. Henry, *Geometric Theory of Semilinear Parabolic Equations*, Springer-Verlag, Berlin, Heidelberg, New York, 1981.
22. A. Vasil'eva, *Mat. Model.*, **2**, 119 (1990) (in Russian). Translated by Wiley, MMCE, **1**, 167 (1993).
23. A. Vasil'eva, *Mat. Model*, **3**, 114 (1991).
24. J. Hale and K. Sakamoto, *Jpn. J. Appl. Math.*, **5**, 367 (1988).
25. Y. Nishiura and H. Fujii, *SIAM J. Math. Anal.*, **18**, 1726 (1987).
26. N. Nefedov, *Dokl. Akad. Nauk Russia* (1996).
27. M. Vishik and L. Lyusternik, *Usp. Mat. Nauk*, **12**, N5, 3 (1957) (in Russian). *Am. Math. Soc. Transl.*, **20**, 239 (1962).
28. V. Trenogin, *Russ. Math. Surveys*, **25**, 119 (1970).
29. V. Butuzov, *Math. USSR-Sb.*, **33**, 403 (1977).
30. V. Butuzov and A. Nikitin, *Differential Equations*, **17**, 1104 (1981).
31. V. Butuzov and L. Kalachev, *Math. Notes*, **38**, 819 (1986).
32. V. Butuzov, *Differential Equations*, **15**, 1318 (1979).
33. V. Butuzov, *Comput. Math. Math. Phys.*, **12**, N3, 14 (1972).
34. N. Emanuel and D. Knorre, *Course of Chemical Kinetics*, Vischaja Shkola, Moscow, 1984.
35. L. Bobisud and C. Christenson, *J. Math. Anal. Appl.*, **74**, 296 (1980).
36. P. Fife and Y. Grinlee, *Usp. Mat. Nauk*, **29**, N4, 103 (1974) (in Russian).
37. N. Nefedov, *Differential'nye Uravneniya*, **31**, N7, 1132 (1995) (in Russian).
38. V. Butuzov and N. Nefedov, *Dokl. Akad. Nauk Russ.* (1996).
39. P. Fife, *Dynamics of Internal Layers and Diffusive Interfaces*, SIAM, Philadelphia, PA, 1988.
40. A. Vol'pert and S. Khudyaev, *Analysis in the Class of Discontinuous Functions and the Equations of Mathematical Physics*, Nauka, Moscow, 1973.
41. I. Gel'fand, *Uspekhi Mat. Nauk*, **14**, N2, 87 (1959).
42. D. Frank-Kamenetskii, *Diffusion and Heat Transfer in Chemical Kinetics*, Plenum, New York, 1969.
43. Yu. Kolesov, *Trans. Moscow Math. Soc.*, **21**, 114 (1970).
44. M. Protter and H. Weinberger, *Maximum Principles in Differential Equations*, Springer-Verlag, New York, Berlin, Heidelberg, Tokyo, 1984.
45. V. Butuzov and L. Kalachev, *Comput. Math. Math. Phys.*, **28**, N4, 39 (1988).
46. D. Sattinger, *Q. Appl. Math.*, **33**, 47 (1975).
47. V. Butuzov and T. Urazgil'dina, *Ukrainskii Matematischeskii Zh.*, **39**, N1, 13 (1987) (in Russian).
48. W. Nowacki, *Dynamic Problems of Thermoelasticity*, PWN-Polish Scientific Publishers, Warsaw, Poland, 1975.
49. V. Butuzov and T. Urazgil'dina, *J. Appl. Math. Mech.*, **51**, 761 (1987).

NUMERICAL METHODS FOR SINGULARLY PERTURBED BOUNDARY VALUE PROBLEMS MODELING DIFFUSION PROCESSES

V. L. KOLMOGOROV and G. I. SHISHKIN

Institute of Engineering Science (V.L.K.) and Institute of Mathematics and Mechanics (G.I.S.), Ural Branch of the Russian Academy of Sciences, 620219 Ekaterinburg, Russia

CONTENTS

Singular Perturbation Problems in Chemical Physics: Analytic and Computational Methods, Edited by John J. H. Miller, Advances in Chemical Physics Series, Vol. XCVII.
ISBN 0-471-11531-2

I. INTRODUCTION

A. Model Problems: Object of the Research

Problems of heat conductivity (diffusion and others), in which the distribution of the temperature (or the concentration of a substance) has a large gradient in thin layers, often appear in practice. Outside of these thin layers the temperature changes only slightly.

For example, at the moment when a car brakes suddenly or an aircraft touches down on a landing strip, the temperature on the wheel surfaces increases sharply because of the friction of the wheels sliding over the supporting surface. It seems that the thickness of the layer, in which the temperature rises, is small. Here the principal question is, What is the temperature of the rubber and what is the temperature distribution as a function of the depth? Also, does the rubber burn away and the aircraft with it? Similar effects appear for all machines and mechanisms where a rapid sliding of their parts takes place.

Many technological processes for the production of inflammable materials involve mixing, which is accompanied by intensive plastic deformation. For some materials local flows may occur under suitable conditions, namely, the relative slipping of parts of the material or cutting. In this case, heat is generated on the shear surface. In thin layers, close to the surface, the temperature rises. Here the main question is the determination of the temperature in the layer, and whether the temperature reaches the point of inflammability.

Similar effects, such as the appearance of cutting and slipping surfaces, are observed, for example, when working metal materials. In this case, the deformation heat can be produced in extremely thin layers. This results in changes in the structural material. For example, in these layers hardening of the metal may occur, which can reduce the quality of the product (see Section V).

It should be emphasized that the main variation of the temperature is observed in rather narrow subregions of the material, or layers. These are boundary layers if the main variations are realized near the material surface, or interior layers if the main variations are inside the material. In most of the material, the temperature varies smoothly, possibly just slightly.

Structural and chemical processes often proceed more intensively at high temperatures and/or high concentrations. Therefore, the temperature and/or concentration distributions in a sample determine the character and completeness of the transformations.

In this chapter, we will consider only such processes that can be accompanied by the appearance of arbitrarily narrow regions having finite variations of the process parameters, such as the temperature and concentration of the material. The temperature distribution u in a material is described by the heat equation expressing heat energy conservation

$$c\rho \frac{du}{dt} = \lambda \nabla^2 u + q(u) \tag{1.1}$$

Here c and λ are the known coefficients of heat capacity and heat conductivity, respectively; ρ is the mass density; q is the known density of distributed heat sources depending, possibly, on the solution;

$$\nabla^2 = \frac{\partial^2}{\partial x_1^2} + \frac{\partial^2}{\partial x_2^2} + \frac{\partial^2}{\partial x_3^2}$$

is the Laplace operator;

$$\frac{d}{dt} = \frac{\partial}{\partial t} + v_1(x,t)\frac{\partial}{\partial x_1} + v_2(x,t)\frac{\partial}{\partial x_2} + v_3(x,t)\frac{\partial}{\partial x_3}$$

$$x = (x_1, x_2, x_3)$$

is the substantive derivative operator, $V(x,t) = [v_1(x,t), v_2(x,t), v_3(x,t)]^T$ is the velocity vector for the particles of the material.

Equation (1.1), which describes the heat conductivity process, is a partial differential equation of the parabolic type. When the coefficients of this equation do not depend on time, the solution can be independent of time as well. In this case, the term with the partial derivative with respect to time is absent from the equation. Such a differential equation is an equation of the elliptic type.

Suppose that the material is fixed and undeformable, and that its thickness is much less than its other linear dimensions. Assume that the temperature varies along the thickness, and that there are no interior sources. Then Eq. (1.1) can be simplified. The temperature distribution is described by the one-dimensional heat equation

$$a \frac{\partial^2}{\partial x^2} u(x,t) - p \frac{\partial}{\partial t} u(x,t) = 0 \qquad 0 < x < d \qquad 0 < t \leq T \tag{1.2}$$

Here a and p are coefficients which, for simplicity, are assumed to be constant. The value d is the material thickness, T is the duration of the process. If the temperature distribution inside the material at the initial instant, as well as on the sample surface at time T, is known

$$\begin{gathered} u(x,0) = \varphi^0(x) \qquad 0 \leq x \leq d \qquad u(0,t) = \varphi_0(t) \\ u(d,t) = \varphi_d(t) \qquad 0 < t \leq T \end{gathered} \tag{1.3}$$

one can find the temperature distribution in the whole sample for the whole time interval by integrating Eq. (1.1) with the boundary conditions (1.3).

The diffusion of a substance in a material can be described by equations similar to Eqs. (1.1)–(1.3). Thus, under the above-mentioned simplified conditions, the distribution of the substance concentrations in the material satisfies the equation

$$D \frac{\partial^2}{\partial x^2} u(x,t) - p \frac{\partial}{\partial t} u(x,t) = 0 \qquad 0 < x < d \qquad 0 < t \leq T \tag{1.4}$$

Here D and p are known constants; D is the diffusion coefficient for the substance; the coefficient p characterizes the mass capacity of the material, that is, the ability of the material to change the concentration of the diffusive substance by varying the quantity of the substance in this material. Note that the quantity

$$P^D(x,t) \equiv D \frac{\partial}{\partial x} u(x,t)$$

is the diffusion flux of the substance, that is, the substance flux caused by the concentration gradient in the material. When the diffusive substance is dissociated over time (in this case we have $q(u) = -Cu$ for the sources),

Eq. (1.4) takes the form

$$D \frac{\partial^2}{\partial x^2} u(x, t) - Cu(x, t) - p \frac{\partial}{\partial t} u(x, t) = 0$$
$$0 < x < d \qquad 0 < t \leq T \tag{1.5}$$

Here the coefficient $C > 0$ characterizes the intensity of substance dissociation (disintegration). It is possible that the sources not only depend on the solution, but also include a part that is a function of the independent variables x, t. For example, the substance under consideration (with its concentration u) can be formed as a by-product during the accompanying reaction. In this case, Eq. (1.5) becomes

$$D \frac{\partial^2}{\partial x^2} u(x, t) - Cu(x, t) - p \frac{\partial}{\partial t} u(x, t) = F(x, t)$$
$$0 < x < d \qquad 0 < t \leq T \tag{1.6}$$

Here $F(x, t)$ is the known intensity of the source of incoming substance. Note that the term $Cu(x, t)$ in Eqs. (1.5) and (1.6) corresponds to a decrease of the substance (for $C > 0$), for example, in the reaction process when the reaction rate satisfies a linear law.

To find the substance distribution in the material at the current time by integrating (1.5), (1.6), it is necessary to know the distribution of the substance at the initial instant and also the substance concentration on the material surface at all times preceding T. In other words, conditions analogous to those in (1.3) must be fulfilled:

$$u(0, t) = \varphi_0(t) \qquad u(d, t) = \varphi_d(t) \qquad 0 < t \leq T$$
$$u(x, 0) = \varphi^0(x) \qquad 0 \leq x \leq d \tag{1.7}$$

The distribution of the substance in the material can be stationary. Then, due to the lack of sources, the process is described by the ordinary differential equation

$$D \frac{d^2}{dx^2} u(x) - Cu(x) = 0 \qquad 0 < x < d \tag{1.8}$$

with the boundary conditions

$$u(0) = \varphi_0 \qquad u(d) = \varphi_d \tag{1.9}$$

We now analyze the influence of the coefficients D and C on the solution of the stationary problem (1.8), (1.9). Let us introduce the parameter ε

$$\varepsilon^2 = DC^{-1}$$

For $d = 1$ and $\varphi_0 = 1$, $\varphi_d = 0$ we arrive at the problem

$$\varepsilon^2 \frac{d^2}{dx^2} u(x) - u(x) = 0 \qquad 0 < x < 1 \tag{1.10}$$

$$u(0) = 1 \qquad u(1) = 0 \tag{1.11}$$

the solution of which is

$$u(x) = [1 - \exp(-2\varepsilon^{-1})]^{-1}\{\exp(-\varepsilon^{-1}x) - \exp[-\varepsilon^{-1}(2 - x)]\} \tag{1.12}$$

Its graph is given in Fig. 1 for various values of ε.

We observe that as the parameter value decreases the solution of problem (1.10), (1.11) tends to zero in the whole interval [0, 1] except for some sufficiently small neighborhood of the point $x = 0$. The size of this neighborhood tends to zero as the parameter tends to zero. In this case, the solution of the problem is said to have boundary layer. The region of sharp variation of the solution (temperature or concentration) in the

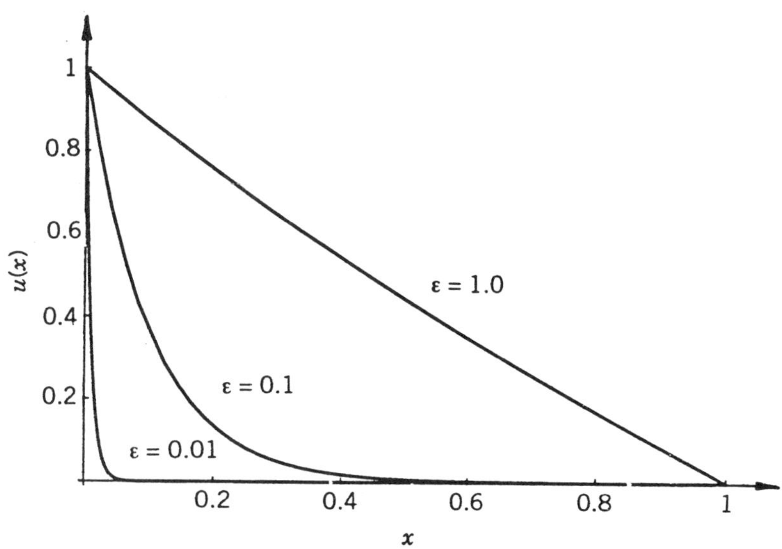

Figure 1. Solution of problem (1.10), (1.11) for $\varepsilon = 1.0$, 0.1, and 0.01.

neighborhood of the boundary is called the boundary layer. Note that the thickness of the boundary layer is equal to ε up to a constant multiplier.

Returning to problem (1.8), (1.9) for the stationary diffusion of the dissociating substance, we arrive at the following conclusion. The concentration distribution of the substance depends essentially on the value of the parameter ε: $u(x) \equiv u(x; \varepsilon)$, where $\varepsilon = \varepsilon(D, C)$. For small values of the parameter ε (i.e., when the diffusion coefficient is much less than the intensity of dissociation, or decay, of the substance) there are small regions in the neighborhood of the material surface where the concentration is finite, and outside of these regions the concentration is practically equal to zero. In this case, it is said that the concentration has a boundary layer.

Let us give some definitions. The parameter ε in Eq. (1.10), taking values close to zero, can be considered as a perturbation parameter, and Eq. (1.10) can be viewed as a perturbed differential equation. With the parameter tending to zero, there appear narrow regions in the solution of the boundary value problem where the solution varies by a finite value. Moreover, the solution derivatives increase without bound as the parameter tends to zero, that is, the solution of the problem has singularities. Taking into account this dependence of the solution on the parameter ε, Eq. (1.10), that is an equation with a small parameter multiplying the highest derivatives, is called a singularly perturbed differential equation. Thus, the boundary value problem (1.10), (1.11) involves a singularly perturbed ordinary differential equation.

In the case when the parameter λ, multiplying the highest derivative in Eq. (1.1), can take arbitrarily small values, Eq. (1.1) is said to be a singularly perturbed equation of the parabolic type (or a singularly perturbed parabolic equation). Singularly perturbed elliptic equations are obtained in a similar way.

In the case when the substance concentration at both ends of the segment [0, 1] differs from zero, boundary layers appear in the neighborhood of both ends of the interval. The qualitative behavior of the solution for the stationary boundary value problem with distributed sources

$$L_{(1.13)}u(x) \equiv \varepsilon^2 \frac{d^2}{dx^2} u(x) - c(x)u(x) = f(x) \qquad 0 < x < 1 \tag{1.13a}$$

$$u(0) = \varphi_0 \qquad u(1) = \varphi_1 \tag{1.13b}$$

where $c(x) > 0$ for $x \in [0, 1]$, $\varepsilon \in (0, 1]$, is depicted in Fig. 2. Here and below the notations $L_{(j.k)}$, $L^i_{(j.k)}$ [or $f_{(j.k)}(x)$, $f^i_{(j.k)}(x)$] indicate that these

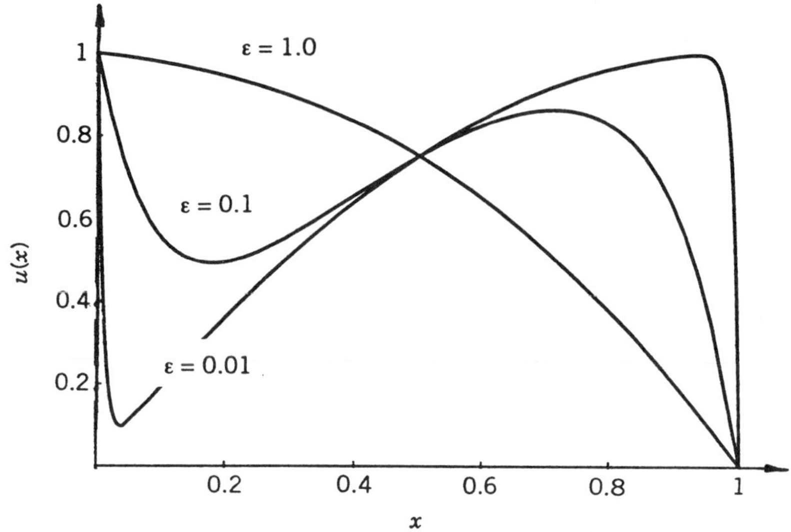

Figure 2. Solution of problem (1.13) for $\varepsilon = 1.0$, 0.1, and 0.01.

operators (or functions) were introduced first in the formula with the number $(j.k)$.

The boundary value problem (1.13) describes the stationary distribution of the concentration of the diffusive substance subjected to decay, in the presence of substance sources in the interior of the material. The parameter ε characterizes the ratio of the diffusion coefficient D of the substance and some constant C representing the intensity of the substance decay, $\varepsilon = \varepsilon(D, C)$, $\varepsilon^2 = DC^{-1}$. Here $c(x) = C(x)C^{-1}$, $f(x) = F(x)C^{-1}$; $F(x)$ is the strength of the interior sources in the material.

We are interested in boundary value problems whose solutions have singularities of the boundary layer type. In particular, for problem (1.13) we are interested in solutions for arbitrary values of the parameter ε in the interval $(0, 1]$ and especially in solutions for small values of ε, i.e. for $\varepsilon \ll 1$.

Solutions of heat and diffusion nonstationary problems can be written in an analytical form, convenient for computations, only in very infrequent cases. Usually in practice the solutions of these problems are found by means of numerical computations. For this purpose, finite difference schemes are often used.

Let us write a finite difference scheme for the boundary value problem (1.13). The function $u(x)$, which is the solution of the problem on the segment $\bar{D} = [0, 1]$, satisfies Eq. (1.13a) in the interval D [where $D =$

(0, 1)] and boundary conditions (1.13b) on the boundary. The solution of the boundary value problem is supposed to be sufficiently smooth (i.e., the solution has the necessary number of derivatives required in the research).

On the segment D we introduce the grid

$$\bar{D}_h = \bar{\omega}_1 \equiv \{x^i : x^0 = 0, \ldots, x^{i-1}, x^i, x^{i+1}, \ldots, x^N = 1\}$$

which is a set of nodes x^i, $i = 0, 1, \ldots, N$ on $\bar{D}$, where $N+1$ is the number of nodes. The nodes x^0 and x^N are called the boundary ones, and other nodes are called the interior ones. When the distance between the nodes (or the step size) is constant, the grid is said to be uniform.

The continuous function $u(x)$, $x \in \bar{D}$ is replaced by the grid function $z(x)$ defined on the grid $\bar{D}_h$.

The second-order derivative in Eq. (1.13a) is approximated by the second-order difference derivative on the set $D_h = D \cap \bar{D}_h$, which is a set of interior nodes in the grid $\bar{D}_h$, in the following way. On the grid $\bar{D}_h$ we put the first-order derivative $(d/dx)u(x)$ into correspondence with the first-order difference derivatives $\delta_x z(x)$ and $\delta_{\bar{x}} z(x)$. Here $\delta_x z(x)$ and $\delta_{\bar{x}} z(x)$ are, respectively, right (or forward) and left (or backward) difference derivatives determined by the relations

$$\delta_x z(x) \equiv z_x(x) = (x^{i+1} - x^i)^{-1}[z(x^{i+1}) - z(x^i)]$$

$$\delta_{\bar{x}} z(x) \equiv z_{\bar{x}}(x) = (x^i - x^{i-1})^{-1}[z(x^i) - z(x^{i-1})]$$

$$x = x^i \qquad x^{i-1}, x^i, x^{i+1} \in \bar{D}_h$$

We put the second-order difference derivative $\delta_{\bar{x}\hat{x}} z(x)$, determined by the relation

$$\delta_{\bar{x}\hat{x}} z(x) \equiv z_{\bar{x}\hat{x}}(x) = 2(x^{i+1} - x^{i-1})^{-1}[\delta_x z(x) - \delta_{\bar{x}} z(x)]$$

$$x = x^i \qquad x^i \in D_h$$

into correspondence with the second-order derivative $(d^2/dx^2)u(x)$. On the uniform grid $\bar{D}_h$ we have

$$\delta_{\bar{x}\hat{x}} z(x) = \delta_{x\bar{x}} z(x) \equiv z_{x\bar{x}}(x) = h^{-2}[z(x-h) - 2z(x) + z(x+h)] \qquad x \in D_h$$

where h is the step size of the uniform grid $\bar{D}_h$, $h = N^{-1}$.

For the sake of simplicity, in this section we will consider uniform grids

$$\bar{D}_h = \bar{D}_{h(1.14)} \tag{1.14}$$

For the boundary value problem (1.13) we obtain a difference scheme (i.e., the combination of the grid sets $\bar{D}_h$, depending on the value of N, and the system of difference equations defined on these sets). The system of finite difference equations is as follows:

$$\Lambda_{(1.15)} z(x) \equiv \varepsilon^2 \delta_{x\bar{x}} z(x) - c(x)z(x) = f(x) \qquad x \in D_h \tag{1.15a}$$

$$z(0) = \varphi_0 \qquad z(1) = \varphi_1 \tag{1.15b}$$

Thus, the boundary value problem (1.13) is replaced by a system of algebraic equations. The grid function $z(x)$, $x \in \bar{D}_h$ is the solution of this system.

The closeness of the grid function $z(x)$, $x \in \bar{D}_h$ to the continuous function $u(x)$, $x \in \bar{D}$ is measured by the quantity

$$\max_{\bar{D}_h} |u(x) - z(x)|$$

that is, by the maximum of the absolute values of the difference of the functions $u(x)$ and $z(x)$ at the nodes of the grid $\bar{D}_h$. This quantity is called the distance between the functions $u(x)$, $x \in \bar{D}$ and $z(x)$, $x \in \bar{D}_h$ in the uniform grid norm.

We emphasize that $z(x)$, which is the solution of the difference problem on uniform grids, depends essentially on the two parameters ε and N, where the parameter ε characterizes the physical essence of the boundary value problem, while N (or h) characterizes the applied numerical method (or the computational resources of our computer).

Let us discuss the behavior of the grid solution for various values of the parameter ε and the number of grid nodes N.

Note that the boundary layer type solution of the boundary value problem varies by a finite value in the boundary layer, that is, in an interval with width of order ε. For fixed values of the parameter ε and as $N \to \infty$ (or as the step size h tends to zero) the step size of the grid becomes much less than ε. In this case, many nodes of the grid are situated in the boundary layer region. Moreover, the grid norm $|u(x) - z(x)|$, $x \in \bar{D}_h$ decreases as N increases, that is, the solution of the grid problem [or difference scheme (1.15), (1.14)] converges to the solution of the boundary value problem. We wonder what will happen if the number of the grid nodes N is large while the parameter ε takes sufficiently small values? It can happen that, for a chosen large number N (or small h), the value of ε becomes commensurable with the value of the grid step. In this case, the step size is commensurable with the width of the boundary layer. So, it is intuitively clear that the grid solution will poorly approximate the solution of the boundary value problem in the vicinity of the boundary layer, and, hence, in the grid norm. This situation appears when one numerically solves the problem on a computer. The power of a given computer defines the maximum possible value of N and, consequently, the smallest possible grid step size. However, in the process under study the width of the boundary layer can become commensurable with (or smaller than) the grid step size.

B. Posing a Computational Problem with Simple Examples

Now we analyze the quality of the approximation of the boundary value problem solution by the grid solution. By using a simple example of an ordinary differential equation, we demonstrate some problems arising from its numerical solution.

Suppose that it is required to find the solution of the following boundary value problem

$$L_{(1.16)}u(x) \equiv \varepsilon^2 \frac{d^2}{dx^2} u(x) - u(x) = -1 \qquad x \in D \tag{1.16a}$$

$$u(0) = u(1) = 0 \tag{1.16b}$$

where $\bar{D} = [0, 1]$ and the parameter ε takes arbitrary values from the interval $(0, 1]$. To solve this problem we apply classical numerical methods, for example, the finite difference method.

Problem (1.16) is approximated by the difference scheme

$$\begin{aligned} \Lambda_{(1.17)}z(x) \equiv \{e^2\delta_{x\bar{x}} - 1\}z(x) = -1 \qquad x \in D_h \\ z(0) = z(1) = 0 \end{aligned} \tag{1.17}$$

where

$$\bar{D}_h = \bar{D}_{h(1.14)} \tag{1.18}$$

As is known (see, e.g., [1, 2]), the approximate solution error (in the uniform grid norm) depends on the values of the parameter ε and the grid step size h in the following way:

$$|u(x) - z(x)| \leq Q(\varepsilon)h^2 \qquad x \in \bar{D}_h \tag{1.19}$$

That is, the maximum nodal (in the grid $\bar{D}_h$ nodes) deflection of the approximate solution from the exact one, in the uniform grid norm, is h^2 within the factor $Q(\varepsilon)$. Thus, the solution of the difference scheme (1.17), (1.18) at $N \to \infty$ (or $h \to 0$) converges to the solution of the boundary value problem (1.16) for each fixed value of the parameter ε, and it converges with second-order accuracy in h. However, the factor $Q(\varepsilon)$ depends essentially on the parameter value.

In fact, for sufficiently small values of the parameter, namely, for $\varepsilon \equiv \varepsilon(h) = h$, this error exceeds some positive constant [3, 4]

$$\max_{\bar{D}_h} |u(x) - z(x)| \geq m_{(1.20)} > 0 \qquad \text{for} \qquad h \to 0 \qquad (1.20)$$

where $u(x) = u(x; \varepsilon)$, $z(x) = z(x; \varepsilon, h)$, $\varepsilon = h$. That is, for any sufficiently small step size of the grid and an arbitrary value of the parameter ε, $\varepsilon \in (0, 1]$, a value of ε can be found (or can occur in practice) such that the error in the approximate solution exceeds or equals a positive constant. In this case, the solution of the difference scheme is said not to converge uniformly with respect to the parameter ε (or, in short, it does not converge ε-uniformly).

It would be desirable to have numerical methods for which the error in the approximate solution tends to zero independently of the parameter ε as the number of grid nodes increases (i.e., at $N \to \infty$). For $N \to \infty$ this approximate solution is said to converge uniformly with respect to the parameter ε (or, in short, it converges ε uniformly).

It follows from estimate (1.20) that the solution of the classical difference scheme (1.17), (1.18) (the scheme constructed with the approximation of the differential derivatives by the difference ones) does not converge ε-uniformly to the solution of problem (1.16). In this connection, the following theoretical problem appears: To construct special numerical methods giving approximate solutions that converge ε uniformly. The present chapter is devoted to this problem.

Let us show by a numerical experiment, the severity of this problem, in particular, in the case of the boundary value problem (1.16). The solution of problem (1.16) is as follows:

$$u(x) = u(x; \varepsilon)$$
$$= 1 - [1 + \exp(-\varepsilon^{-1})]^{-1}\{\exp(-\varepsilon^{-1}x) + \exp[-\varepsilon^{-1}(1-x)]\} \qquad x \in \bar{D}$$

The function $u(x)$ satisfies the following relations:

$$0 \leq u(x) < 1 \qquad x \in \bar{D}$$
$$\max_{\bar{D}} u(x) = u(0.5) \qquad \lim_{\varepsilon \to 0} u(0.5) = 1$$

In Fig. 3 the graphs of the function $u(x) = u(x; \varepsilon)$ are given for some

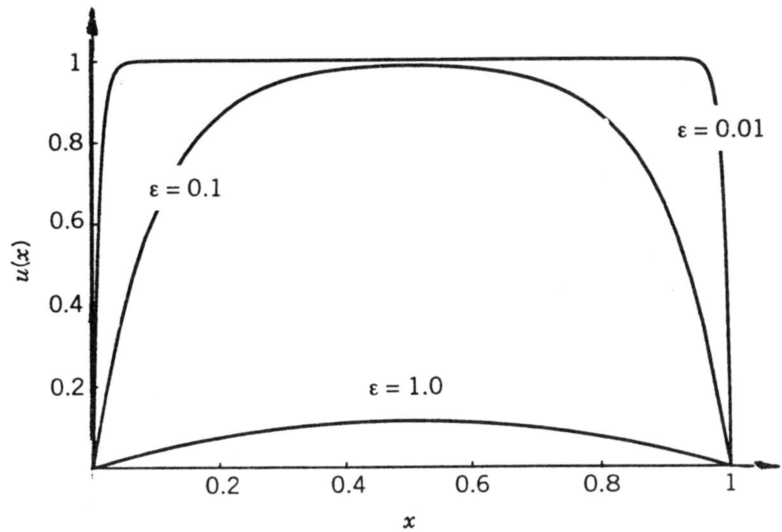

Figure 3. Solution of problem (1.16) for $\varepsilon = 1.0$, 0.1, and 0.01.

values of the parameter ε. In Fig. 4 the graphs of the function $z(x; \varepsilon, N)$ are given for some values of the parameter ε and for the value $N = 16$.

After visual comparison of the graphs in Figs. 3 and 4 one can see that the difference solution approximates the solution of the boundary value problem fairly well.

In Table I we give the results of computing the quantity $E(\varepsilon, N)$

$$E(\varepsilon, N) = \max_{\bar{D}_h} |u(x) - z(x)| = \max_{\bar{D}_h} |u(x; \varepsilon) - z(x; \varepsilon, N)| \qquad (1.21)$$

for various values of ε and N, where $u(x) = u(x; \varepsilon)$ is the solution of problem (1.16) and $z(x) = z(x; \varepsilon, N)$ is the solution of problem (1.17), (1.18), $\varepsilon = 2^{-12} \div 1.0$, $N = 4 \div 1024$. Here the values of $\bar{E}(N)$ are also given. With respect to ε (for $\varepsilon = 2^{-12} \div 1.0$), the value of

$$\bar{E}(N) = \max_{\varepsilon} E(\varepsilon, N) \qquad \varepsilon = 2^{-12} \div 1.0$$

is the largest error in the approximate solution for a fixed value of N, $N = 4 \div 1024$. The quantity $\bar{E}(N)$ defines the best guaranteed accuracy that one can obtain using scheme (1.17), (1.18) to solve problem (1.16) for a given fixed N and various values of the parameter ε. On the other hand, the quantity $\bar{E}(N)$ is also the worst realizable (for some value of ε, $\varepsilon = 2^{-12} \div 1.0$) error of the approximate solution.

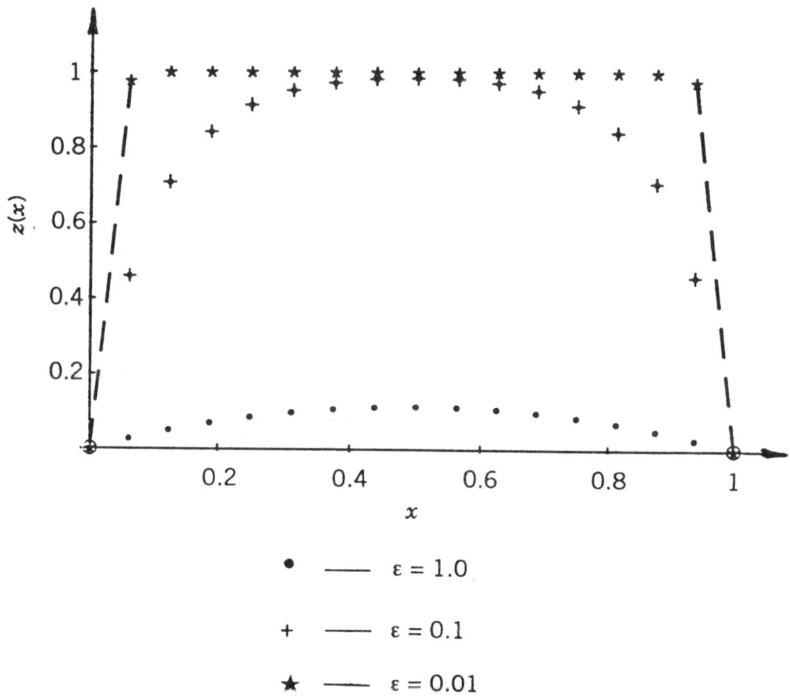

Figure 4. Solution of the classical scheme (1.17), (1.18) for problem (1.16) for $\varepsilon = 1.0$, 0.1, 0.01, and $N = 16$.

Naturally, it would be expected that the error $E(\varepsilon, N)$ behaves correctly (in other words, its behavior is regular). This means that, as the number of nodes N increases, this error should decrease for a fixed value of the parameter ε. But it does not.

One can see from Table I that the errors $E(\varepsilon, N)$ in the table are divided into two parts by the diagonal $N = 4^{-1}\varepsilon^{-1}$. In one part of the table (for $N > 4^{-1}\varepsilon^{-1}$) the error decreases when N increases, while in the other part (for $N < 4^{-1}\varepsilon^{-1}$) it increases with increasing N. We see that the maximum error $E(\varepsilon, N)$ is achieved for $\varepsilon = 2^{-4}$, $N = 4$, and it is equal to 3.76×10^{-2}. The error becomes stable along the diagonals as ε decreases: it takes constant (the same) values for $\varepsilon N = \text{const}$ (i.e., on the diagonals). We see that the errors $E(\varepsilon, N)$ illustrate the theoretical estimate (1.20). For a fixed value of the parameter ε, the solution of the difference scheme (1.17), (1.18) converges with the growth of N (for $N > 4^{-1}\varepsilon^{-1}$). However, the approximate solution does not converge ε-uniformly. Indeed, for a fixed, arbitrarily large value of N one can find

TABLE I
Table of Errors $E(\varepsilon, N)$ for the Classical Scheme (1.17), (1.18) in the Case of Problem (1.16)

$\varepsilon \backslash N$	4	16	64	256	1024
1	5.29e − 4	3.33e − 5	2.08e − 6	1.30e − 7	8.14e − 9
2^{-2}	1.99e − 2	1.33e − 3	8.34e − 5	5.21e − 6	3.26e − 7
2^{-4}	3.76e − 2	1.41e − 2	9.53e − 4	5.99e − 5	3.74e − 6
2^{-6}	3.88e − 3	3.74e − 2	1.41e − 2	9.53e − 4	5.99e − 5
2^{-8}	2.44e − 4	3.88e − 3	3.74e − 2	1.41e − 2	9.53e − 4
2^{-10}	1.53e − 5	2.44e − 4	3.88e − 3	3.74e − 2	1.41e − 2
2^{-12}	9.54e − 7	1.53e − 5	2.44e − 4	3.88e − 3	3.74e − 2
$\bar{E}(N)$	3.76e − 2	3.74e − 2	3.74e − 2	3.74e − 2	3.74e − 2

a value of the parameter ε, for example, $\varepsilon = \varepsilon(N) = 4^{-1}N^{-1}$, such that the error of the approximate solution is no less than a positive constant, namely, this error is equal to 3.74×10^{-2}. For any large N, we cannot guarantee an accuracy better than 3.74×10^{-2} (for $\varepsilon = 2^{-12} \div 1.0$). Thus, for the worst realizable error $\bar{E}(N)$ the lower bound

$$\bar{E}(N) \geq 3.74 \times 10^{-2}$$

is valid. From here and from estimate (1.20) we see that the constant $m_{(1.20)}$ is no larger than 3.74×10^{-2}. We can compute the relative worst realizable error for a fixed N:

$$\delta(N) \equiv \bar{E}(N)[\max_{\bar{D}} |u(x)|]^{-1} \tag{1.22}$$

Since $\max_{\bar{D}}|u(x)| < 1$, the quantity $\delta(N)$ is estimated from below by

$$\delta(N) \geq \bar{E}(N)$$

which is independent of N and no less than 3.74%. On the other hand, for $\delta(N)$ the following upper bound holds:

$$\delta(N) \leq 3.76\%$$

when $N = 4^n$, $\varepsilon = 2^{-2k}$, $n = 1, 2, \ldots, 5$, $k = 0, 1, 2, \ldots, 6$, that is, $\delta(N)$ is independent of N and is no more than 3.76%.

Consequently, for the relative worst realizable error in the case of the difference scheme (1.17), (1.18), the estimate

$$3.74\% \leq \delta(N) \leq 3.76\% \tag{1.23}$$

is valid at $N = 4^n$, $n = 1, 2, \ldots, 5$, for $\varepsilon = 2^{-2k}$, $k = 0, 1, 2, \ldots, 6$. Taking into account the stabilized behavior of the errors $E(\varepsilon, N)$ for small ε and large N, we see that the estimate (1.23) is fulfilled for all $N = 4^n$, $n = 1, 2, \ldots$ and $\varepsilon = 2^{-2k}$, $k = 0, 1, 2, \ldots$.

Thus, although the approximate solution does not converge ε uniformly, the accuracy offered by relation (1.23) can be acceptable in some cases. The computed solution qualitatively represents the behavior of the exact solution for all the values of the parameter ε. Dissatisfaction can arise only from the fact that the increase in the number of grid nodes does not guarantee a better result.

The accuracy problem is more serious in those boundary value problems where, besides the solution (the values of the unknown function), it is also required to find the solution gradient. For example, suppose that in the case of problem (1.16) we want to find the gradient of the function $u(x)$ on the boundary of the segment $\bar{D}$. A problem like this can arise if we are interested in the diffusion or heat flux on the material surface.

The derivative $(d/dx)u(x)$ of the solution for the boundary value problem (1.16) is determined by the formula

$$\frac{d}{dx}u(x) = \frac{d}{dx}u(x;\varepsilon)$$
$$= \varepsilon^{-1}[1 + \exp(-\varepsilon^{-1})]^{-1}\{\exp(-\varepsilon^{-1}x) - \exp[-\varepsilon^{-1}(1-x)]\} \quad x \in \bar{D}$$

The graph of the function $(d/dx)u(x)$ for some values of the parameter is given in Fig. 5.

The modulus of the derivative $(d/dx)u(x)$ attains its largest values on the boundary of D. In particular, at the ends of the interval $\bar{D}$ we have

$$\frac{d}{dx}u(0) = -\frac{d}{dx}u(1) = \varepsilon^{-1}[1 + \exp(-\varepsilon^{-1})]^{-1}[1 - \exp(-\varepsilon^{-1})]$$

Note that the derivative increases unboundedly on the boundary as the parameter tends to zero. However, the function

$$P(x) \equiv \varepsilon \frac{d}{dx}u(x)$$

remains bounded for all the values of the parameter ε (i.e., it remains bounded ε-uniformly). In contrast to the quantity $P^D(x) = D(d/dx)u(x)$, which is the diffusion flux of the substance, the function $P(x)$ is called the normalized diffusion flux or, in short, the normalized flux. For the

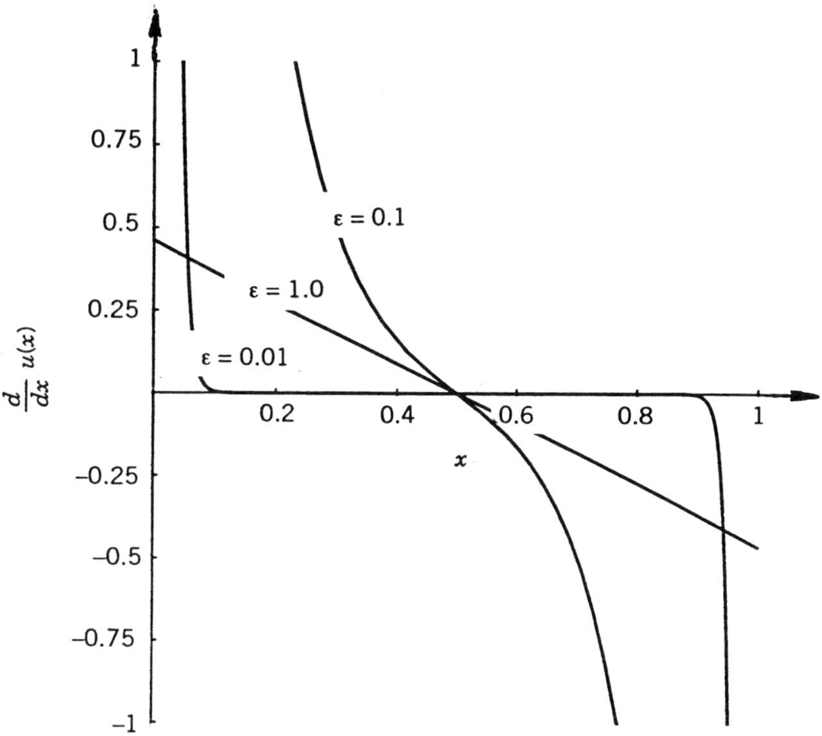

Figure 5. Derivative of the solution of problem (1.16) for $\varepsilon = 1.0$, 0.1, and 0.01.

boundary value problem (1.16) we obtain

$$P(x) = [1 + \exp(-\varepsilon^{-1})]^{-1}\{\exp(-\varepsilon^{-1}x) - \exp[-\varepsilon^{-1}(1-x)]\} \qquad x \in \bar{D}$$

The graph of the function $P(x)$ is given in Fig. 6.

Therefore it is natural to consider the following problem:

Find the solution $u(x)$ of the boundary value problem and the normalized diffusion flux $P(x)$ for $x \in \bar{D}$

or the simpler, particular problem:

Find the solution $u(x)$ of the boundary value problem for $x \in \bar{D}$ and the normalized diffusion flux $P(x)$ on the boundary Γ (1.24)

where $u(x)$ is the solution of problem (1.16), $\Gamma = \bar{D} \backslash D$.

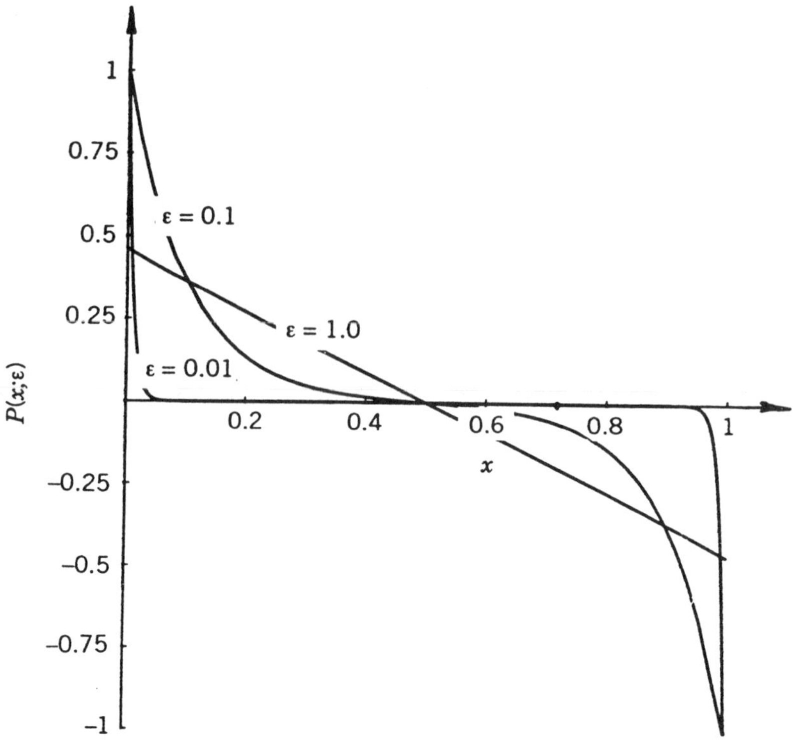

Figure 6. Normalized flux $P(x;\varepsilon)$ for problem (1.16) for $\varepsilon = 1.0$, 0.1, and 0.01.

Note that the function $P(x) = P(x;\varepsilon)$ satisfies the relations

$$\max_{\bar{D}} |P(x)| < 1 \qquad P(x=0;\varepsilon) = -P(x=1;\varepsilon) > 0$$

$$\lim_{\varepsilon \to 0} P(x=0;\varepsilon) = 1$$

To solve problem (1.16), (1.24), we apply the difference scheme (1.17), (1.18).

The quantity $\delta_{(1.22)}(N)$ satisfies the estimate (1.23) in the case of problem (1.16), that is, the computed solution approximates the exact solution for all values of the parameter qualitatively (visually) well.

The function $P(x) = P(x;\varepsilon)$ for $x = 0$ is approximated by the function $P^{h+}(x) \equiv P^{h+}(x;\varepsilon, N)$, where

$$P^{h+}(x) \equiv \varepsilon \delta_x z(x) \qquad x = 0 \tag{1.25}$$

$P^{h+}(x)$, $x = 0$ is the approximation of the normalized diffusion flux at the point $x = 0$. The quantity $P^{h+}(0)$ will be called the computed normalized diffusion flux (or, in short, the computed normalized flux) at the point $x = 0$.

The graphs of the functions $P(0; \varepsilon)$ and $P^{h+}(0; \varepsilon, N)$ are given in Figs. 7 and 8 respectively. From Figs. 7 and 8 it is seen that the approximate solution approaches the exact solution only for those values of the parameter which are considerably larger than the step size of the applied grid.

In Table II we give the results of the computation of the quantity

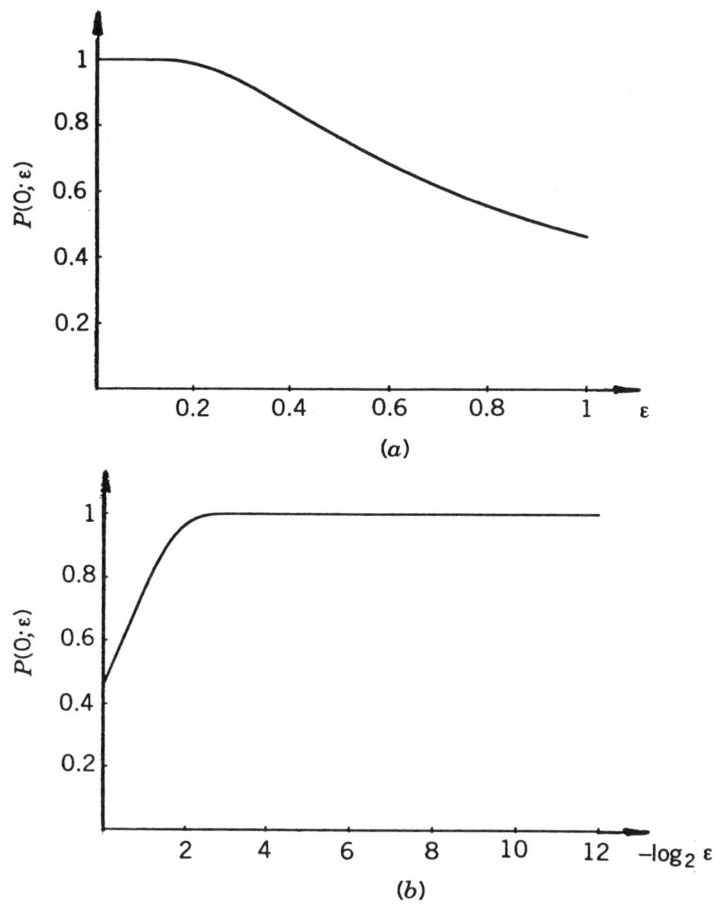

Figure 7. Normalized flux $P(x; \varepsilon)$ for problem (1.16) at $x = 0$ with (a) natural (linear) scale and (b) log_2-scale.

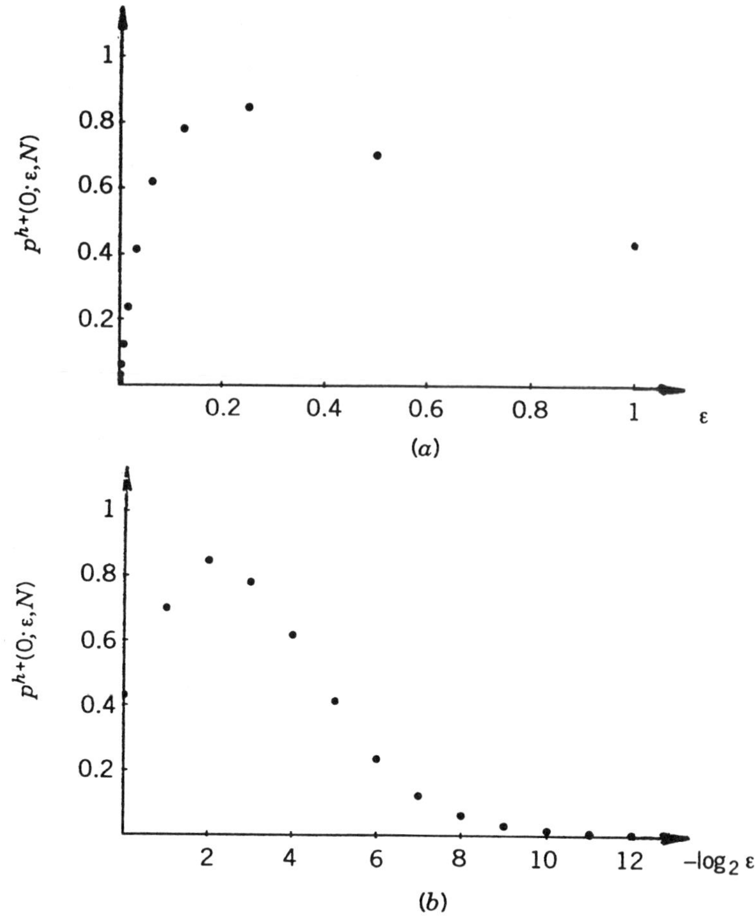

Figure 8. Computed normalized flux $P^{h+}(x;\varepsilon, N)$ for scheme (1.17), (1.18) at $x = 0$ and $N = 16$ with (a) natural (linear) scale and (b) log_2-scale.

$Q(\varepsilon, N)$:

$$Q(\varepsilon, N) = |P(x = 0; \varepsilon) - P^{h+}(x = 0; \varepsilon, N)|$$

which is the error in the normalized flux on the boundary $x = 0$ for various values of ε and N, $\varepsilon = 2^{-12} \div 1.0$, $N = 4 \div 1024$. Here the values of $\bar{Q}(N)$ are also given, where

$$\bar{Q}(N) = \max_{\varepsilon} Q(\varepsilon, N) \qquad \varepsilon = 2^{-12} \div 1.0$$

TABLE II
Table of Errors of the Normalized Flux $Q(\varepsilon, N)$ for the Classical Scheme (1.17), (1.18) in the Case of Problem (1.16)

$\varepsilon \backslash N$	4	16	64	256	1024
1	1.22e − 1	3.11e − 2	7.80e − 3	1.95e − 3	4.88e − 4
2^{-2}	3.93e − 1	1.18e − 1	3.08e − 2	7.78e − 3	1.95e − 3
2^{-4}	7.64e − 1	3.82e − 1	1.17e − 1	3.08e − 2	7.78e − 3
2^{-6}	9.38e − 1	7.64e − 1	3.82e − 1	1.17e − 1	3.08e − 2
2^{-8}	9.84e − 1	9.38e − 1	7.64e − 1	3.82e − 1	1.17e − 1
2^{-10}	9.96e − 1	9.84e − 1	9.38e − 1	7.64e − 1	3.82e − 1
2^{-12}	9.99e − 1	9.96e − 1	9.84e − 1	9.38e − 1	7.64e − 1
$\bar{Q}(N)$	9.99e − 1	9.96e − 1	9.84e − 1	9.38e − 1	7.64e − 1

The quantity $\bar{Q}(N)$ characterizes the largest (with respect to ε) error in the computed normalized flux for $x = 0$; this value defines the best guaranteed accuracy that one can obtain by scheme (1.17), (1.18), (1.25) to solve problem (1.16), (1.24) for a given N and various values of the parameter ε, $\varepsilon = 2^{-12} \div 1.0$.

It follows from Table II that, for a fixed value of the parameter ε, the quantity $P^{h+}(0) = P^{h+}(x = 0; \varepsilon, N)$, which is the computed normalized flux for $x = 0$, converges to the quantity $P(0; \varepsilon)$ as N increases. However, the computed flux $P^{h+}(0)$ does not converge ε-uniformly. In fact, for any given value of N one can find a value of the parameter $\varepsilon = \varepsilon(N)$ such that the error $Q(\varepsilon, N)$ is no less than some positive constant, for example, $\varepsilon(N) = \text{const}\, N^{-1}$.

From the table, one can see that the error $Q(\varepsilon, N)$ can be commensurable with the computed value of the normalized flux itself. For example, when the number of the grid nodes is $N = 1024$, the error is larger than 75% of the real value of the flux, even when the parameter ε does not exceed $2^{-12} = 1/4096$. Moreover, for any fixed N, the error $Q(\varepsilon, N)$ tends to the value P_0 as ε decreases, where

$$P_0 \equiv \lim_{\varepsilon \to 0} P(0; \varepsilon) = 1$$

is the limiting (for $\varepsilon \to 0$) normalized flux on the boundary $x = 0$. Thus, the error $Q(\varepsilon, N)$ becomes arbitrarily close to the value of the normalized flux for sufficiently small values of the parameter and for a large, but fixed, value of N.

Note that, for the quantity $\lambda(\varepsilon, N)$, where

$$\lambda(\varepsilon, N) \equiv P(0; \varepsilon)[P^{h+}(0; \varepsilon, N)]^{-1}$$

is the ratio of the exact normalized flux and the computed flux for $x = 0$, the relation

$$\lambda(\varepsilon, N) = \frac{d}{dx} u(0; \varepsilon)[\delta_x z(0; \varepsilon, N)]^{-1}$$

is valid, that is, the quantity $\lambda(\varepsilon, N)$ also characterizes the ratio of the derivative of the boundary value problem solution to the difference derivative of the grid problem solution for $x = 0$.

In Table III the values of $\lambda(\varepsilon, N)$ and $\bar{\lambda}(N)$ are given. Here

$$\bar{\lambda}(N) = \max_{\varepsilon} \lambda(\varepsilon, N) \qquad \varepsilon = 2^{-12} \div 1.0$$

The values of $\lambda(\varepsilon, N)$ tend to unity with $N \to \infty$ for a fixed value of the parameter ε. The ratio of fluxes unboundedly increases for a fixed value of N when the parameter tends to zero. This ratio sharply increases with increasing N, whereas the product εN decreases (i.e., for $\varepsilon \ll N^{-1}$). For example, the real diffusion flux exceeds the computed one by no more than 15% if $\varepsilon \leq 4^{-1}$ and $\varepsilon N \geq 4$ (i.e., for $h \leq 4^{-1}\varepsilon$, $\varepsilon \leq 4^{-1}$). However, even for the number of grid nodes $N = 1024$ and for ε not exceeding $2^{-12} = 1/4096$, the real diffusion flux is four times as small as the computed diffusion flux. The real flux will be more than 1000 times the computed flux for ε, N such that $\varepsilon N \leq 1/1024$.

Thus, in the case of the boundary value problem (1.16), (1.24), the use of the scheme (1.17), (1.18), (1.25) leads, for small values of the parameter ε, to sharp underestimation of the computed normalized flux (and also of the solution gradient) on the boundary. Therefore, even qualitatively, the normalized flux cannot be approximated by the computed flux ε-uniformly.

Table III
Table of the Ratios of the Normalized Fluxes $\lambda(\varepsilon, N)$ for the Classical Scheme (1.17), (1.18) in the Case of Problem (1.16)

$\varepsilon \backslash N$	4	16	64	256	1024
1	1.36e + 0	1.07e + 0	1.02e + 0	1.00e + 0	1.00e + 0
2^{-2}	1.69e + 0	1.14e + 0	1.03e + 0	1.01e + 0	1.00e + 0
2^{-4}	4.24e + 0	1.62e + 0	1.13e + 0	1.03e + 0	1.01e + 0
2^{-6}	1.61e + 1	4.24e + 0	1.62e + 0	1.13e + 0	1.03e + 0
2^{-8}	6.40e + 1	1.61e + 1	4.24e + 0	1.62e + 0	1.13e + 0
2^{-10}	2.56e + 2	6.40e + 1	1.61e + 1	4.24e + 0	1.62e + 0
2^{-12}	1.02e + 3	2.56e + 2	6.40e + 1	1.61e + 1	4.24e + 0
$\bar{\lambda}(N)$	1.02e + 3	2.56e + 2	6.40e + 1	1.61e + 1	4.24e + 0

It is known that, in the case of singularly perturbed elliptic equations for which (as the parameter ε equals zero) the equation does not contain any derivatives with respect to the space variable, the principal term in the singular part of the solution is described by an ordinary differential equation similar to Eq. (1.16a) (see, e.g., [3–6]). Thus, it can be expected that, when solving singularly perturbed elliptic and parabolic equations using classical difference schemes, one faces computational problems similar to the computational problems for the boundary value problem (1.16).

We pay special attention to one additional type of problem, for which the accuracy problem is very severe. Suppose that on the segment D it is required to find the solution of the differential equation

$$L_{(1.16)}u(x) = 0 \qquad x \in D \tag{1.26a}$$

The solution satisfies the boundary conditions

$$\varepsilon \frac{d}{dx} u(0) = -1 \qquad \varepsilon \frac{d}{dx} u(1) = 1 \tag{1.26b}$$

that is, the diffusion (normalized) flux is given at the ends of the segment. The solution of problem (1.26) is as follows:

$$u(x) \equiv u(x; \varepsilon) = [1 - \exp(-\varepsilon^{-1})]^{-1}\{\exp(-\varepsilon^{-1}x) + \exp[-\varepsilon^{-1}(1 - x)]\}$$
$$x \in \bar{D}$$

The graphs of the function $u(x)$ for different values of the parameter ε are shown in Fig. 9. For small values of the parameter ε a boundary layer appears in a neighborhood of each of the segment ends.

Note that the function $u(x)$ satisfies the relations

$$u(x) > 0 \qquad x \in D$$
$$\max_{\bar{D}} u(x) = u(0) = u(1) = [1 - \exp(-\varepsilon^{-1})]^{-1}[1 + \exp(-\varepsilon^{-1})] > 1$$
$$\lim_{\varepsilon \to 0} u(0; \varepsilon) = 1$$

To solve problem (1.26), we apply the classical difference scheme on uniform grids

$$\begin{aligned} &\Lambda_{(1.17)}z(x) = 0 \qquad x \in D_h \\ &\varepsilon\delta_x z(0) = -1 \qquad \varepsilon\delta_{\bar{x}} z(1) = 1 \end{aligned} \tag{1.27}$$

where $\bar{D}_h = \bar{D}_{h(1.18)}$.

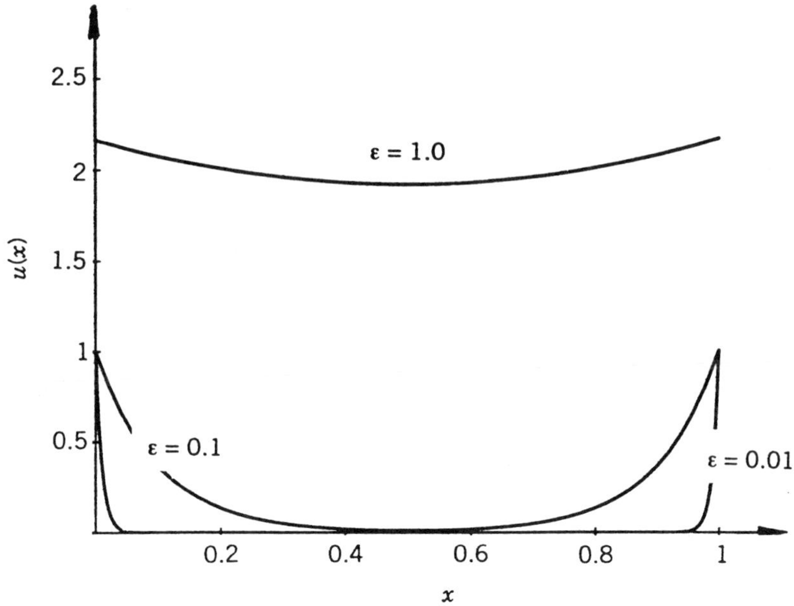

Figure 9. Solution of problem (1.26) for $\varepsilon = 1.0$, 0.1, and 0.01.

The graphs of the function $z(x)$, for different values of the parameter ε with $N = 16$, are given in Fig. 10. Comparing the graphs of the functions $u(x)$ and $z(x)$, we see that the approximate solutions unsatisfactorily approach the exact solution, and that the accuracy is much worse if the value of ε is small.

In Table IV we give the results of the computation of the errors $E_{(1.21)}(\varepsilon, N)$ for $u(x) = u_{(1.26)}(x)$ and $z(x) = z_{(1.27)}(x)$, which are the solutions of problems (1.26) and (1.27), (1.18), respectively.

From Table IV we see that, for a fixed value of the parameter ε, the error in the approximate solution decreases as N increases, that is, the solution of the grid problem converges for a fixed value of the parameter. However, the solution of the grid problem does not converge ε-uniformly. Moreover, the error in the approximate solution increases unboundedly when the product εN decreases to zero. For example, the error becomes more than three times as large as the real solution, even for $\varepsilon N \le 4^{-1}$, and it is more than 1000 times as large for $\varepsilon N \le 1/1024$.

Note that the computed normalized diffusion flux does not approximate the real normalized diffusion flux ε-uniformly.

Thus, for the boundary value problem (1.26) the use of the finite

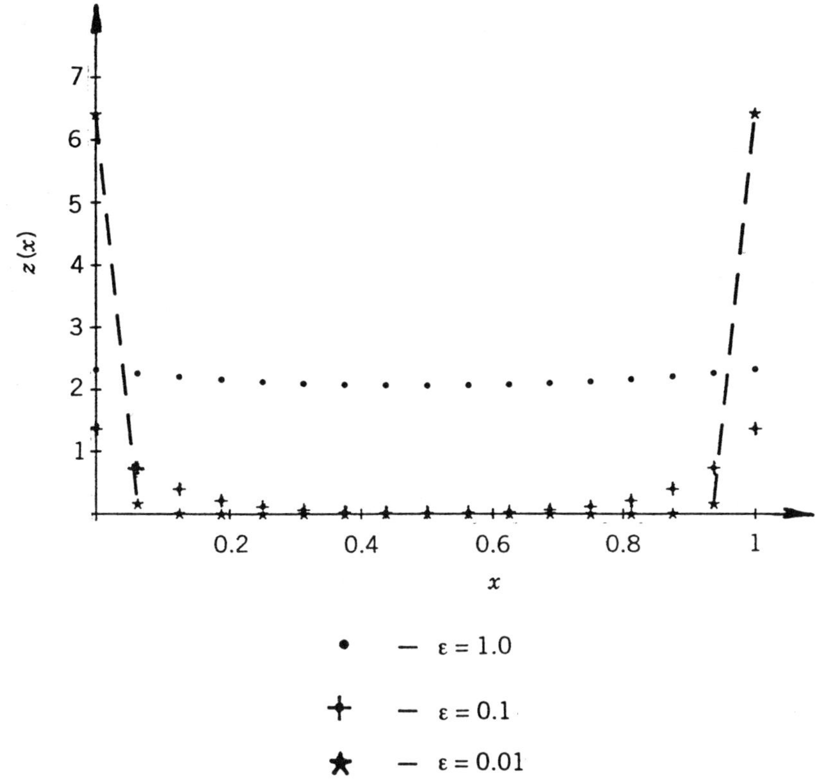

Figure 10. Solution of the classical scheme (1.27), (1.18) for problem (1.26) for $\varepsilon = 1.0$, 0.1, 0.01, and $N = 16$.

TABLE IV
Table of Errors $E(\varepsilon, N)$ for the Classical Scheme (1.27), (1.18) in the Case of Problem (1.26)

$\varepsilon \backslash N$	4	16	64	256	1024
1	7.80e − 1	1.56e − 1	3.72e − 2	9.18e − 3	2.29e − 3
2^{-2}	7.13e − 1	1.45e − 1	3.42e − 2	8.45e − 3	2.10e − 3
2^{-4}	3.24e + 0	6.18e − 1	1.33e − 1	3.17e − 2	7.84e − 3
2^{-6}	1.51e + 1	3.24e + 0	6.18e − 1	1.33e − 1	3.17e − 2
2^{-8}	6.30e + 1	1.51e + 1	3.24e + 0	6.18e − 1	1.33e − 1
2^{-10}	2.55e + 2	6.30e + 1	1.51e + 1	3.24e + 0	6.18e − 1
2^{-12}	1.02e + 3	2.55e + 2	6.30e + 1	1.51e + 1	3.24e + 0
$\bar{E}(N)$	1.02e + 3	2.55e + 2	6.30e + 1	1.51e + 1	3.24e + 0

difference scheme (1.27), (1.18) leads (for small values of the parameter ε) to sharp overestimation of the computed solution. The error in the approximate solution exceeds the required exact solution considerably. The approximate solution does not approach the exact solution ε-uniformly, even in qualitative terms.

Similar computational problems arise when one solves singularly perturbed elliptic and parabolic equations when the flux is given on the domain boundary.

The above examples illustrate the fact that, in the case of singularly perturbed elliptic and parabolic equations, the use of classical finite difference schemes does not enable us to find the approximate solutions and the normalized diffusion fluxes with ε-uniform accuracy. To find approximate solutions and normalized fluxes that converge ε-uniformly, it is necessary to develop special numerical methods, in particular, special finite difference schemes.

In conclusion we may state that, if we apply classical numerical methods to the solution of the above problems, we cannot expect to obtain even the qualitative characteristics of the processes accompanied by heat transfer and/or diffusion, if boundary or interior layers (i.e., heat or diffusion layers) appear. Such processes often take place, for example, in catalytic reactions, burning, detonation and so on. To study these problems, it is necessary to use numerical methods that allow us to approximate both the solutions of the problems and the normalized diffusion fluxes uniformly with respect to the perturbation parameter.

Now we are well motivated to study more precisely classical finite difference schemes for the diffusion equation and to try to develop some new finite difference schemes with the desired properties.

II. NUMERICAL SOLUTIONS OF THE DIFFUSION EQUATION WITH PRESCRIBED VALUES ON THE BOUNDARY

In Section I we obtained an intuitive impression of the numerical problems appearing when one uses classical finite difference schemes to solve singularly perturbed boundary value problems for ordinary differential equations. In this section, for a parabolic equation, we study the nature of the errors in the approximate solution and the normalized diffusion flux for a classical finite difference scheme on a uniform grid and also on a grid with an arbitrary distribution of nodes in space. We find distributions of the grid nodes for which the solution of the finite difference scheme approximates the exact one uniformly with respect to the parameter. The efficiency of the new scheme for finding the approximate solution will be demonstrated with numerical examples.

A. Mathematical Formulation of the Physical Processes

We consider the simplest meaningful example leading to singularly perturbed boundary value problems. Suppose that we want to find the distributions of concentration C of a substance in a homogeneous material or in a solid material layer with thickness L. Suppose that the quantity C depends only on the variable y, which is the distance to one side of the material, and that generally speaking, the quantity C varies in time τ, that is, $C = C(y, \tau)$. Assume also that inside the material the distributed sources of the substance have a density $F(y, \tau)$. Suppose that the diffusion coefficient D is constant. In this case, the distribution of the substance in a material layer is described by the diffusion equation

$$D \frac{\partial^2}{\partial y^2} C(y, \tau) - \frac{\partial}{\partial \tau} C(y, \tau) = -F(y, \tau) \qquad 0 < y < L \qquad 0 < \tau \leq \vartheta \tag{2.1}$$

Here ϑ is the characteristic duration of the process. Observe that the diffusion flux is defined by the value

$$P^D(y, \tau) \equiv D \frac{\partial}{\partial y} C(y, \tau)$$

Let the concentration of the substance in a material be known at an initial instant

$$C(y, 0) = C^0(y) \qquad 0 \leq y \leq L \tag{2.2}$$

In addition, in some problems, the concentration of the substance is given on the material boundary

$$C(0, \tau) = C_0(\tau) \qquad C(L, \tau) = C_L(\tau) \qquad 0 < \tau \leq \vartheta \tag{2.3}$$

In other problems, on the material boundary (or on one side of the material) the diffusion flux is given

$$D \frac{\partial}{\partial y} C(0, \tau) = \Psi_0(\tau) \qquad D \frac{\partial}{\partial y} C(L, \tau) = \Psi_L(\tau) \qquad 0 < \tau \leq \vartheta \tag{2.4}$$

In similar situations, but for the heat transfer process, the distribution of temperature in a material is described by Eq. (2.1), where C is temperature, D is the coefficient of temperature conductivity

$$D = \lambda c^{-1} \rho^{-1}$$

λ and c are heat conduction and heat capacity coefficients, ρ is the material density. The heat flux is defined by the formula

$$P^H(y,\tau) \equiv \lambda \frac{\partial}{\partial y} C(y,\tau)$$

Equations (2.2) and (2.3) are valid for the heat transfer problem while Eqs. (2.4) take the form

$$\lambda \frac{\partial}{\partial y} C(0,\tau) = \Psi_0(\tau) \qquad \lambda \frac{\partial}{\partial y} C(L,\tau) = \Psi_L(\tau) \qquad 0 < \tau \leq \vartheta \quad (2.5)$$

When we study problems (2.1)–(2.3) [or problems (2.1), (2.2), and (2.4)], for convenience, we change to new variables in which the differential equations are simpler. Using the variables

$$x = L^{-1} y \qquad t = \vartheta^{-1} \tau \quad (2.6)$$

and writing

$$\varepsilon^2 = D\vartheta L^{-2} \qquad u(x,t) = C_*^{-1} C(y(x), \tau(t))$$
$$f(x,t) = -\vartheta C_*^{-1} F(y(x), \tau(t)) \quad (2.7)$$

we obtain the following equation

$$\varepsilon^2 \frac{\partial^2}{\partial x^2} u(x,t) - \frac{\partial}{\partial t} u(x,t) = f(x,t) \qquad 0 < x < 1 \qquad 0 < t \leq 1 \quad (2.8)$$

Here C_* is constant, which is a typical concentration of the substance in the process under study, for example, the maximal concentration observed during the diffusion period in the material or in part of it. The boundary conditions (2.2) and (2.3) take the form

$$u(x,0) = \varphi^0(x) \qquad 0 \leq x \leq 1 \quad (2.9)$$

$$u(0,t) = \varphi_0(t) \qquad u(1,t) = \varphi_1(t) \qquad 0 < t \leq 1 \quad (2.10)$$

where

$$\varphi^0(x) = C_*^{-1} C^0[y(x)] \qquad \varphi_0(t) = C_*^{-1} C_0[\tau(t)]$$
$$\varphi_1(t) = C_*^{-1} C_L[\tau(t)]$$

The diffusion flux of the substance in the new variables is defined by the formula

$$P^D(y,\tau) = D\frac{\partial}{\partial y}C(y,\tau) = C_*DL^{-1}\frac{\partial}{\partial x}u(x,t)$$
$$= C_*D^{1/2}\vartheta^{-1/2}\varepsilon\frac{\partial}{\partial x}u(x,t)$$

and the heat flux is defined by the formula

$$P^H(y,\tau) = \lambda\frac{\partial}{\partial y}C(y,\tau)$$
$$= C_*\lambda L^{-1}\frac{\partial}{\partial x}u(x,t) = C_*\lambda^{1/2}c^{1/2}\rho^{1/2}\vartheta^{-1/2}\varepsilon\frac{\partial}{\partial x}u(x,t)$$

where

$$x = x(y) \qquad t = t(\tau)$$

In the case of boundary conditions (2.4), we have

$$\varepsilon\frac{\partial}{\partial x}u(0,t) = \psi_0(t) \qquad \varepsilon\frac{\partial}{\partial x}u(1,t) = \psi_1(t) \qquad 0<t\leq 1 \qquad (2.11)$$

where the functions $\psi_0(t)$ and $\psi_1(t)$ are as follows:

$$\psi_0(t) = C_*^{-1}D^{-1/2}\vartheta^{1/2}\Psi_0[\tau(t)] \qquad \psi_1(t) = C_*^{-1}D^{-1/2}\vartheta^{1/2}\Psi_L[\tau(t)]$$

for diffusion processes and

$$\psi_0(t) = C_*^{-1}\lambda^{-1/2}c^{-1/2}\rho^{-1/2}\vartheta^{1/2}\Psi_0[\tau(t)]$$
$$\psi_1(t) = C_*^{-1}\lambda^{-1/2}c^{-1/2}\rho^{-1/2}\vartheta^{1/2}\Psi_L[\tau(t)]$$

for heat processes. Thus, problem (2.1)–(2.3) in the new variables (2.6), (2.7) transforms into problem (2.8)–(2.10), and problem (2.1), (2.2), (2.4) transforms into problem (2.8), (2.9), (2.11). These new problems are considered on a unit-length segment and on a unit-time interval. Note that the quantity ε^2 is dimensionless. This quantity is the diffusion Fourier number in the diffusion problem and the heat Fourier number in the heat transfer problem.

Problems (2.1)–(2.3) and (2.8)–(2.10), in which the unknown functions $C(y,\tau)$ and $u(x,t)$ are given on the boundary of the domain D, are called Dirichlet problems (or first boundary value problems). Problems (2.1), (2.2), (2.4) and (2.8), (2.9), (2.11), in which the derivatives of the unknown functions $C(y, \tau)$ and $u(x, t)$ are given on the boundary of the domain D, are called Neumann problems (or second boundary value problems). Note that, on the domain boundary a linear combination of the values of the unknown function and its derivatives can be given (these problems are called Robin problems, or third boundary value problems). On different parts of the boundary, different types of boundary conditions may be given (these problems are called mixed-boundary value problems).

For rapid processes, the value of ϑ can be very small. Generally speaking, the diffusion coefficient varies over a wide range for different media (from 10^{-5} m^2/s for gases to 10^{-14} m^2/s for solid materials). The coefficient of temperature conductivity also varies over a wide range and can take very small values in the case of molecular heat transfer in gases. Therefore, if the sample thickness is not very small, the parameter ε can take very small values. Note that we cannot say in advance what lower bound the parameter ε has, that is, the parameter ε can be arbitrarily small.

For the two-dimensional case, in the presence of distributed sources (sinks) of the diffusing substance, the concentration distribution is described under simplified conditions by the following equation of parabolic type

$$L_{(2.12)}u(x,t) \equiv \left\{ \varepsilon^2 \sum_{s=1,2} a_s(x,t) \frac{\partial^2}{\partial x_s^2} - p(x,t)\frac{\partial}{\partial t} - c(x,t) \right\} u(x,t) = f(x,t)$$

$$(x,t) \in G \qquad \text{(2.12a)}$$

Here

$$G = D \times (0, T] \qquad \text{(2.13a)}$$

$$D = \{x \colon d_{0s} < x_s < d_{1s},\, s = 1, 2\} \qquad \text{(2.13b)}$$

where D is a rectangular domain.

In Dirichlet problems the function $u(x, t)$ is given on the boundary S $(S = \bar{G} \setminus G)$

$$u(x,t) = \varphi(x,t) \qquad (x,t) \in S \qquad \text{(2.12b)}$$

The coefficients and the right-hand side of the differential equation are sufficiently smooth (i.e., they have high-order derivatives) and satisfy the conditions

$$a_1(x,t), a_2(x,t) \geq a_0 \qquad p(x,t) \geq p_0 \qquad c(x,t) \geq 0$$
$$(x,t) \in \bar{G} \qquad a_0, p_0 > 0$$

In the one-dimensional case, under simplified conditions, diffusion can be described by the equations

$$L_{(2.14)}u(x,t) \equiv \left\{ \varepsilon^2 a(x,t) \frac{\partial^2}{\partial x^2} - p(x,t)\frac{\partial}{\partial t} - c(x,t) \right\} u(x,t) = f(x,t) \qquad (x,t) \in G \tag{2.14a}$$

$$u(x,t) = \varphi(x,t) \qquad (x,t) \in S \tag{2.14b}$$

Here

$$G = D \times (0, T] \qquad D = \{x \colon d_0 < x < d_1\} \tag{2.15}$$

B. Numerical Experiments with the Classical Finite Difference Scheme

In this section, we illustrate computational problems that arise in the numerical solution of singularly perturbed problems. The decrease of the perturbation parameter (i.e., the parameter multiplying the highest derivatives) to zero leads to the appearance of boundary layers, that is, small regions in the domain of definition of the solution where it varies sharply. Their width is of the same order as the parameter ε. When the step size of the space grid (or the distance between the nodes) is comparable to the parameter value, the exact solution of the problem varies significantly between neighboring nodes. This situation causes large errors in the approximate solution, which is typical when the step size of the uniform grid and the parameter value are comparable. As is known, according to the natural formulation of the problem, we are interested in the solution corresponding to an arbitrary value of the parameter in the interval $(0, 1]$.

Let us formulate the problem for which we study a classical finite difference scheme and also newly constructed ones.

Suppose that it is required to find the solution of the Dirichlet problem for the singularly perturbed heat equation

$$\begin{gathered} L_{(2.16)}u(x,t) \equiv \varepsilon^2 \frac{\partial^2}{\partial x^2} u(x,t) - \frac{\partial}{\partial t} u(x,t) = f(x,t) \qquad (x,t) \in G \\ u(x,t) = \varphi(x,t) \qquad (x,t) \in S \end{gathered} \tag{2.16}$$

Here the domain of definition of Eq. (2.16) is as follows:

$$G = D \times (0, T] \tag{2.17}$$

where $D = (0, 1)$, that is, $G = \{(x, t)\colon\ x \in (0, 1),\ t \in (0, T]\}$, S is the boundary of the set G, $S = \bar{G} \setminus G$, $f(x, t)$, $(x, t) \in \bar{G}$, $\varphi(x, t)$, $(x, t) \in S$ are given functions. Let us define these functions by the relations

$$\begin{aligned} f(x, t) &= -2t \qquad (x, t) \in \bar{G} \\ \varphi(x, t) &= \sin(\pi x) \qquad x \in \bar{D} \\ \varphi(0, t) = t^2 + t \qquad \varphi(1, t) &= t^2 \qquad 0 < t \leq T \qquad T = 1 \end{aligned} \tag{2.18}$$

Now we discuss some properties of this problem. Problem (2.16), (2.18) is linear. According to the theory of partial differential equations, the solution can be represented as a sum of two functions

$$u(x, t) = U(x, t) + V(x, t) \qquad (x, t) \in \bar{G} \tag{2.19}$$

Here, together with its partial derivatives with respect to both x and t, the function $U(x, t)$ is bounded ε-uniformly, while the other function $V(x, t)$ is bounded ε-uniformly together with its partial derivatives only with respect to t. However, the derivatives with respect to the space variable x are not bounded ε-uniformly. Therefore the functions $U(x, t)$ and $V(x, t)$ are called the regular and singular parts of the solution, respectively. The functions $U(x, t)$ and $V(x, t)$ are the solutions of the problems

$$\begin{aligned} L_{(2.16)} U(x, t) &= -2t \qquad (x, t) \in G \\ U(x, t) &= \begin{cases} \sin(\pi x) & x \in \bar{D} \qquad t = 0 \\ t^2 & x = 0, 1 \qquad 0 \leq t \leq 1 \end{cases} \end{aligned} \tag{2.20}$$

$$\begin{aligned} L_{(2.16)} V(x, t) &= 0 \qquad (x, t) \in G \\ V(x, t) &= \begin{cases} t & x = 0 \qquad 0 \leq t \leq 1 \\ 0 & (x, t) \in S \qquad x \neq 0 \end{cases} \end{aligned} \tag{2.21}$$

The solution of problem (2.20) is as follows:

$$U(x, t) = t^2 + \sin(\pi x) \exp(-\varepsilon^2 \pi^2 t) \qquad (x, t) \in \bar{G} \tag{2.22}$$

the graph being given in Fig. 11 for $\varepsilon = 1.0$, 0.1, and 0.01. The graph of the solution of problem (2.21) is given in Fig. 12 for $\varepsilon = 1.0$, 0.1, and 0.01.

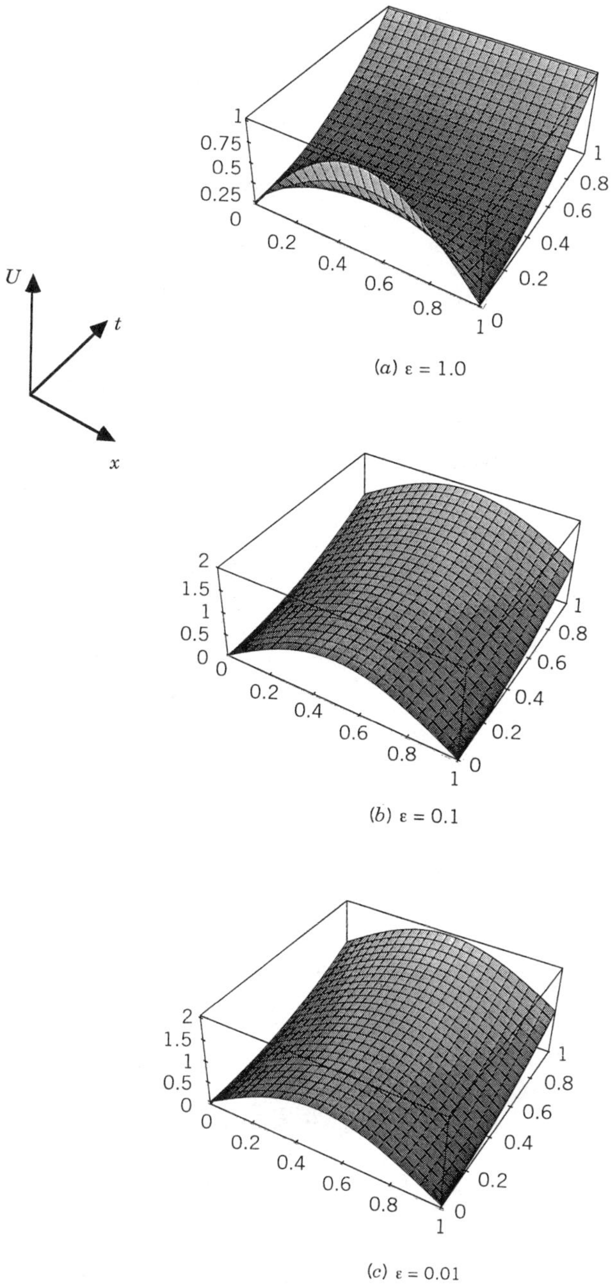

Figure 11. Regular part of the solution $U(x, t)$ of problem (2.16), (2.18) for $\varepsilon = 1.0$, 0.1, and 0.01.

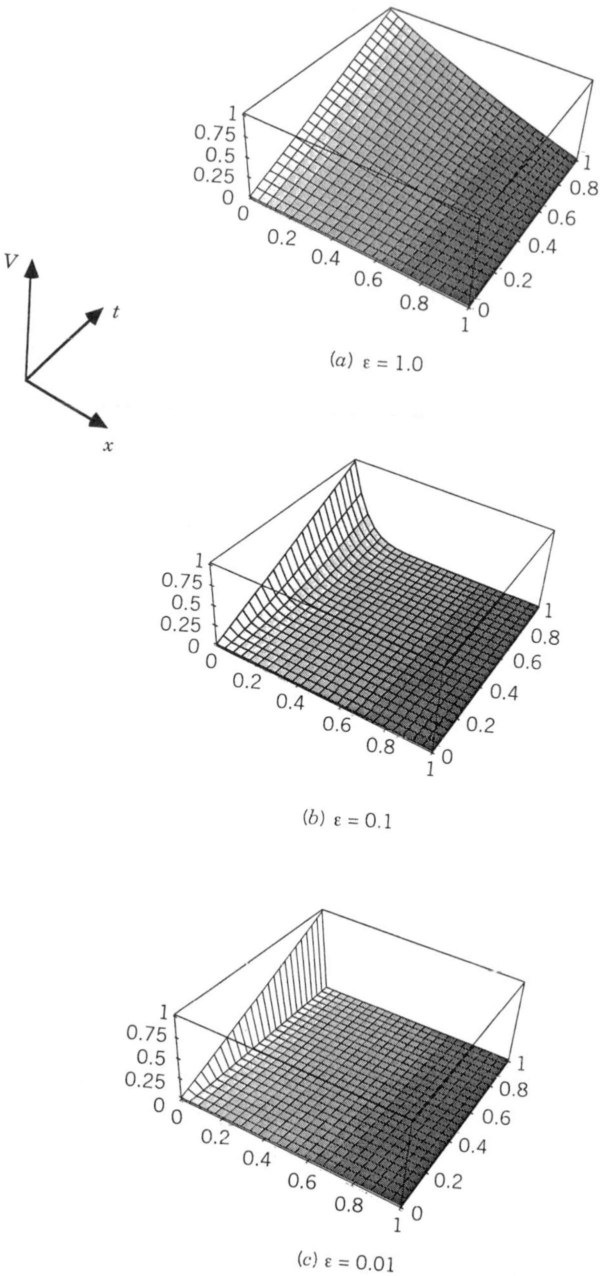

Figure 12. Singular part of the solution $V(x, t)$ of problem (2.16), (2.18) for $\varepsilon = 1.0$, 0.1, and 0.01.

It is convenient for the research to represent the function $V(x, t)$ as the sum of the functions

$$V(x, t) = W(x, t) + v(x, t) \qquad (x, t) \in \bar{G}$$

where

$$W(x, t) = \left(\frac{x^2}{2\varepsilon^2} + t\right)\operatorname{erfc}\left(\frac{x}{2\varepsilon\sqrt{t}}\right) - \frac{1}{\sqrt{\pi}}\frac{x\sqrt{t}}{\varepsilon}\exp\left(-\frac{x^2}{4\varepsilon^2 t}\right)$$
$$0 \le x < \infty \qquad t \ge 0 \tag{2.23}$$

The function $W(x, t)$ is called the main term of the singular part of the solution, and the function $v(x, t)$ is called the remainder term of the solution expansion. It is known from the theory [4] that for the function $v(x, t)$ the estimate

$$\max_{\bar{G}} |v(x, t)| \le M\varepsilon^n$$

is valid, where the integer n can be chosen very large, as large as desired, and the constant M, which is independent of x, t, does not depend on the parameter ε. A similar estimate holds for the derivative $(\partial/\partial x)v(x, t)$

$$\max_{\bar{G}} \left|\frac{\partial}{\partial x} v(x, t)\right| \le M\varepsilon^n$$

Thus, the function $V(x, t)$ approximates the function $W(x, t)$ to an accuracy of $M\varepsilon^n$. As one can easily verify, the function $W(x, t)$ is the solution of the problem

$$L_{(2.16)} W(x, t) = 0 \qquad 0 < x < \infty \qquad t \ge 0$$
$$W(x, 0) = 0 \qquad 0 \le x < \infty \qquad W(0, t) = t \qquad t \ge 0$$

and it satisfies the relations

$$\max_{\substack{0 \le x < \infty, \\ 0 \le \vartheta \le t}} W(x, \vartheta) \le t \qquad \frac{\partial}{\partial x} W(x, t) \le 0 \qquad 0 \le x < \infty \qquad t \ge 0$$

$$\max_{0 \le x < \infty} \varepsilon \left|\frac{\partial}{\partial x} W(x, t)\right| \le \varepsilon \left|\frac{\partial}{\partial x} W(0, t)\right| = 2\pi^{-1/2} t^{1/2} \qquad t \ge 0$$

For $x \ge x_0 > m$, $0 \le t \le 1$, as ε decreases, the function $W(x, t)$ tends to zero more rapidly than any power of the parameter ε

$$|W(x, t)| \le M\varepsilon^n \qquad x \ge x_0 > m \qquad 0 \le t \le 1$$

Taking into account the analysis given above, we arrive at the following representation of the function $u_{(2.16)}(x, t)$

$$u(x, t) = U(x, t) + W(x, t) + v(x, t) \qquad (x, t) \in \bar{G} \tag{2.24}$$

We have explicit analytical representations for the functions $U(x, t)$, $W(x, t)$, that is, for the regular and singular parts of the solution. These functions are defined by the formulas (2.22) and (2.23). The function $v(x, t)$ and its derivatives are small for small values of the parameter. The function $U(x, t)$ is the solution of problem (2.20), the function $W(x, t)$ is the solution of the problem

$$\begin{gathered} L_{(2.16)}W(x, t) = 0 \qquad (x, t) \in G \\ W(x, t) = \left\{\begin{array}{lll} 0 & x \in \bar{D} & t = 0 \\ t & x = 0 & 0 < t \leq 1 \\ W(1, t) & x = 1 & 0 < t \leq 1 \end{array}\right\} \qquad (x, t) \in S \end{gathered} \tag{2.25}$$

and the function $v(x, t)$ is the solution of the problem

$$\begin{gathered} L_{(2.16)}v(x, t) = 0 \qquad (x, t) \in G \\ v(x, t) = \left\{\begin{array}{lll} 0 & x \in \bar{D} & t = 0 \\ 0 & x = 0 & 0 < t \leq 1 \\ -W(1, t) & x = 1 & 0 < t \leq 1 \end{array}\right\} \qquad (x, t) \in S \end{gathered} \tag{2.26}$$

Here the function $W(1, t)$, $0 < t \leq 1$ is defined by formula (2.23).

Recall that we are interested in the behavior of the error in the approximate solution for various values of the parameter ε. To compute the solution of the boundary value problem (2.16), (2.18), we use a classical finite difference scheme. We now describe this scheme. On the set $\bar{G}$ the uniform rectangular grid

$$\bar{G}_h = \bar{\omega}_1 \times \bar{\omega}_0 \tag{2.27a}$$

is constructed. Here $\bar{\omega}_1$ and $\bar{\omega}_0$ are uniform grids on the interval $\bar{D}$ on the axis x and on the interval $[0, T]$ on the axis t, respectively. Suppose that $N + 1$ and $N_0 + 1$ are the number of nodes in the grids $\bar{\omega}_1$ and $\bar{\omega}_0$, respectively, and $h = N^{-1}$, $h_0 = N_0^{-1}$ are the step sizes of the grids, that is,

$$\bar{G}_h = \bar{G}_h(N, N_0) \tag{2.27b}$$

On the grid $\bar{G}_h$, the problem (2.16), (2.18) is approximated by the system of finite difference equations

$$\begin{aligned} \Lambda_{(2.28)} z(x,t) \equiv \varepsilon^2 \delta_{x\bar{x}} z(x,t) - \delta_{\bar{t}} z(x,t) = f(x,t) \qquad (x,t) \in G_h \\ z(x,t) = \varphi(x,t) \qquad (x,t) \in S_h \end{aligned} \tag{2.28}$$

Here $G_h = G \cap \bar{G}_h$, $S_h = S \cap \bar{G}_h$, $\delta_{x\bar{x}} z(x,t)$ and $\delta_{\bar{t}} z(x,t)$ are, respectively, the central finite difference derivative on the grid ω_1 and the first backward finite difference derivative on the grid ω_0. As is known, the combination of the grid (2.27) and Eqs. (2.28) on these grids is called the finite difference scheme (2.28), (2.27).

The scheme (2.28), (2.27) allows us to find the grid function $z(x,t)$, $(x,t) \in \bar{G}_h$, which is the approximate solution of the original problem (2.16), (2.18). We are interested in examining the quantity

$$\max_{\bar{G}_h} |u(x,t) - z(x,t)|$$

which is the error in the approximate solution. Moreover, we are interested in examining the behavior of the quantity

$$E(\varepsilon, N, N_0) \equiv \max_{\bar{G}_h} |u(x,t) - z(x,t)| = \max_{\bar{G}_h} |u(x,t;\varepsilon) - z(x,t;\varepsilon,N,N_0)| \tag{2.29}$$

depending on the value of the parameter ε for N, $N_0 \to \infty$.

It is known (see, e.g., [1]) from the theory of finite difference schemes that for N, $N_0 \to \infty$, and for a fixed value of the parameter ε, the solution of the finite difference scheme (2.28), (2.27) [or, as is usually said in short, the finite difference scheme (2.28), (2.27)] converges to the solution of the boundary value problem. In the case of sufficiently smooth solutions for problem (2.16), the following estimate is fulfilled

$$\max_{\bar{G}_h} |u(x,t) - z(x,t)| \le Q(\varepsilon)[N^{-2} + N_0^{-1}]$$

where the constant $Q(\varepsilon)$, a quantity depending on the smoothness of the solution (i.e., depending on the maximum of the absolute values of the solution derivatives), increases when ε decreases.

To show the influence of ε on the behavior of the quantity $E(\varepsilon, N, N_0)$, we perform a numerical experiment. The quantity $E(\varepsilon, N, N_0)$ depends on three parameters that make the analysis more difficult. Therefore, we set $N = N_0$ and we consider the quantities $E(\varepsilon, N)$ and $\bar{E}(N)$ where

$$E(\varepsilon, N) \equiv E(\varepsilon, N, N) \qquad \bar{E}(N) \equiv \bar{E}(N, N) \tag{2.30}$$

and

$$\bar{E}(N, N_0) = \max_{\varepsilon} E(\varepsilon, N, N_0)$$

is the maximum (with respect to the parameter ε) of the error $E(\varepsilon, N, N_0)$.

Unfortunately, we have no analytical representation of the function $u_{(2.16)}(x, t)$ to compute the value $E(\varepsilon, N, N_0; u_{(2.16)}(\cdot))$ directly. In the representation (2.24), we know only the formulas for the functions $U(x, t)$ and $W(x, t)$, but not for the function $v(x, t)$.

Therefore, to analyze the quantity $E(\varepsilon, N, N_0; u_{(2.16)}(\cdot))$, we use the technique described below.

For the function $u_{(2.16)}(x, t)$ we apply the representation in the form of Eq. (2.24). Then we use the fact that the error in the approximate solution for problem (2.16), (2.18) does not exceed the sum of the errors in the approximate solutions for problems (2.20), (2.25) and (2.26):

$$\begin{aligned} E(\varepsilon, N, N_0; u_{(2.16)}(\cdot)) \le E(\varepsilon, N, N_0; U_{(2.20)}(\cdot)) + E(\varepsilon, N, N_0; W_{(2.25)}(\cdot)) \\ + E(\varepsilon, N, N_0; v_{(2.26)}(\cdot)) \end{aligned} \tag{2.31}$$

Here $E(\varepsilon, N, N_0; u_{(2.16)}(\cdot))$ is determined by relation (2.29), where $u(x, t) = u_{(2.16)}$, and $z(x, t)$ is the solution of the corresponding difference problem (2.28), (2.27). In a similar way the quantities

$$E(\varepsilon, N, N_0; U_{(2.20)}(\cdot)) \qquad E(\varepsilon, N, N_0; W_{(2.25)}(\cdot))$$
$$\text{and} \qquad E(\varepsilon, N, N_0; v_{(2.26)}(\cdot))$$

can be determined. It is obvious that the following finite difference schemes correspond to the boundary value problems (2.20), (2.25) and (2.26), respectively:

$$\begin{aligned} \Lambda_{(2.28)} z(x, t) &= f(x, t) \qquad (x, t) \in G_h \\ z(x, t) &= U(x, t) \qquad (x, t) \in S_h \end{aligned} \tag{2.32}$$

$$\begin{aligned} \Lambda_{(2.28)} z(x, t) &= 0 \qquad (x, t) \in G_h \\ z(x, t) &= W(x, t) \qquad (x, t) \in S_h \end{aligned} \tag{2.33}$$

TABLE V
Table of Errors $E(\varepsilon, N)$ for the Classical Scheme (2.32), (2.27) in the Case of Problem (2.20)

$\varepsilon \backslash N$	4	16	64	256	1024
1	2.360e − 1	9.698e − 2	2.795e − 2	7.288e − 3	1.842e − 3
2^{-2}	2.423e − 1	6.182e − 2	1.546e − 2	3.864e − 3	9.659e − 4
2^{-4}	2.515e − 1	6.266e − 2	1.564e − 2	3.910e − 3	9.773e − 4
2^{-6}	2.501e − 1	6.251e − 2	1.563e − 2	3.906e − 3	9.766e − 4
2^{-8}	2.500e − 1	6.250e − 2	1.563e − 2	3.906e − 3	9.766e − 4
2^{-10}	2.500e − 1	6.250e − 2	1.563e − 2	3.906e − 3	9.766e − 4
2^{-12}	2.500e − 1	6.250e − 2	1.563e − 2	3.906e − 3	9.766e − 4
$\bar{E}(N)$	2.515e − 1	9.698e − 2	2.795e − 2	7.288e − 3	1.842e − 3

$$\begin{aligned} \Lambda_{(2.28)} z(x,t) &= 0 \qquad (x,t) \in G_h \\ z(x,t) &= v(x,t) \qquad (x,t) \in S_h \end{aligned} \tag{2.34}$$

where $\bar{G}_h = \bar{G}_{h(2.27)}$.

In Tables V and VI, the values of $E(\varepsilon, N)$, $\bar{E}(N)$, which are calculated for problems (2.20) and (2.25) for various values of ε and N, with $N_0 = N$, are given. To solve the problems, we use the finite difference schemes (2.32) and (2.33) on the grid (2.27).

From Table V, one can see that, as N increases, the error $E(\varepsilon, N)$ vanishes both for a fixed value of the parameter ε and for all the values of ε. Letting $\bar{E}(N)$ denote the maximum of the error $E(\varepsilon, N)$ for ε varying from 10^{-12} to 1.0, we see also from Table V that $\bar{E}(N)$ decreases as $N_0 = N \to \infty$. Thus, as N and $N_0 = N$ increase, the approximate solution $z_{(2.32)}(x,t)$ converges to the function $U_{(2.20)}(x,t)$, which is the solution of problem (2.20), both for a fixed value of the parameter ε and uniformly in ε.

TABLE VI
Table of Errors $E(\varepsilon, N)$ for the Classical Scheme (2.33), (2.27) in the Case of Problem (2.25)

$\varepsilon \backslash N$	4	16	64	256	1024
1	1.630e − 2	6.144e − 3	1.780e − 3	4.651e − 4	1.176e − 4
2^{-2}	4.374e − 2	8.624e − 3	1.960e − 3	4.769e − 4	1.184e − 4
2^{-4}	3.601e − 2	2.558e − 2	3.131e − 3	5.507e − 4	1.230e − 4
2^{-6}	2.432e − 3	3.095e − 2	2.061e − 2	1.728e − 3	1.968e − 4
2^{-8}	1.526e − 4	2.069e − 3	2.966e − 2	1.934e − 2	1.376e − 3
2^{-10}	9.537e − 6	1.297e − 4	1.978e − 3	2.934e − 2	1.902e − 2
2^{-12}	5.960e − 7	8.106e − 6	1.240e − 4	1.956e − 3	2.926e − 2
$\bar{E}(N)$	4.374e − 2	3.095e − 2	2.966e − 2	2.934e − 2	2.926e − 2

TABLE VII
Table of Errors $E(\varepsilon, N, N_0)$ for the Classical Scheme (2.32), (2.27) in the Case of Problem (2.20) at $\varepsilon = 2^{-4}$

$N_0 \backslash N$	4	16	64	256	1024
1	2.515e − 1	2.503e − 1	2.502e − 1	2.502e − 1	2.502e − 1
16	6.432e − 2	6.266e − 2	6.255e − 2	6.255e − 2	6.254e − 2
64	1.748e − 2	1.576e − 2	1.564e − 2	1.564e − 2	1.564e − 2
256	5.774e − 3	4.028e − 3	3.916e − 3	3.910e − 3	3.909e − 3
1024	2.846e − 3	1.096e − 3	9.847e − 4	9.777e − 4	9.773e − 4

In Table VII, the values of $E(\varepsilon, N, N_0)$ for problem (2.20) and the finite difference scheme (2.32), (2.27) are given for $\varepsilon = 2^{-4}$ and different values of N and N_0. From the table, we see that the errors vanish as the values of both N and N_0 increase. Similar behavior of the error $E(\varepsilon, N, N_0)$ is observed for other values of the parameter ε. Consequently,

$$E(\varepsilon, N, N_0) \leq E(\varepsilon, N) \qquad \text{at} \qquad N_0 \geq N$$

It follows from this relation and from the results in Table V that the approximate solution $z_{(2.32)}(x, t)$ converges to the function $U_{(2.20)}(x, t)$, as N and N_0 increase, both for a fixed value of the parameter ε and ε-uniformly.

As is obvious from Table VI, the behavior of the error $E(\varepsilon, N)$ as a function of ε and N is rather complicated. In the case of problem (2.25), and the corresponding scheme (2.33), (2.27), the behavior of the error $E(\varepsilon, N)$ is similar to that in the case of problem (1.16) and scheme (1.17), (1.18) (see Table I). When N increases, the error $E(\varepsilon, N)$ decreases only for $N \geq 4^{-1}\varepsilon^{-1}$; and it increases, as N increases, for $N < 4^{-1}\varepsilon^{-1}$. We can see that the approximate solution $z_{(2.33)}(x, t)$ converges to the solution of problem (2.25) for a fixed value of the parameter ε. However, the approximate solution does not converge ε uniformly, while the value of $\bar{E}(N)$ is bounded from below

$$\bar{E}(N) \geq 2.92 \times 10^{-2}$$

that is, the value of $\delta(N)$, where

$$\delta(N) = \bar{E}(N)[\max_{\bar{G}} |u(x, t)|]^{-1}$$

is the worst realizable relative error for fixed N and $N_0 = N$, is guaranteed to be no smaller than 2.9%.

This means that, for any large N, N_0, one can find a value of the parameter ε such that the difference is $100\% - \delta(N)$, that is, the guaranteed accuracy of the approximate solution, does not exceed 97.1%.

One should bear in mind that we cannot compute the values of $E(\varepsilon, N, N_0; v(\cdot))$ and $\bar{E}(N, N_0; v(\cdot))$, since we have no analytical representation of the function $v(x, t)$. Note that the behavior of the functions $v(x, t)$ and $U(x, t)$ is roughly similar. Consequently, the function $v(x, t)$ and its partial derivatives, as well as the function $U(x, t)$, are bounded ε-uniformly. Moreover, the maximum of the absolute value of the function $v(x, t)$ and its derivatives on the set $\bar{G}$ tends to zero as $\varepsilon \to 0$. In this case, it follows from the theory of finite difference schemes that the solution of the finite difference scheme (2.34), (2.27) converges for $N, N_0 \to \infty$ to the function $v(x, t)$ [the solution of the boundary value problem (2.26)] ε-uniformly. Therefore, the quantities $E(\varepsilon, N, N_0; v(\cdot))$ exhibit the natural behavior, that is, they tend to zero with the growth of N, N_0.

Thus, in order to evaluate $E(\varepsilon, N, N_0; v(\cdot))$ and $\bar{E}(N, N_0; v(\cdot))$ for the function $v(x, t)$, instead of these values we compute the quantities

$$E_{N^*, N_0^*}(\varepsilon, N, N_0) = \max_{(x,t) \in \bar{G}_h} |\bar{z}_{N^*, N_0^*}(x, t) - z(x, t)|$$

$$\bar{E}_{N^*, N_0^*}(N, N_0) = \max_{\varepsilon} E_{N^*, N_0^*}(\varepsilon, N, N_0)$$

Here $\bar{z}_{N^*, N_0^*}(x, t)$ is a continuous function defined on the set $\bar{G}$. The function $\bar{z}_{N^*, N_0^*}(x, t)$ is constructed by linear interpolation with respect to the variables x and t from the values of the grid function $z_{N^*, N_0^*}(x, t)$.

The function $z_{N^*, N_0^*}(x, t)$ is the solution of the finite difference scheme (2.34) on the grid

$$\bar{G}_h(N^*, N_0^*)$$

where $\bar{G}_h(N^*, N_0^*) = \bar{G}_{h(2.27)}(N^*, N_0^*)$, provided that the values N^* and N_0^* are chosen so that they considerably exceed N and N_0, respectively. We set

$$E_{N^*, N_0^*}(\varepsilon, N) = E_{N^*, N_0^*}(\varepsilon, N, N) \qquad \bar{E}_{N^*, N_0^*}(N) = \bar{E}_{N^*, N_0^*}(N, N) \quad (2.35)$$

In Table VIII, the values of $E_{1024,1024}(\varepsilon, N)$, $\bar{E}_{1024,1024}(N)$ are given. They are calculated for the finite difference scheme (2.34), (2.27) corresponding to the boundary value problem (2.26) for different values of ε and N, with $N_0 = N$. For values of ε smaller than 2^{-4} the errors $E_{1024,1024}(\varepsilon, N)$ become equal to zero up to the computer accuracy. From

TABLE VIII

Table of Errors $E_{1024,1024}(\varepsilon, N)$ for the Classical Scheme (2.34), (2.27) in the Case of Problem (2.26)

$\varepsilon\backslash N$	4	16	64	256
1	4.609e − 03	1.273e − 03	3.114e − 04	6.275e − 05
2^{-2}	9.645e − 05	3.058e − 05	6.804e − 06	1.330e − 06
2^{-4}	2.561e − 33	9.477e − 33	1.616e − 32	3.497e − 33
2^{-6}	0.000e − 00	0.000e − 00	0.000e − 00	0.000e − 00
$\bar{E}(N)$	4.609e − 03	1.273e − 03	3.114e − 04	6.275e − 05

the table one can see that, as was predicted above, the error $E(s, N)$ tends to zero, when N increases, both for a fixed value of the parameter ε and ε uniformly.

Note that for the solutions $v(x, t)$, $z_{(2.34,2.27)}(x, t)$ of the boundary value problem (2.26), and the finite difference scheme (2.34), (2.27), the following estimate is fulfilled:

$$\max_{\bar{G}} |v(x, t)|, \max_{\bar{G}_h} |z_{(2.34,2.27)}(x, t)| = \max_{0 \le t \le T} W(1, t) \le W(1, T)$$

$$\max_{\bar{G}} |P(x, t; v(\cdot))| \le |P(1, T; W(\cdot))|$$

where $\quad P(x, t; v(\cdot)) = \varepsilon \dfrac{\partial}{\partial x} v(x, t) \qquad P(x, t; W(\cdot)) = \varepsilon \dfrac{\partial}{\partial x} W(x, t)$

In Table IX, the quantities $W(1, T) = W(1, T; \varepsilon)$ and $P(1, T) = P(1, T; W(\cdot))$ are presented for different values of the parameter ε. The values of $W(1, T; \varepsilon)$ and $P(1, T; W(\cdot))$ tend to zero rapidly as the parameter tends to zero.

From Tables V, VI, and VIII we see that the function $W(x, t)$, that is, the singular part of the solution, makes the greatest contribution to the error when one solves the boundary value problem (2.16), (2.18) with the use of the difference scheme (2.28), (2.27). From the results in Tables V,

TABLE IX

Table of Values $W(1, T; \varepsilon)$ and $P(1, T; W(\cdot))$ for Various Values of the Parameter ε; $T = 1$

ε	1	2^{-2}	2^{-4}	2^{-6}	2^{-8}
$W(1, T = 1, \varepsilon)$	2.799e − 01	5.679e − 02	7.656e − 04	8.401e − 10	1.690e − 31
$P(1, T = 1, W(\cdot))$	−3.993e − 01	−1.005e − 01	−1.956e − 03	−3.644e − 09	−1.382e − 30

VI, IX, and the inequality

$$E(\varepsilon, N, N_0; u_{(2.16)}(\cdot)) \le E(\varepsilon, N, N_0; W(\cdot)) + E(\varepsilon, N, N_0; U(\cdot)) + \max_{\bar{G}} |v(x, t)| + \max_{\bar{G}_h} |z_{(2.34)}(x, t)|$$

it follows that the solution of the finite difference scheme (2.28), (2.27) converges for a fixed value of the parameter ε when N and N_0 increase. However, it follows from the inequality

$$E(\varepsilon, N, N_0; u_{(2.16)}(\cdot)) \ge E(\varepsilon, N, N_0; W(\cdot)) - E(\varepsilon, N, N_0; U(\cdot)) - \max_{\bar{G}} |v(x, t)| - \max_{\bar{G}_h} |z_{(2.34)}(x, t)|$$

and from the results in Tables V, VI, and IX that the solution of the difference scheme (2.28), (2.27) does not converge ε-uniformly.

The error $E(\varepsilon, N; u_{(2.16)}(\cdot))$, obtained from (2.31) as the sum of the errors of the components in the representation (2.24), can be overestimated. Let us point out a recipe for computing a more accurate estimate for the quantity $E(\varepsilon, N; u_{(2.16)}(\cdot))$.

We determine the quantities $E^0_{N^*,N_0^*}(\varepsilon, N, N_0; U_{(2.16)}(\cdot))$ and $E_{N^*,N_0^*}(N, N_0; u_{(2.16)}(\cdot))$ by the relations

$$E^0_{N^*,N_0^*}(\varepsilon, N, N_0; u_{(2.16)}(\cdot)) = \max_{\bar{G}} |u(x, t) - z_{(2.28)}(x, t)|$$

$$\bar{E}^0_{N^*,N_0^*}(N, N_0; u_{(2.16)}(\cdot)) = \max_{\varepsilon} E^0_{N^*,N_0^*}(\varepsilon, N, N_0; u_{(2.16)}(\cdot))$$

where

$$\begin{gathered} u(x, t) = \bar{u}(x, t) \\ \bar{u}(x, t) = U(x, t) + W(x, t) + \bar{z}_{N^*,N_0^*}(x, t) \qquad (x, t) \in \bar{G} \end{gathered} \tag{2.36}$$

Further, we assume that

$$E^0_{N^*,N_0^*}(\varepsilon, N; u_{(2.16)}(\cdot)) = \bar{E}^0_{N^*,N_0^*}(\varepsilon, N, N; u_{(2.16)}(\cdot))$$

$$\bar{E}^0_{N^*,N_0^*}(N; u_{(2.16)}(\cdot)) = \bar{E}^0_{N^*,N_0^*}(N, N; u_{(2.16)}(\cdot))$$

The quantities $E^0_{N^*,N_0^*}(\varepsilon, N; u_{(2.16)}(\cdot))$ and $\bar{E}^0_{N^*,N_0^*}(N; u_{(2.16)}(\cdot))$ approach the quantities $E(\varepsilon, N; u_{(2.16)}(\cdot))$ and $\bar{E}(N; u_{(2.16)}(\cdot))$ fairly well when N^*, N_0^* exceed the number N.

In Table X we give the values of $E^0_{1024,1024}(\varepsilon, N)$, $\bar{E}^0_{1024,1024}(N)$

TABLE X
Table of Errors $E^0_{1024,1024}(\varepsilon, N)$ for the Classical Scheme (2.28), (2.27) in the Case of Problem (2.16)

$\varepsilon \backslash N$	4	16	64	256
1	2.485e − 1	1.010e − 1	2.772e − 2	5.782e − 3
2^{-2}	2.783e − 1	6.816e − 2	1.612e − 2	3.215e − 3
2^{-4}	2.773e − 1	7.816e − 2	1.635e − 2	3.140e − 3
2^{-6}	2.509e − 1	9.048e − 2	3.064e − 2	4.069e − 3
2^{-8}	2.491e − 1	6.346e − 2	4.382e − 2	2.066e − 2
2^{-10}	2.490e − 1	6.165e − 2	1.660e − 2	3.214e − 2
2^{-12}	2.490e − 1	6.153e − 2	1.477e − 2	4.878e − 3
$\bar{E}(N)$	2.783e − 1	1.010e − 1	4.382e − 2	3.214e − 2

computed for different values of the parameter ε and the number of nodes N. From the table it is seen that, as N increases, the errors $E^0_{1024,1024}(\varepsilon, N)$ converge for fixed values of the parameter ε and do not converge ε-uniformly.

In conclusion, following the results of Tables V and VI, we may state that, for $N, N_0 \to \infty$, the approximate solution converges ε-uniformly in the case of the smooth solution of the boundary value problem, for example, for $u(x,t) = U_{(2.20)}(x,t)$. When the solution of the boundary value problem is singular, for example, $u(x,t) = V_{(2.21)}(x,t)$, the approximate solution converges for a fixed value of the parameter ε, however, it does not converge ε-uniformly. Therefore, the solution of problem (2.28), (2.27) does not converge ε-uniformly either.

Now we analyze the behavior of the error when the derivative with respect to x is being approximated. We consider the function

$$P(x,t) \equiv \varepsilon \frac{\partial}{\partial x} u(x,t) \qquad (x,t) \in \bar{G}$$

called the normalized diffusion flux (or, in short, the normalized flux, or the flux). According to the analytical form of the function $U(x,t)$, we see that the quantity

$$P(x,t; U_{(2.20)}(\cdot)) = \varepsilon \frac{\partial}{\partial x} U_{(2.20)}(x,t)$$

satisfies the relations

$$\max_{\bar{G},\varepsilon} |P(x,t; U_{(2.20)}(\cdot))| \le M$$

$$\lim_{\varepsilon \to 0} \max_{\bar{G}} |P(x,t; U_{(2.20)}(\cdot))| = 0$$

Taking into account the estimates for the functions $W(x, t)$ and $v(x, t)$ and for their derivatives with respect to x, we come to the relations

$$\max_{\bar{G},\varepsilon} |P(x, t; V_{(2.21)}(\cdot))| \leq M$$

$$\lim_{\varepsilon\to 0} \max_{\bar{G}} |P(x, t; V_{(2.21)}(\cdot))| = |P(x = 0, t = 1, W(\cdot))| = 2\pi^{-1/2}$$

Hence, for the function $P(x, t; u_{(2.16)}(\cdot))$ the following relations hold

$$\max_{\bar{G},\varepsilon} |P(x, t; u_{(2.16)}(\cdot))| \leq M$$

$$\lim_{\varepsilon\to 0} \max_{\bar{G}} |P(x, t; u_{(2.16)}(\cdot))| = |P(x = 0, t = 1, W(\cdot))|$$

In Figs. 13 and 14 the graphs of the functions $P(x, t; U(\cdot))$ and $P(x, t; V(\cdot))$ are given.

We approximate the normalized diffusion flux by the function

$$P^{h+}(x, t) \equiv \varepsilon\delta_x z(x, t) \qquad (x, t) \in \bar{G}_h^-$$

which is the computed normalized diffusion flux (or, in short, the computed normalized flux, or the computed flux), where $\bar{G}_h^- = \{(x, t): (x, t) \in \bar{G}_h, x < 1\}$ and $\delta_{\underline{x}} z(x, t)$ is the first (forward) difference derivative on the grid $\bar{G}_h^-$, $\bar{G}_h = \bar{G}_{h(2.27)}$.

To evaluate how the function $P^{h+}(x, t)$ approximates the function $P(x, t)$, we use the quantities $Q(\varepsilon, N, N_0)$ and $\bar{Q}(N, N_0)$

$$Q(\varepsilon, N, N_0) \equiv \max_{\bar{G}_h^-} |P(x, t) - P^{h+}(x, t)|$$

$$= \max_{\bar{G}_h^-} |P(x, t; \varepsilon) - P^{h+}(x, t; \varepsilon, N, N_0)|$$

$$\bar{Q}(N, N_0) \equiv \max_{\varepsilon} Q(\varepsilon, N, N_0)$$

which are the errors in the normalized fluxes. We define

$$Q(\varepsilon, N) = Q(\varepsilon, N, N) \qquad \bar{Q}(N) = \bar{Q}(N, N)$$

Since the analytical representation of the function $u_{(2.16)}(x, t)$ is unknown, and, consequently, the flux $P(x, t; u_{(2.16)}(\cdot))$ is unknown, in order to analyze the quantities $Q(\varepsilon, N; u_{(2.16)}\cdot))$ and $\bar{Q}(N; u_{(2.16)}(\cdot))$ we use the same method that we used to analyze $E(\varepsilon, N; u_{(2.16)}\cdot))$ and $\bar{E}(N; u_{(2.16)}(\cdot))$. In this connection we start from an inequality similar to

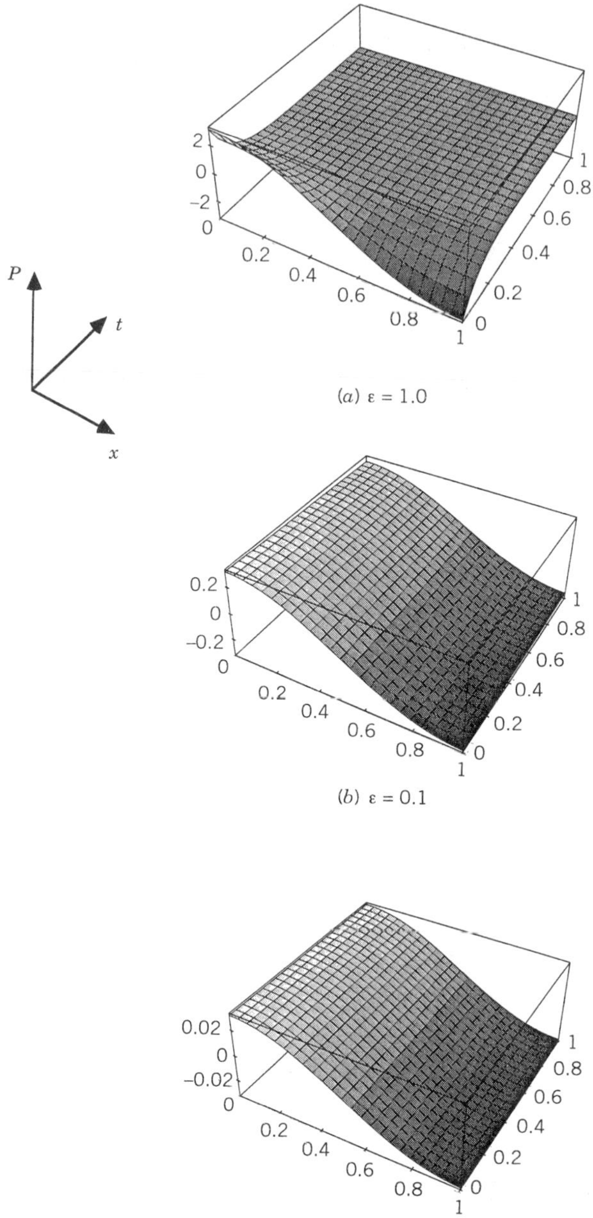

(a) $\varepsilon = 1.0$

(b) $\varepsilon = 0.1$

(c) $\varepsilon = 0.01$

Figure 13. Normalized flux $P(x, t; U(\cdot))$ for problem (2.16), (2.18) for $\varepsilon = 1.0$, 0.1, and 0.01.

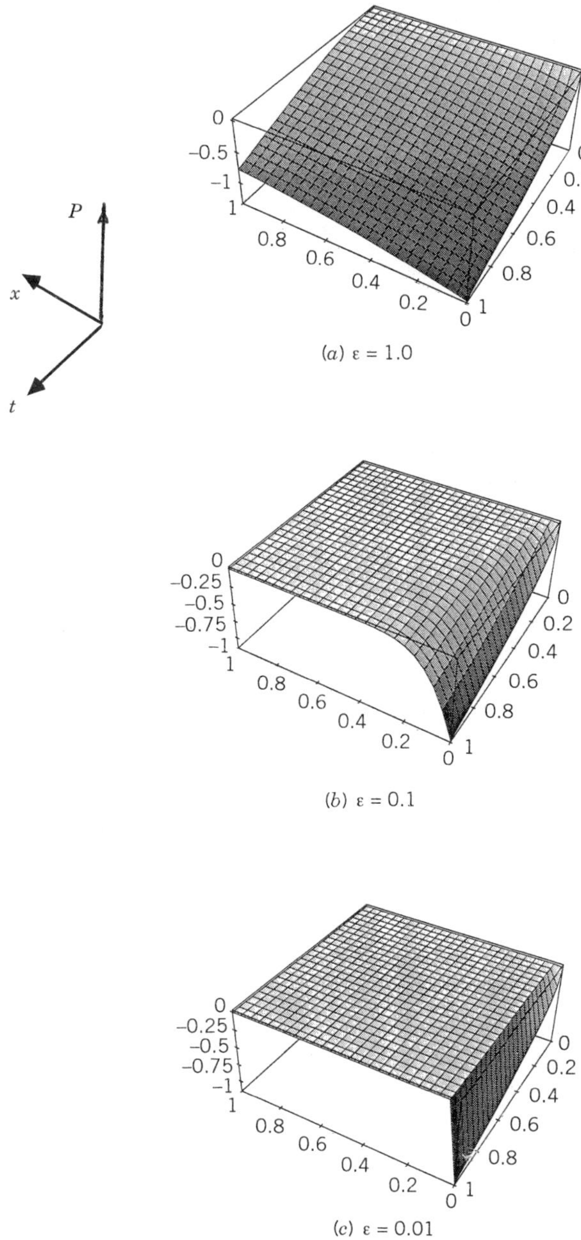

Figure 14. Normalized flux $P(x, t; V(\cdot))$ for problem (2.16), (2.18) for $\varepsilon = 1.0$, 0.1, and 0.01.

TABLE XI
Table of Errors of the Normalized Flux $Q(\varepsilon, N)$ for the Classical Scheme (2.32), (2.27) in the Case of Problem (2.20)

$\varepsilon \backslash N$	4	16	64	256	1024
1	7.240e − 1	3.218e − 1	9.282e − 2	2.413e − 2	6.095e − 3
2^{-2}	2.729e − 1	7.402e − 2	1.909e − 2	4.807e − 3	1.204e − 3
2^{-4}	9.706e − 2	4.590e − 2	1.587e − 2	4.295e − 3	1.095e − 3
2^{-6}	2.505e − 2	1.576e − 2	1.098e − 2	3.937e − 3	1.072e − 3
2^{-8}	6.276e − 3	4.055e − 3	3.797e − 3	2.740e − 3	9.841e − 4
2^{-10}	1.569e − 3	1.016e − 3	9.771e − 4	9.473e − 4	6.853e − 4
2^{-12}	3.923e − 4	2.540e − 4	2.447e − 4	2.437e − 4	2.368e − 4
$\bar{Q}(N)$	7.240e − 1	3.218e − 1	9.282e − 2	2.413e − 2	6.095e − 3

inequality (2.31)

$$Q(\varepsilon, N, N_0; u_{(2.16)}(\cdot)) \le Q(\varepsilon, N, N_0; U_{(2.20)}(\cdot)) + Q(\varepsilon, N, N_0; W_{(2.25)}(\cdot)) + Q(\varepsilon, N, N_0; v_{(2.26)}(\cdot))$$

In Tables XI and XII the results of the computed values of $Q(\varepsilon, N)$ and $\bar{Q}(N)$ for problems (2.20) and (2.25) are presented.

From the results in Table XI, it is obvious that, as N increases, provided that $N_0 = N$, the computed normalized flux converges to the exact normalized flux $P(x, t; U_{(2.20)}(\cdot))$ both for fixed values of the parameter and ε-uniformly.

The behavior of the errors in the normalized flux is more complicated in the case of problem (2.25). It is similar to the behavior of the errors in the normalized diffusion flux in the case of boundary value problem (1.16) for an ordinary differential equation, when the difference scheme

TABLE XII
Table of Errors of the Normalized Flux $Q(\varepsilon, N)$ for the Classical Scheme (2.33), (2.27) in the Case of Problem (2.25)

$\varepsilon \backslash N$	4	16	64	256	1024
1	1.620e − 1	6.123e − 2	2.362e − 2	9.946e − 3	4.496e − 3
2^{-2}	4.516e − 1	1.328e − 1	4.345e − 2	1.535e − 2	5.904e − 3
2^{-4}	8.876e − 1	4.332e − 1	1.228e − 1	3.321e − 2	1.086e − 2
2^{-6}	1.066e + 0	8.863e − 1	4.282e − 1	1.211e − 1	3.111e − 2
2^{-8}	1.113e + 0	1.066e + 0	8.860e − 1	4.270e − 1	1.207e − 1
2^{-10}	1.124e + 0	1.113e + 0	1.066e + 0	8.859e − 1	4.267e − 1
2^{-12}	1.127e + 0	1.124e + 0	1.113e + 0	1.066e − 0	8.859e − 1
$\bar{Q}(N)$	1.127e + 0	1.124e + 0	1.113e + 0	1.066e + 0	8.859e − 1

(1.17), (1.18) is used (see Table II). From the results in Table XII one can see that, as N increases, the computed normalized flux converges to the real flux for a fixed value of the parameter ε. However, the computed normalized flux does not converge ε-uniformly. Moreover, the error in the computed flux tends to the maximum modulus of the unknown function $P(x, t; \varepsilon, V(\cdot))$ on the set $\bar{G}$, as the parameter ε decreases, that is, it tends to a constant greater than 1

$$P(x = 0, t = 1; \varepsilon, W(\cdot)) = 2\pi^{-1/2} \approx 1.128$$

We know that we are not able to compute the quantities $Q(\varepsilon, N; v(\cdot))$ and $\bar{Q}(N; v(\cdot))$ because an analytical formula for $v(x, t)$, which is the solution of problem (2.26), is unknown. Since the derivatives of the function $v(x, t)$ decrease rapidly with $\varepsilon \to 0$, the errors $Q(\varepsilon, N, N_0; v(\cdot))$ and $\bar{Q}(N, N_0; v(\cdot))$ are well approximated by the corresponding values of

$$Q_{N^*, N_0^*}(\varepsilon, N, N_0) = \max_{(x,t) \in \bar{G}_h^-} |P_{N^*, N_0^*}(x, t) - P^{h+}(x, t)|$$

$$\bar{Q}_{N^*, N_0^*}(N, N_0) = \max_{\varepsilon} Q_{N^*, N_0^*}(\varepsilon, N, N_0)$$

where

$$P_{N^*, N_0^*}(x, t) \equiv \varepsilon \frac{\partial}{\partial x} \bar{z}_{N^*, N_0^*}(x', t) \qquad (x', t) \in G_h(N^*, N_0^*)$$

$$x \in [x^*, x^* + h^*) \qquad h^* = (N^*)^{-1}$$

We define

$$Q_{N^*, N_0^*}(\varepsilon, N) = Q_{N^*, N_0^*}(\varepsilon, N, N)$$

$$\bar{Q}_{N^*, N_0^*}(N) = \bar{Q}_{N^*, N_0^*}(N, N)$$

In Table XIII we give the errors in the normalized diffusion fluxes $Q_{1024,1024}(\varepsilon, N)$ and $\bar{Q}_{1024,1024}(N)$ for the classical finite difference scheme (2.34), (2.27) related to problem (2.26). These errors are computed for different values of the parameter ε and the number of nodes N. For values of ε smaller than 2^{-4} the errors $Q_{1024,1024}(\varepsilon, N)$ are equal to zero up to the computer accuracy. From Table XIII it follows that, as N increases, the error $Q_{1024,1024}(\varepsilon, N)$ tends to zero for a fixed value of the parameter ε and ε-uniformly.

From the results in Tables XI–XIII we see that the greatest error in computing the normalized flux for problem (2.16) is the error in the singular part of the solution of the boundary value problem.

TABLE XIII
Table of Errors of the Normalized Flux $Q_{1024,1024}(\varepsilon, N)$ for the Classical Scheme (2.34), (2.27) in the Case of Problem (2.26)

$\varepsilon \backslash N$	4	16	64	256
1	3.934e − 02	1.176e − 02	3.268e − 03	1.033e − 03
2^{-2}	4.777e − 04	3.054e − 04	1.066e − 04	3.701e − 05
2^{-4}	4.161e − 32	1.591e − 31	3.514e − 31	2.499e − 31
2^{-6}	0.000e − 00	0.000e − 00	0.000e − 00	0.000e − 00
$\bar{Q}(N)$	3.934e − 02	1.176e − 02	3.268e − 03	1.033e − 03

Having analyzed Tables XI–XIII, we conclude that, for a fixed value of the parameter, the function $P^{h+}(x, t; z_{(2.28)}(\cdot))$ approximates the function $P(x, t; u_{(2.16)}(\cdot))$ with $N, N_0 \to \infty$. Nevertheless, the computed normalized flux does not approach the real one ε-uniformly, even qualitatively.

Thus, by numerical experiments we verify that the approximate solution of the Dirichlet problem (2.16), found by the classical finite difference scheme (2.28), (2.27), and the computed normalized diffusion flux converge for N, $N_0 \to \infty$, respectively, to the solution of the boundary value problem and the real normalized diffusion flux for fixed ε. However, we can also see that they do not converge ε-uniformly. The solution of the grid problem approaches the solution of the boundary value problem uniformly in ε qualitatively well. The normalized flux computed according to the solution of the difference problem does not approach ε-uniformly the real normalized flux (i.e., the flux related to the solution of the boundary value problem) even qualitatively. Nevertheless, if the solution of the singularly perturbed boundary value problem is smooth and ε-uniformly bounded, the approximate solution and the computed normalized flux converge ε-uniformly (when $N, N_0 \to \infty$) to the exact solution and flux.

The computational example described in this section is not far-fetched. The power of a computer is characterized by the value N. From the numerical experiment under consideration it follows that, for any large N and $N_0 = N$, a value of the parameter ε (corresponding to the physical process under study) can be found such that the error of the approximate solution is no smaller than 2.9%. Although the maximum error is not so large (at most 4.4%), the situation is unsatisfactory when we cannot guarantee the error to be less than 3.0% for arbitrarily large N and $N_0 = N$. But, it is obvious from the computational results that, for any large N and $N_0 = N$, a value of ε can be found for which the magnitude of the real normalized diffusion flux on the material surface exceeds many times the magnitude of the computed flux.

C. Principles for Constructing Special Finite Difference Schemes

In this section, using an example of a boundary value problem for a singularly perturbed ordinary differential equation, we discuss some principles for constructing special finite difference schemes. In Section II.D, these principles will be applied to the construction of special schemes for singularly perturbed equations of the parabolic type.

Now we consider some ideas that will be applied to the construction of ε-uniformly convergent finite difference schemes. From the results of Section II.B we obtain the following. When the parameter ε is of the order of unity, the solution of the boundary value problem varies smoothly in the whole domain of integration. Therefore, partial derivatives in the differential equations are well approximated by the corresponding difference derivatives, and the finite difference equation approximates the differential one. Here grids with an arbitrary distribution of nodes can be used (in particular, uniform grids). It is important that the domain does not contain parts where grid nodes are absent. It is clear that the use of uniform grids is preferable here because these grids are the simplest, and a better approximation of the differential derivatives by finite differences can be achieved on them.

As the parameter ε tends to zero (and in this case the thickness of the boundary layer tends to zero), parts of the domain with sharp variations of the function appear in a neighborhood of the boundary. Their width is comparable to the parameter ε. The solution varies sufficiently smoothly outside of these parts. The partial derivatives are well approximated by the difference derivatives on regions of smooth variation of the solution. On these regions, the use of grids with an arbitrary distribution of grid nodes is admitted. As a consequence, the differential equation is well approximated by the finite difference equation on the regions of smooth variation of the solution.

However, on the parts with sharp variation of the solution, that is, in the neighborhood of the boundary layer, the following situation is realized. The parameter ε can take arbitrarily small values. Therefore, for the chosen grid with very condensed nodes (or for the chosen step size in the case of uniform grids) a value of the parameter ε can be found that is comparable to the distance between neighboring nodes in the boundary layer. In this case, the exact solution of the problem varies considerably between the nodes mentioned. With this argument we cannot expect that the finite difference equation and the grid solution approximate well the differential equation and the exact solution, respectively, at these nodes.

Thus, here, the accuracy of the approximate solution is lost for small values of the parameter ε.

We recall that the interval with sharp variation of the solution, where we refine the grid, decreases as ε vanishes. We may try to use grids with a small step size in a neighborhood of the boundary layer. The following question arises: is it possible to redistribute a given number of nodes N so that the nodes in the boundary layer are sufficiently densely distributed, while outside the boundary layer the nodes are not too sparse, and, as a result, the finite difference equation and the grid solution on those grids will approach well the differential equation and the exact solution, respectively? Note that such grids with a special distribution of nodes satisfying the above-mentioned conditions and providing ε-uniform convergence of the approximate solution for increasing N do exist. A rigorous justification of this statement is given below.

If the reader is not interested in the detailed justification, he can pass on to the description of the special piecewise uniform grid $\bar{D}^{sp}_{h(2.59)}$ at the end of this section.

We consider the Dirichlet problem for the ordinary differential equation (1.13a) on the unit segment D, $\bar{D} = [0, 1]$:

$$L_{(2.37)}u(x) \equiv \varepsilon^2 \frac{d^2}{dx^2} u(x) - c(x)u(x) = f(x) \qquad x \in D \tag{2.37a}$$

$$u(x) = \varphi(x) \qquad x \in \Gamma \tag{2.37b}$$

where $L_{(2.37)} = L_{(1.13)}$, the functions $c(x)$, $f(x)$, $x \in \bar{D}$ are assumed to be sufficiently smooth, moreover, $c(x) \geq c_0 > 0$, $x \in \bar{D}$; the function $f(x)$ corresponds to the density of the substance sources in the diffusion problem, and the function $c(x)$ corresponds to the substance dissociation rate.

For this problem, we give a finite difference scheme on a grid with an arbitrary distribution of grid nodes for a fixed number of nodes $N + 1$. Then we analyze the behavior of the error in the grid solution corresponding to the distribution of the grid nodes. Further, we show that the nodes can be distributed in such a way that the errors in the grid solution tend to zero ε-uniformly when N tends to infinity.

First we estimate the solution of problem (2.37). Intuitively, it is clear that for $f(x) \equiv 0$, $x \in \bar{D}$ the following inequality is fulfilled

$$\min_{\Gamma} \varphi(\xi) \leq u(x) \leq \max_{\Gamma} \varphi(\xi) \qquad x \in \bar{D} \tag{2.38}$$

that is, due to the lack of sources the substance concentration inside the sample is no smaller than the minimal concentration and no greater than the maximal concentration on the boundary Γ of the set D (or, equivalently, on the sample surface). It is also clear that for $\varphi(x) \equiv 0$, $x \in \Gamma$ the solution of problem (2.37) satisfies the inequality

$$\min_{\bar{D}} [-c^{-1}(\xi)f(\xi)] \le u(x) \le \max_{\bar{D}} [-c^{-1}(\xi)f(\xi)] \qquad x \in \bar{D} \quad (2.39)$$

This means that, indeed, when the concentration on the sample surface is equal to zero and there is no diffusion (i.e., for $\varepsilon = 0$), the concentration $u_0(x)$ inside the sample is defined by the relations

$$u_0(x) = -c^{-1}(x)f(x) \qquad x \in D$$

$$\min_{\bar{D}} [-c^{-1}(\xi)f(\xi)] \le u_0(x) \le \max_{\bar{D}} [-c^{-1}(\xi)f(\xi)] \qquad x \in D$$

In this case, in the presence of diffusion, relation (2.39) is valid. Similar arguments lead to the estimate

$$\begin{aligned} &\min[\min_{\bar{D}} [-c^{-1}(\xi)f(\xi)], \min_{\Gamma} \varphi(\xi)] \\ &\quad \le u(x) \le \max[\max_{\bar{D}} [-c^{-1}(\xi)f(\xi)], \max_{\Gamma} \varphi(\xi)] \end{aligned}$$

for $x \in \bar{D}$. Consequently, for the solution of problem (2.37) the following estimate is true:

$$\max_{\bar{D}} |u(x)| \le M_{(2.40)} \qquad (2.40)$$

where

$$M_{(2.40)} = \max_{\Gamma} |\varphi(\xi)| + (c_0)^{-1} \max_{\bar{D}} |f(\xi)|$$

That is, the solution of the problem is bounded by the quantity $M_{(2.40)}$ independent of the parameter ε. Note that the smaller the values of $\max_{\Gamma}|\varphi(\xi)|$ and $\max_{\bar{D}}|f(\xi)|$, the smaller is the value of the quantity $M_{(2.40)}$; the solution $u(x)$ is equal to zero if $\varphi(x) \equiv 0$, $x \in \Gamma$ and $f(x) \equiv 0$, $x \in \bar{D}$.

Here and below, we denote by M (and by m) sufficiently large (and, respectively, small) positive constants independent of the small parameter ε. Generally speaking, such constants appearing in different formulas are different.

As is known from the theory of differential equations [7, 8], the solution of the boundary value problem (2.37) can be represented as a sum of two functions

$$u(x) = U(x) + V(x) \qquad x \in \bar{D} \tag{2.41}$$

Here the function $U(x)$ (called the regular, or smooth, part of the solution) and its derivatives are bounded by a constant independent of the value of the parameter ε (alternatively, they are bounded ε-uniformly)

$$\left| \frac{d^k}{dx^k} U(x) \right| \leq M \qquad x \in \bar{D} \qquad 0 \leq k \leq 4 \tag{2.42}$$

The function $V(x)$ (the singular part of the solution) is bounded ε-uniformly; however, its derivatives increase unboundedly in the neighborhood of the ends of the segment $\bar{D}$ when the parameter ε tends to zero.

When the point x moves away from the ends of the segment $\bar{D}$, the function $V(x)$ itself and its derivatives decrease rapidly for small values of the parameter. For the function $V(x)$ and its derivatives the following estimate is fulfilled:

$$\left| \frac{d^k}{dx^k} V(x) \right| \leq M\varepsilon^{-k} \exp(-m_0 \varepsilon^{-1} r(x)) \qquad x \in \bar{D} \qquad 0 \leq k \leq 4 \tag{2.43}$$

where $r(x)$ is the distance from the point x to the ends of the segment $\bar{D}$, $m_0 < m_{(2.43)}$, $m_{(2.43)} \equiv (c_0)^{1/2}$.

For example, in the case of problem (1.16) we have

$$U(x) = 1 \qquad V(x) = -[1 + \exp(-\varepsilon^{-1})]^{-1}[\exp(-\varepsilon^{-1}x) + \exp(-\varepsilon^{-1}(1-x))]$$
$$x \in \bar{D}$$

For problem (2.37), we write the classical difference scheme on a grid with an arbitrary distribution of nodes. On the set $\bar{D}$ we introduce the grid

$$\bar{D}_h = \bar{\omega}_1 \tag{2.44}$$

where $\bar{\omega}_1 = \{x^i : x^0 = 0, x^i < x^{i+1}, x^N = 1, i = 0, 1, \ldots, N\}$ is a grid with an arbitrary distribution of nodes. Set $h = \max_i h^i$, where $h^i = x^{i+1} - x^i$, x^i, $x^{i+1} \in \bar{\omega}_1$. Let h be the maximum step size of the grid satisfying the condition

$$h \leq MN^{-1} \tag{2.45}$$

On the set $D_h = D \cap \bar{D}_h$ (interior nodes of the grid) the second-order derivative $(d^2/dx^2)u(x)$ is approximated by the second difference derivative $\delta_{\bar{x}\hat{x}}u(x)$

$$\delta_{\bar{x}\hat{x}}u(x) = 2(h^{i-1} + h^i)^{-1}[\delta_x u(x) - \delta_{\bar{x}}u(x)] \qquad x = x^i \in D_h$$

Here $\delta_x u(x)$ and $\delta_{\bar{x}}u(x)$ are the first difference derivatives, forward and backward, respectively

$$\delta_x u(x) = (h^i)^{-1}[u(x^{i+1}) - u(x^i)]$$

$$\delta_{\bar{x}}u(x) = (h^{i-1})^{-1}[u(x^i) - u(x^{i-1})] \qquad x = x^i,\ x^{i-1},\ x^i,\ x^{i+1} \in \bar{D}_h$$

Note that the quantity

$$\eta(x) = \eta(x; u(\cdot\,)) \equiv \delta_{\bar{x}\hat{x}}u(x) - \frac{d^2}{dx^2}u(x) \qquad x \in D_h$$

which is the approximation error for the second-order derivative by the second difference, depends on the value of the parameter ε and on the node distribution in the grid $\bar{D}_h$.

Now we discuss the behavior of $\eta(x)$. For the function $\eta(x)$ we have the estimate

$$|\eta(x)| \le 3^{-1} \max_{x^{i-1} \le \xi \le x^{i+1}} \left| \frac{d^3}{dx^3} u(\xi) \right| (x^{i+1} - x^{i-1}) \qquad x = x^i \in D_h$$

On the uniform grid $\bar{D}_h$ we obtain

$$|\eta(x)| \le 12^{-1} \max_{x^{i-1} \le \xi \le x^{i+1}} \left| \frac{d^4}{dx^4} u(\xi) \right| h^2 \qquad x = x^i \in D_h$$

where $h = N^{-1}$ is the step size of the grid $\bar{D}_h$.

When the parameter ε is not too small, for example, $\varepsilon \ge m$, it follows from estimates (2.42) and (2.43) and the inequality

$$\max_{x^{i-1} \le \xi \le x^{i+1}} \left| \frac{d^3}{dx^3} u(\xi) \right| \le \max_{x^{i-1} \le \xi \le x^{i+1}} \left| \frac{d^3}{dx^3} U(\xi) \right| + \max_{x^{i-1} \le \xi \le x^{i+1}} \left| \frac{d^3}{dx^3} V(\xi) \right| \qquad x = x^i \in \bar{D}_h$$

that the quantity $\eta(x)$ is small for a sufficiently large N

$$\max_{D_h} |\eta(x)| \leq MN^{-1}$$

However, the function $\eta(x)$ may take large values as ε decreases. In particular, on the uniform grid $\bar{D}_h$ [e.g., when $\varepsilon = \varepsilon(N) = h$] for problem (1.16) we have

$$|\eta(x)| = |\eta(x; u_{(1.16)}(\cdot))| \geq mN^2 \qquad x = x^1 = h$$

On the grid $\bar{D}_h$ we consider the finite difference scheme

$$\Lambda_{(2.46)} z(x) \equiv \varepsilon^2 \delta_{\bar{x}\hat{x}} z(x) - c(x)z(x) = f(x) \qquad x \in D_h \tag{2.46a}$$

$$z(x) = \varphi(x) \qquad x \in \Gamma_h \tag{2.46b}$$

We are interested in the behavior of the function

$$w(x) = z(x) - u(x) \qquad x \in \bar{D}_h$$

which is the error in the approximate solution [or, in other words, the error in the solution of the finite difference scheme (2.46), (2.44)]. We would like to construct a difference scheme the solution of which converges to the solution of the boundary value problem ε-uniformly.

In the case of the boundary value problem (2.37) the solution of the finite difference scheme (2.46), (2.44) converges ε-uniformly for $N \to \infty$, if the estimate

$$\max_{\bar{D}_h} |z(x) - u(x)| \leq \lambda(N) \tag{2.47}$$

is valid for the difference of the functions $z(x) - u(x) = w(x)$, where the function $\lambda(N)$ does not depend on the parameter ε and tends to zero for $N \to \infty$.

Now we construct an ε-uniformly convergent finite difference scheme.

The arguments that lead to estimate (2.40) for the solution of the boundary value problem (2.37) make it possible to find a similar estimate for the solution of grid problem (2.46), (2.44)

$$\max_{\bar{D}_h} |z(x)| \leq M_{(2.48)} \tag{2.48}$$

where

$$M_{(2.48)} = \max_{\Gamma_h} |\varphi(\xi)| + (c_0)^{-1} \max_{\bar{D}_h} |f(\xi)|$$

This estimate means that the solution of grid problem (2.46), (2.44), as well as the solution of the differential problem (2.37), is bounded ε-uniformly.

One can easily see that

$$\Lambda_{(2.46)}u(x) - L_{(2.37)}u(x) = \varepsilon^2\eta(x) \qquad x \in D_h \tag{2.49}$$

Alternatively, we have

$$\Lambda_{(2.46)}u(x) - L_{(2.37)}u(x) = -\Lambda_{(2.46)}w(x) = \Lambda_{(2.46)}u(x) - f(x) \equiv -\beta(x)$$
$$x \in D_h \tag{2.50}$$

The quantity $\beta(x)$ is called the approximation error of Eq. (2.46a) for the solution $u(x)$ of Eq. (2.37a). From Eqs. (2.49) and (2.50) it follows that

$$\beta(x) = -\varepsilon^2\eta(x) \qquad x \in D_h$$

Thus, the function $w(x)$ is the solution for the following problem:

$$\begin{aligned} \Lambda_{(2.46)}w(x) &= -\varepsilon^2\eta(x) \qquad x \in D_h \\ w(x) &= 0 \qquad x \in \Gamma_h \end{aligned} \tag{2.51}$$

For the function $w(x)$, $x \in \bar{D}_h$, which is the solution of grid problem (2.51), (2.44), the following estimate is valid

$$\max_{\bar{D}_h} |w(x)| \leq M_{(2.52)} \tag{2.52}$$

where

$$M_{(2.52)} = (c_0)^{-1}\varepsilon^2 \max_{D_h} |\eta(\xi)|$$

Thus, in order that the solution of the finite difference scheme (2.46), (2.44) converges ε-uniformly, it suffices that the function $\eta(x)$ satisfies the inequality

$$\varepsilon^2 \max_{D_h} |\eta(x)| \leq M\lambda(N) \tag{2.53}$$

From the results in Section I we know that even when the coefficient $c(x)$ and the right-hand side $f(x)$ are constant, the solution of the finite difference scheme on a uniform grid [namely, scheme (1.17), (1.18)] does not converge ε-uniformly to the solution of the boundary value problem (1.16). Therefore, we will try to find a distribution of the nodes of the grid $\bar{D}_{h(2.44)}$ so that inequality (2.53) is valid for the function $\eta(x)$.

For the function $\eta(x) = \eta(x; u_{(2.37)}(\cdot))$ the following estimate is fulfilled

$$|\eta(x; u_{(2.37)}(\cdot))| \le |\eta(x; U(\cdot))| + |\eta(x; V(\cdot))| \qquad x \in D_h$$

According to estimate (2.42), for the function $\eta(x; U(\cdot))$ we have

$$|\eta(x; U(\cdot))| \le 3^{-1} \max_{x^{i-1} \le \xi \le x^{i+1}} \left| \frac{d^3}{dx^3} U(\xi) \right| (x^{i+1} - x^{i-1}) \le MN^{-1}$$
$$x = x^i \in D_h$$

where $\bar{D}_h = \bar{D}_{h(2.44)}$. Consequently,

$$\varepsilon^2 \max_{D_h} |\eta(x; U(\cdot))| \le MN^{-1}$$

that is, in the case of the smooth part of the solution (in short, the smooth solution) inequality (2.53), where $\lambda(N) = N^{-1}$, is satisfied for any distribution of the nodes in the grid $\bar{D}_{h(2.44)}$.

According to estimate (2.43), for the function $\eta(x; V(\cdot))$ we have

$$\begin{aligned}|\eta(x; V(\cdot))| &\le M\varepsilon^{-3} \max_{x^{i-1} \le \xi \le x^{i+1}} \exp(-m_0\varepsilon^{-1} r(\xi))(x^{i+1} - x^{i-1}) \\ &= M\varepsilon^{-3}[\exp(-m_0\varepsilon^{-1}x^{i-1}) + \exp(-m_0\varepsilon^{-1}(1 - x^{i+1}))](x^{i+1} - x^{i-1})\end{aligned}$$
$$x = x^i \in D_h$$

It can be easily verified that

$$\max_{x^{i-1} \le \xi \le x^{i+1}} \exp(-m_0\varepsilon^{-1} r(\xi)) = O(1) \tag{2.54a}$$

in a neighborhood of the ends of the segment $\bar{D}$, and the quantity

$$\max_{x^{i-1} \le \xi \le x^{i+1}} \exp(-m_0\varepsilon^{-1} r(\xi))$$

decreases rapidly (exponentially) when the nodes x^{i-1}, x^{i+1} are shifted far from the ends of the segment. Thus, we obtain the inequality

$$\max_{x^{i-1}\leq\xi\leq x^{i+1}} \exp(-m_0\varepsilon^{-1}r(\xi)) \leq M\varepsilon^n \qquad \text{for} \qquad r(x^{i-1}), r(x^{i+1}) \geq m$$
$$x^i \in D_h \tag{2.54b}$$

where the integer n can be chosen sufficiently large, $M = M(n)$, $\bar{D}_h = \bar{D}_{h(2.44)}$.

In the case of the boundary value problem (1.16), we have a lower bound for the function $\eta(x; V(\cdot))$ on the uniform grid $\bar{D}_h$

$$|\eta(x; V(\cdot))| = \left|\frac{d^2}{dx^2}V(x) - \delta_{x\bar{x}}V(x)\right| \geq 24^{-1}\left|\frac{d^4}{dx^4}V(x)\right|N^{-2}$$
$$= 24^{-1}\varepsilon^{-4}N^{-2}\exp(-\varepsilon^{-1}x) \qquad x = x^i \in D_h \qquad \bar{D}_h = \bar{D}_{h(2.44)} \tag{2.55}$$

On piecewise uniform grids, which are uniform in neighborhoods of the ends of the segment $\bar{D}$, we obtain an estimate similar to (2.55)

$$|\eta(x; V(\cdot))| \geq m\varepsilon^{-4}h_0^2\exp(-\varepsilon^{-1}x) \qquad x = x^i \in D_h$$

where h_0 is the step size in the neighborhoods of the segment ends.

Suppose that the node x^i of the grid $\bar{D}_h$ satisfies the relation

$$\exp(-\varepsilon^{-1}x^i) = O(1) \qquad x^i \in D_h$$

Then, for the inequality

$$\varepsilon^2|\eta(x^i; V(\cdot))| \leq M\lambda(N) \qquad x^i \in D_h$$

to be valid for some function $\lambda(N)$, the following condition should be fulfilled:

$$\varepsilon^{-2}h_0^2 \leq M\lambda(N)$$

that is, the step size between the nodes should satisfy the condition

$$h_0 = \varepsilon o(N^{-1}) \tag{2.56}$$

that is, the step size h_0 should be much smaller than the parameter ε.

If at the point x^i near the segment ends the step size satisfies a condition similar to (2.56)

$$x^{i+1} - x^i, x^i - x^{i-1} = \varepsilon o(N^{-1}) \tag{2.57}$$

then, at the node x^i, in virtue of relation (2.54a), the following inequality holds for a certain function $\lambda_1(N)$

$$\varepsilon^2 |\eta(x^i; V(\cdot))| \le M\lambda_1(N)$$

According to inequality (2.54b), in the nodes $x^i \in D_h$ for which $r(x^{i-1})$, $r(x^{i+1}) \ge m$ we have the inequality

$$\varepsilon^2 |\eta(x^i; V(\cdot))| \le M\lambda(N)$$

where $\lambda(N) = N^{-1}$. In this case no restrictions connected with the value of ε are imposed on the values $h^i = x^{i+1} - x^i$, $h^{i-1} = x^i - x^{i-1}$. Thus, the grid step size should satisfy the condition (or restriction) (2.57) in a small neighborhood of the ends of the segment $\bar{D}$; outside of this m-neighborhood of the ends of $\bar{D}$ no restrictions are imposed on the distribution of the nodes in the grid $\bar{D}_{h(2.44)}$.

It would be interesting to find a grid that satisfies the above mentioned conditions in a class of the simplest grid, namely, in the class of piecewise uniform grids.

We now describe such a grid. Suppose that

$$\bar{\omega}_*(\sigma, N_*, N, d_1, d_2) \tag{2.58a}$$

is a piecewise uniform grid on the segment $[d_1, d_2]$ condensing in neighborhoods of both the ends of this segment. The grid $\bar{\omega}_*$ is uniform on each of the subintervals $[d_1, d_1 + \sigma]$, $[d_1 + \sigma, d_2 - \sigma]$, $[d_2 - \sigma, d_2]$ with the grid-step size $h_{(1)}$ on the segments $[d_1, d_1 + \sigma]$ and $[d_2 - \sigma, d_2]$ with $N_* + 1$ nodes on each of them and with the grid-step size $h_{(2)}$ on the segment $[d_1 + \sigma, d_2 - \sigma]$; $N + 1$ is the overall number of the $\bar{\omega}_*$-grid nodes and $\sigma \le 4^{-1}(d_2 - d_1)$. The value σ is determined by the relation

$$\sigma = \sigma(\varepsilon, N, d_1, d_2, m) = \min[4^{-1}(d_2 - d_1), m^{-1}\varepsilon \ln N] \tag{2.58b}$$

where $m = m_{(2.58)} < m_{(2.43)}$. Suppose that

$$\bar{D}_h^{sp} = \bar{\omega}_*(\sigma_{(2.58)}(\varepsilon, N, 0, 1, m), N_*, N, 0, 1) \tag{2.59}$$

where $N_* = 4^{-1}N$. This is the required special grid $\bar{D}_h^{sp}$.

In the case of the grid $\bar{D}_h^{sp}$ for the function $\eta(x)$ we have the estimate

$$\varepsilon^2 \max_{D_h} |\eta(x; u_{(2.37)}(\cdot))| \leq MN^{-1}\ln N \qquad \bar{D}_h = \bar{D}_h^{sp} \tag{2.60}$$

The proof of this estimate is given in [4].

By virtue of estimate (2.52), we obtain

$$\max_{\bar{D}_h} |u(x) - z(x)| \leq MN^{-1}\ln N \tag{2.61}$$

where $\bar{D}_h = \bar{D}_h^{sp}$ and $z(x)$ is the solution of scheme (2.46), (2.59). That is, the solution of the finite difference scheme (2.46), (2.59) for $N \to \infty$ converges ε-uniformly to the solution of the boundary value problem (2.37). Moreover, the function $\lambda(N)$ in the right-hand side of estimate (2.47) is defined by the relation $\lambda(N) = M_{(2.61)}N^{-1}\ln N$.

A rigorous analysis of convergence for the finite difference scheme (2.46), (2.59) leads to the estimate

$$\max_{\bar{D}_h^{sp}} |u(x) - z(x)| \leq MN^{-2}\ln^2 N \tag{2.62}$$

if the following condition holds

$$m_{(2.58)} < 2^{-1} m_{(2.43)} \tag{2.63}$$

Thus, we have constructed the grid $\bar{D}_h^{sp}$ on which the solution of the finite difference scheme (2.46), (2.59) converges ε-uniformly. For the approximate solution the estimate (2.61) for $m_{(2.58)} < m_{(2.43)}$ and the estimate (2.62) for $m_{(2.58)} < 2^{-1}m_{(2.43)}$ are valid.

As we know, for the solution of the grid equations (2.46) on the uniform grid $\bar{D}_h$ the following estimate is true

$$\max_{\bar{D}_h} |u(x) - z(x)| \leq Q(\varepsilon)N^{-2} \tag{2.64}$$

This inequality means that, for a fixed value of the parameter ε, the classical finite difference scheme on the uniform grid converges with increasing N with practically the same order of convergence (with respect to N) as the special finite difference scheme (2.46), (2.59), (2.63). Note that in the case of the classical scheme on a uniform grid the rate of convergence [more precisely, the constant $Q(\varepsilon)$ in estimate (2.64) called the convergence constant] depends essentially on the value of the parameter ε (see the results in Section II.B). However, in the case of the special finite difference scheme, the rate of convergence [namely, the convergence constant M in inequality (2.62)] is independent of the value of the parameter ε.

Note that in [9] a special grid is constructed where the distance between the neighboring nodes depends on the parameter ε and on the number of nodes N, and varies gradually. The approximate solution on that grid satisfies the estimate

$$\max_{\bar{D}_h} |u(x) - z(x)| \leq MN^{-2} \tag{2.65}$$

The right-hand sides of estimates (2.62) and (2.65) differ only in the multiplier $\ln^2 N$ and constant coefficients M (we remember that the values of M in these estimates are different). But piecewise uniform grids are considerably simpler to use than grids with a gradually varying step size. It is obvious that for large N the value $\ln^2 N$ is small in comparison with the value N^ν, where $\nu > 0$ is an arbitrarily small quantity. Hence, it follows that the orders of convergence in estimates (2.62) and (2.65) are close.

D. Special Finite Difference Schemes for Problems (2.12), (2.13) and (2.14), (2.15): Numerical Experiments with the Special Difference Scheme

First, we give a special finite difference scheme for the one-dimensional boundary value problem (2.14), (2.15). A special piecewise uniform grid for problem (2.14), (2.15) is constructed, which is rectangular and condensing with respect to the variable x in a neighborhood of the boundary layers, that is similar to the grid for problem (2.37). The grid is uniform with respect to the time variable. In the case of the two-dimensional boundary value problem (2.12), (2.13) the special grids are rectangular, piecewise uniform with respect to the variables x_1 and x_2 and uniform with respect to the time variable. The distribution of nodes with respect to x for a one-dimensional problem, and with respect to both x_1 and x_2 for a two-dimensional problem, is similar to that for an ordinary differential equation. We describe special grids for problems (2.12), (2.13) and (2.14), (2.15), and then, we give theoretical estimates for the approximate solutions.

On the set $\bar{G}_{(2.15)}$, we introduce the rectangular grid

$$\bar{G}_h = \bar{\omega}_1 \times \bar{\omega}_0 \tag{2.66}$$

where $\bar{\omega}_1$ is a grid on the segment $\bar{D}_{(2.15)}$, generally speaking, a nonuniform grid, $\bar{\omega}_0$ us a uniform grid on the segment $[0, T]$. We denote the number of nodes in the grids $\bar{\omega}_1$ and $\bar{\omega}_0$ by $N+1$ and N_0+1, respectively. We set $h = \max_i h^i$, $h^i = x^{i+1} - x^i$, x^i, $x^{i+1} \in \bar{\omega}_1$, $h \leq MN^{-1}$,

h_0 is the step size of the grid $\bar{\omega}_0$, $h_0 = TN_0^{-1}$. On the grid $\bar{G}_h$ we consider the difference equation corresponding to differential equation (2.14a)

$$\Lambda_{(2.67)} z(x,t) \equiv \{\varepsilon^2 a(x,t)\delta_{\bar{x}\hat{x}} - p(x,t)\delta_{\bar{t}} - c(x,t)\} z(x,t) = f(x,t) \tag{2.67a}$$
$$(x,t) \in G_h$$

The function $z(x,t)$ takes the given values on the boundary S_h

$$z(x,t) = \varphi(x,t) \qquad (x,t) \in S_h \tag{2.67b}$$

Here $G_h = G \cap \bar{G}_h$, $S_h = S \cap \bar{G}_h$, $\delta_{\bar{x}\hat{x}} z(x,t)$, $\delta_{\bar{t}} z(x,t)$ are the second and first (backward) difference derivatives.

We want to know the behavior of the function

$$w(x,t) = z(x,t) - u(x,t) \qquad (x,t) \in \bar{G}_h$$

with N, $N_0 \to \infty$ for different values of the parameter ε; the function $w(x,t)$ is the error in the approximate solution.

According to the definition, the solution of finite difference scheme (2.67), (2.66) with N, $N_0 \to \infty$ converges to the solution of the boundary value problem (2.14) ε-uniformly if for the difference of the functions $z(x,t) - u(x,t) = w(x,t)$ the estimate

$$\max_{\bar{G}_h} |z(x,t) - u(x,t)| \le \lambda(N, N_0)$$

is satisfied, where the function $\lambda(N, N_0)$ does not depend on the parameter ε and tends to zero for N, $N_0 \to \infty$.

The function $w(x,t)$ is the solution for the grid problem

$$\begin{aligned} \Lambda_{(2.67)} w(x,t) &= \beta(x,t) \qquad (x,t) \in G_h \\ w(x,t) &= 0 \qquad (x,t) \in S_h \end{aligned} \tag{2.68}$$

where $\beta(x,t)$ is the approximation error of Eq. (2.67a) for the solution $u(x,t)$ of Eq. (2.14a)

$$\begin{aligned} \beta(x,t) &= L_{(2.14)} u(x,t) - \Lambda_{(2.67)} u(x,t) \\ &= \varepsilon^2 a(x,t) \left[\frac{\partial^2}{\partial x^2} u(x,t) - \delta_{\bar{x}\hat{x}} u(x,t) \right] + p(x,t) \left[\delta_{\bar{t}} u(x,t) - \frac{\partial}{\partial t} u(x,t) \right] \\ &\equiv \beta_1(x,t) + \beta_2(x,t) \end{aligned}$$

It is known that, for the solution of the grid problem (2.67), (2.66), the estimate

$$\max_{\bar{G}_h} |z(x, t)| \leq M_{(2.69)} \tag{2.69}$$

is valid, where

$$M_{(2.69)} = \max_{S_h} |\varphi(\xi, \tau)| + (p_0)^{-1} T \max_{G_h} |f(\xi, \tau)|$$

This estimate can be derived similarly to estimate (2.48). Hence, for the function $w(x, t)$ which is the solution of Problem (2.68), we arrive at the estimate

$$\max_{\bar{G}_h} |w(x, t)| \leq M \max_{G_h} |\beta(\xi, \tau)| \tag{2.70a}$$

It is known from the theory of differential equations that the solution of problem (2.14) can be represented as a sum of two functions

$$u(x, t) = U(x, t) + V(x, t) \qquad (x, t) \in \bar{G}$$

For the functions $U(x, t)$ and $V(x, t)$ (i.e., for the regular and singular parts of the solution) and their derivatives the following estimates are fulfilled

$$\left| \frac{\partial^{k+k_0}}{\partial x^k \, \partial t^{k_0}} U(x, t) \right| \leq M \tag{2.70b}$$

$$\left| \frac{\partial^{k+k_0}}{\partial x^k \, \partial t^{k_0}} V(x, t) \right| \leq M\varepsilon^{-k} \exp[-m\varepsilon^{-1} r(x)] \tag{2.70c}$$

$$(x, t) \in \bar{G} \qquad 0 \leq k, k_0 \qquad k_0 \leq 2 \qquad k \leq 4$$

where $r(x)$ is the distance from the point x to the boundary Γ of the set D, $\Gamma = \bar{D} \backslash D$ and m is a sufficiently small arbitrary number.

We see that the partial derivatives with respect to time are bounded ε-uniformly for both of the functions $U(x, t)$ and $V(x, t)$, while the partial derivatives with respect to the space variable x are bounded ε-uniformly only for the function $U(x, t)$.

Hence, we obtain the following inequality for the second-order derivative of the problem solution with respect to time

$$\max_{\bar{G}_h} \left| \frac{\partial^2}{\partial t^2} u(x, t) \right| \leq \max_{\bar{G}_h} \left| \frac{\partial^2}{\partial t^2} U(x, t) \right| + \max_{\bar{G}_h} \left| \frac{\partial^2}{\partial t^2} V(x, t) \right| \leq M$$

and, consequently, for the function $\beta_2(x,t)$ we have

$$\max_{G_h} |\beta_2(x,t)| \le M \max_{G_h} \left| \frac{\partial^2}{\partial t^2} u(\xi,\tau) \right| N_0^{-1} \le M N_0^{-1}$$

Taking into account the latter inequality and estimates (2.70), we arrive at the following conclusion. For the construction of an ε-uniformly convergent finite difference scheme it is enough to construct the grid $\bar{D}_h$ so that the following inequality is satisfied

$$\max_{G_h} |\beta_1(x,t)| \le \lambda(N, N_0) \tag{2.71}$$

We also see from (2.70) that the estimates of the derivatives with respect to x of the functions $U(x,t)$ and $V(x,t)$ are similar to estimates (2.42) and (2.43) of the components in the representation (2.41) of the solution of problem (2.37). To construct the special grid $\bar{D}_h^{sp}$ and, consequently, the grid $\bar{G}_h^{sp}$ where the inequality (2.71) is fulfilled, we take advantage of the results in Section II.C.

We construct the special grid

$$\bar{G}_h^{sp} = \bar{D}_h^{sp} \times \bar{\omega}_0 = \bar{\omega}_1^* \times \bar{\omega}_0 \tag{2.72a}$$

as follows. We define the grid $\bar{D}_h^{sp}$ by the relation

$$\begin{aligned} \bar{D}_h^{sp} &= \bar{D}_h^{sp}(\varepsilon, N, d_0, d_1, m) \\ &= \bar{\omega}_{*(2.58)}[\sigma_{(2.58)}(\varepsilon, N, d_0, d_1, m), N_* = 4^{-1}N, N, d_0, d_1] \end{aligned} \tag{2.72b}$$

where $m = m_{(2.72)}$ is an arbitrary number; $\bar{\omega}_0 = \bar{\omega}_{0(2.66)}$.

On the grid $\bar{G}_h^{sp}$ the function $\beta(\xi,\tau)$ satisfies the inequality

$$|\beta(\xi,\tau)| \le \begin{cases} M[N^{-2}\ln^2 N + N_0^{-1}] & (\xi,\tau) \in G_h^{sp} \quad \xi \ne d_0 + \sigma, d_1 - \sigma \\ M[N^{-1}\ln N + N_0^{-1}] & (\xi,\tau) \in G_h^{sp} \quad \xi = d_0 + \sigma, d_1 - \sigma \end{cases}$$

Hence, by virtue of estimate (2.70a), we obtain the ε-uniform convergence of the finite difference scheme (2.67), (2.72). Thus, we arrive at the error estimate

$$\max_{\bar{G}_h^{sp}} |u(x,t) - z(x,t)| \le M[N^{-2}\ln^2 N + N_0^{-1}]$$

In the case of the boundary value problem (2.12), (2.13) on the set

$\bar{G}_{(2.13)}$, we construct the rectangular grid

$$\bar{G}_h = \bar{D}_h \times \bar{\omega}_0 = \bar{\omega}_1 \times \bar{\omega}_2 \times \bar{\omega}_0 \tag{2.73}$$

where $\bar{\omega}_s$ is a grid on the segment $[d_{0s}, d_{1s}]$ on the axis x_s, $s = 1, 2$, generally speaking, a nonuniform grid; $\bar{\omega}_0$ is a uniform grid; $N_s + 1$ and $N_0 + 1$ are the number of nodes in the grids $\bar{\omega}_s$ and $\bar{\omega}_0$, respectively; $h_0 = TN_0^{-1}$ is the step size of the grid $\bar{\omega}_0$; $h_s = \max_i h_s^i$, $h_s^i = x_s^{i+1} - x_s^i$, x_s^i, $x_s^{i+1} \in \bar{\omega}_s$, $h = \max_s h_s$, $s = 1, 2$; $N = \min_s N_s$, $s = 1, 2$, $h \le MN^{-1}$. For the approximation of problem (2.12), on the grid $\bar{G}_h$, we use the finite difference scheme

$$\begin{aligned} \Lambda_{(2.74)} z(x, t) &= f(x, t) \qquad (x, t) \in G_h \\ z(x, t) &= \varphi(x, t) \qquad (x, t) \in S_h \end{aligned} \tag{2.74}$$

Here, $G_h = G \cap \bar{G}_h$, $S_h = S \cap \bar{G}_h$

$$\Lambda_{(2.74)} z(x, t) \equiv \left\{ \varepsilon^2 \sum_{s=1,2} a_s(x, t) \delta_{\overline{xs}\widehat{xs}} - p(x, t)\delta_{\bar{t}} - c(x, t) \right\} z(x, t)$$

The analysis of convergence for the finite difference scheme (2.74), (2.73) is similar to that for the finite difference scheme (2.67), (2.66). We have the estimate

$$\max_{\bar{G}_h} |u(x, t) - z(x, t)| \le Q(\varepsilon)[N^{-1} + N_0^{-1}]$$

for the solution of the finite difference scheme (2.74), (2.73). Even if the grid

$$\bar{G}_h \tag{2.75}$$

is uniform with respect to all the variables, we have the estimate

$$\max_{\bar{G}_h} |u(x, t) - z(x, t)| \le Q(\varepsilon)[N^{-2} + N_0^{-1}]$$

To construct the special grid so that the solution of the finite difference scheme converges ε-uniformly, we use grids that are piecewise uniform with respect to the variables x_1 and x_2. Remember that the estimates for the derivatives (with respect to the space and time variables) of the regular and singular parts of the solution for problem (2.12) are similar to the estimates of the derivatives for problem (2.14).

Now we describe a special grid for the finite difference scheme (2.74).

With this aim in view we define

$$\bar{G}_h^{sp} = \bar{D}_h^{sp} \times \bar{\omega}_0 = \bar{\omega}_1^* \times \bar{\omega}_2^* \times \bar{\omega}_0 \tag{2.76a}$$

where

$$\bar{\omega}_s^* = \bar{\omega}_{*(2.58)}[\sigma_{(2.58)}(\varepsilon, N_s, d_{0s}, d_{1s}, m), N_* = 4^{-1}N_s, N_s, d_{0s}, d_{1s}] \\ s = 1, 2 \tag{2.76b}$$

m is an arbitrary number.

Suppose that the solution of problem (2.14), (2.13) is sufficiently smooth. In this case, the finite difference scheme (2.74), (2.76) converges ε-uniformly, and we obtain the estimate

$$\max_{\bar{G}_h^{sp}} |u(x, t) - z(x, t)| \leq M[N^{-2}\ln^2 N + N_0^{-1}]$$

We examine the efficiency of special finite difference schemes for the model boundary value problem (2.16), (2.18). On the set $\bar{G}_{(2.17)}$ we construct the special grid

$$\bar{G}_h^{sp} = \bar{D}_h^{sp} \times \bar{\omega}_0 \tag{2.77a}$$

where

$$\bar{D}_h^{sp} = \bar{D}_{h(2.72)}^{sp}(\varepsilon, N, d_0 = 0, d_1 = 1, m) \qquad \bar{\omega}_0 = \bar{\omega}_{0(2.66)} \tag{2.77b}$$

Problem (2.16) is approximated on this grid by the finite difference scheme

$$\begin{aligned} \Lambda_{(2.78)} z(x, t) = f(x, t) \qquad (x, t) \in G_h^{sp} \\ z(x, t) = \varphi(x, t) \qquad (x, t) \in S_h^{sp} \end{aligned} \tag{2.78}$$

Here $f(x, t) = f_{(2.18)}(x, t)$, $\varphi(x, t) = \varphi_{(2.18)}(x, t)$,

$$\Lambda_{(2.78)} z(x, t) \equiv \{\varepsilon^2 \delta_{\bar{x}\hat{x}} - \delta_{\bar{t}}\} z(x, t)$$

We know that the solution of problem (2.16) can be represented in the form (2.24). We used this representation when analyzing the classical finite difference scheme. Here we also use representation (2.24) for the analysis of the special finite difference scheme.

When one studies the errors in the approximate solution obtained by scheme (2.78), (2.77), it is convenient to consider the finite difference

TABLE XIV
Table of Errors $E(\varepsilon, N)$ for the Special Scheme (2.79), (2.77) in the Case of Problem (2.20)

$\varepsilon \backslash N$	4	16	64	256	1024
1	2.360e − 1	9.698e − 2	2.795e − 2	7.288e − 3	1.842e − 3
2^{-2}	2.423e − 1	6.182e − 2	1.546e − 2	3.864e − 3	9.659e − 4
2^{-4}	2.526e − 1	6.266e − 2	1.564e − 2	3.910e − 3	9.773e − 4
2^{-6}	2.504e − 1	6.252e − 2	1.563e − 2	3.906e − 3	9.766e − 4
2^{-8}	2.500e − 1	6.250e − 2	1.563e − 2	3.906e − 3	9.766e − 4
2^{-10}	2.500e − 1	6.250e − 2	1.563e − 2	3.906e − 3	9.766e − 4
2^{-12}	2.500e − 1	6.250e − 2	1.563e − 2	3.906e − 3	9.766e − 4
$\bar{E}(N)$	2.526e − 1	9.698e − 2	2.795e − 2	7.288e − 3	1.842e − 3

schemes

$$\begin{aligned} \Lambda_{(2.78)} z(x, t) &= f(x, t) \qquad (x, t) \in G_h^{sp} \\ z(x, t) &= U(x, t) \qquad (x, t) \in S_h^{sp} \end{aligned} \tag{2.79}$$

$$\begin{aligned} \Lambda_{(2.78)} z(x, t) &= 0 \qquad (x, t) \in G_h^{sp} \\ z(x, t) &= W(x, t) \qquad (x, t) \in S_h^{sp} \end{aligned} \tag{2.80}$$

$$\begin{aligned} &\Lambda_{(2.78)} z(x, t) = 0 \qquad (x, t) \in G_h^{sp} \\ &z(x, t) = \begin{cases} 0 & (x, t) \in S_h^{sp} \quad x < 1 \\ -W(1, t) & x = 1 \quad 0 < t \le 1 \end{cases} \qquad (x, t) \in S_h^{sp} \end{aligned} \tag{2.81}$$

Here $U(x, t)$, $W(x, t)$, $v(x, t)$ are the components from the representation

TABLE XV
Table of Errors $E(\varepsilon, N)$ for the Special Scheme (2.80), (2.77) in the Case of Problem (2.25)

$\varepsilon \backslash N$	4	16	64	256	1024
1	1.630e − 2	6.144e − 3	1.780e − 3	4.651e − 4	1.176e − 4
2^{-2}	4.374e − 2	8.624e − 3	1.960e − 3	4.769e − 4	1.184e − 4
2^{-4}	3.976e − 2	2.558e − 2	3.131e − 3	5.507e − 4	1.230e − 4
2^{-6}	4.494e − 3	4.156e − 2	7.214e − 3	1.076e − 3	1.772e − 4
2^{-8}	9.440e − 3	4.156e − 2	7.214e − 3	1.076e − 3	1.772e − 4
2^{-10}	1.207e − 2	4.156e − 2	7.214e − 3	1.076e − 3	1.772e − 4
2^{-12}	1.273e − 2	4.156e − 2	7.214e − 3	1.076e − 3	1.772e − 4
$\bar{E}(N)$	4.374e − 2	4.156e − 2	7.214e − 3	1.076e − 3	1.772e − 4

TABLE XVI
Table of Errors $E_{1024,1024}(\varepsilon, N)$ for the Special Scheme (2.81), (2.77) in the Case of Problem (2.26)

$\varepsilon \backslash N$	4	16	64	256
1	4.609e − 03	1.273e − 03	3.114e − 04	6.275e − 05
2^{-2}	9.645e − 05	3.058e − 05	6.804e − 06	1.330e − 06
2^{-4}	3.683e − 33	9.477e − 33	1.616e − 32	3.497e − 33
2^{-6}	0.000e − 00	0.000e − 00	0.000e − 00	0.000e − 00
$\bar{E}(N)$	4.609e − 03	1.273e − 03	3.114e − 04	6.275e − 05

(2.24) of the solution of problem (2.16). The finite difference schemes (2.79), (2.77); (2.80), (2.77), and (2.81), (2.77) [we denote their solutions by $z_{(2.79)}(x, t)$, $z_{(2.80)}(x, t)$, and $z_{(2.81)}(x, t)$, respectively] approximate the boundary value problems (2.20), (2.25), and (2.26).

In Tables XIV and XV we give the values of $E_{(2.30)}(\varepsilon, N)$, $\bar{E}_{(2.30)}(N)$ for finite difference schemes (2.79), (2.77), and (2.80), (2.77) corresponding to problems (2.20) and (2.25), respectively, which are computed for different values of ε and N with $N_0 = N$. From the tables one can see that, as N increases, the functions $z_{(2.79)}(x, t)$ and $z_{(2.80)}(x, t)$ converge ε-uniformly to the functions $U(x, t)$ and $W(x, t)$, respectively.

To analyze the convergence of the function $z_{(2.81)}(x, t)$ to the function $v(x, t)$, we have computed the values of $E_{1024,1024(2.35)}(\varepsilon, N)$ and $\bar{E}_{1024,1024(2.35)}(N)$. The results are given in Table XVI. Tables XVI and VIII differ only in the value of $E_{1024,1024}(\varepsilon, N)$ for $\varepsilon = 2^{-4}$, $N = 4$ because, for the other ε and N, the special grid is uniform. From Table XVI we can see that the error $E_{1024,1024}(\varepsilon, N)$ converges to zero ε-uniformly.

From the results given in Tables XIV–XVI it follows that the error in the solution of the finite difference scheme (2.78), (2.77) does not exceed the sum of the errors in the solutions of the finite difference schemes (2.79), (2.77); (2.80), (2.77), and (2.81), (2.77). Therefore, the solution of the special scheme (2.78), (2.77) converges ε-uniformly to the solution of the boundary value problem (2.16). Thus, numerical experiments illustrate the efficiency of the constructed finite difference scheme.

In Section III we will see that the new finite difference schemes allow us to find the normalized diffusion fluxes with an ε-uniform accuracy.

III. NUMERICAL SOLUTIONS OF THE DIFFUSION EQUATION WITH PRESCRIBED DIFFUSION FLUXES ON THE BOUNDARY

In Section II.A, the problem of the diffusion of a substance in a layer of some material was considered. The determination of the substance concentration was reduced to the solution of a one-dimensional singularly perturbed diffusion equation (2.8). The substance concentration at the initial instant is known, that is, the unknown function is given by condition (2.9). On the domain boundary, depending on the physical conditions, one of two sharply differing laws for mass exchange between the boundary and the surrounding medium is imposed. In the first case, the substance concentration on the material surface is known for all times, that is, the unknown function takes given values, and this function satisfies condition (2.10). In the second case, the diffusion flux on the surfacc is known for the current time, that is, the space derivative (or gradient of the unknown function) is prescribed, and this derivative satisfies condition (2.11). As we will see later, the approximate solutions obtained with classical finite difference schemes differ sharply in the first and second cases.

In Section II (see Sections II.B, II.D) we considered finite difference schemes in the case when the unknown function takes given values on the boundary. The boundary value problem for the singularly perturbed parabolic equation on a rectangle, that is, a two-dimensional problem, is described by Eqs. (2.12), while the boundary value problem on a segment, that is, a one-dimensional problem, is described by equations (2.14). In Section II.B classical finite difference schemes were analyzed. It was shown that the error in the approximate solution, as a function of the perturbation parameter, is comparable to the required solution for any fine grid. For the above mentioned problems special finite difference schemes were constructed. The error in the approximate solution obtained by the new scheme does not depend on the parameter value and tends to zero as the number of grid nodes increases.

In this section, we consider singularly perturbed diffusion equations when the diffusion flux is given on the domain boundary. We show (see Section III.B) that the error in the approximate solution obtained by a classical finite difference scheme, depending on the parameter value, can be many times greater than the magnitude of the exact solution. For the boundary value problems under study we construct special finite difference schemes (see Sections III.C and III.D), which allow us to find the solution and diffusion flux. The errors in the approximate solution for these schemes and the computed diffusion flux are independent of the parameter value and depend only on the number of nodes in the grid.

A. Mathematical Formulation of the Problems

On the set

$$G = D \times (0, T] \tag{3.1}$$

where D is the segment $[d_0, d_1]$ on the x axis, we consider the singularly perturbed equation of parabolic type [see Eq. (2.14a)]

$$L_{(3.2)}u(x,t) \equiv \left\{ \varepsilon^2 a(x,t) \frac{\partial^2}{\partial x^2} - p(x,t) \frac{\partial}{\partial t} - c(x,t) \right\} u(x,t) = f(x,t) \qquad (x,t) \in G \tag{3.2a}$$

Here $a(x,t) \geq a_0 > 0$, $c(x,t) \geq 0$, $p(x,t) \geq p_0 > 0$, $f(x,t)$ are sufficiently smooth functions on $\bar{G}$; the parameter ε takes arbitrary values in the interval $(0, 1]$. The following condition is given on the lateral boundary $S^1 = \{x: x = d_0, d_1\} \times (0, T]$ of the set G

$$l_{(3.2)}u(x,t) \equiv \varepsilon \frac{\partial}{\partial n} u(x,t) = \psi(x,t) \qquad (x,t) \in S^1 \tag{3.2b}$$

Here $\partial / \partial n$ is the derivative with respect to the outward normal to the set G

$$\frac{\partial}{\partial n} u(x,t) = -\frac{\partial}{\partial x} u(x,t) \qquad \text{for} \qquad x = d_0$$

$$\frac{\partial}{\partial n} u(x,t) = \frac{\partial}{\partial x} u(x,t) \qquad \text{for} \qquad x = d_1$$

and $\psi(x,t)$ is a sufficiently smooth function. Thus, the first space derivative of the unknown function is given on the domain boundary by condition (3.2b). At the initial instant, that is, on the set $S^0 = \bar{D} \times \{t = 0\}$, which is the lower base of the set G, the function $u(x,t)$ is known

$$u(x,t) = \varphi(x) \qquad (x,t) \in S^0 \tag{3.2c}$$

Here $\varphi(x)$ is a sufficiently smooth function.

We want to find the solution of problem (3.2), (3.1) (a Neumann boundary value problem), that is, the function $u(x,t)$, $(x,t) \in \bar{G}$, which satisfies Eq. (3.2a) on the set G and the boundary conditions (3.2b), (3.2c) on the boundary $S = S^0 \cup S^1$.

We discuss the motivation for including the parameter ε in the boundary condition (3.2b). For the Dirichlet problem (2.14), when the parameter tends to zero, boundary layers appear in a neighborhood of the lateral boundary. In this connection, the derivatives with respect to x increase unboundedly [see, e.g., estimates (2.70c)]. However, the product $\varepsilon(\partial/\partial x)u(x,t)$ is bounded uniformly with respect to the parameter ε for all $\varepsilon \in (0, 1]$. Therefore, when studying the Neumann problem (3.2) and considering the derivatives $(\partial/\partial x)u(x,t)$ (e.g., when analyzing the diffusion fluxes), we consider it natural to introduce the function $P(x,t) = \varepsilon(\partial/\partial x)u(x,t)$. Note that, if the parameter ε does not appear in the boundary condition (3.2b), then (3.2b) is practically equivalent to the absence of flux exchange between the surface and the surrounding medium for small values of the parameter. A boundary layer does not appear in practice in this case.

The function

$$P(x,t) = \varepsilon \frac{\partial}{\partial x} u(x,t)$$

in contrast to the function

$$P^D(y,\tau) = D \frac{\partial}{\partial y} C(y,\tau)$$

which is the diffusion flux considered in problems (2.1)–(2.3) and (2.1), (2.2), (2.4), is called the normalized diffusion flux (or, in short, the normalized flux, or simply, the flux). The variable y is the distance from the point x to the material surface, x is a dimensionless space variable, $x = x(y)$, t is a dimensionless time variable, $t = t(\tau)$; the functions $P(x,t)$ and $P^D(y,\tau)$ are related to each other as follows:

$$P^D(y,\tau) = C_* D^{1/2} \vartheta^{-1/2} P(x,t) \qquad x = x(y) \qquad t = t(\tau)$$

For details see, for example, Section II.A.

Problem (3.2) for $a(x,t) = p(x,t) = 1$, $c(x,t) = 0$ and $d_0 = 0$, $d_1 = T = 1$ reduces to problem (2.8), (2.9), (2.11).

It should be noted that, generally speaking, the function $P(x,t)$ is not continuous on the set $\bar{G}$. It can have discontinuities on the set $S_0^1 = \Gamma \times \{t = 0\}$. For example, since the function $u(x,t)$ is known at the initial instant

$$u(x,0) = \varphi(x) \qquad x \in \bar{D}$$

we can find the function

$$\varepsilon \frac{\partial}{\partial n} \varphi(x) \qquad (x,t) \in S_0^1$$

for $t = 0$, that is, the normalized diffusion flux at the initial instant. Generally speaking,

$$\varepsilon \frac{\partial}{\partial n} \varphi(x) \neq \psi(x,t) \qquad (x,t) \in S_0^1$$

that is, the normalized flux found at the initial instant is not equal to the normalized flux prescribed on the boundary S^1 for $t = 0$. Therefore, we consider the function $P(x, t)$ on the set $\bar{G}_0 = \bar{G} \backslash S_0^1$. The function $P(x, t)$ is continuous on $\bar{G}$ if the data of problem (3.2) [more precisely, the functions $f(x, t)$, $\varphi(x)$, $\psi(x, t)$] satisfy a special condition of compatibility on the set S_0^1 (see, e.g., [10]). This compatibility condition guarantees sufficient smoothness of the solution of the problem on $\bar{G}$. For example, in the case of problem (3.2), (3.1), in order that the function $P(x, t)$ should be continuous on $\bar{G}$, the following condition is sufficient

$$\varepsilon \frac{\partial}{\partial n} \varphi(x) = \psi(x, t) \qquad (x, t) \in S_0^1$$

As the parameter ε tends to zero in Eqs. (3.2a), (3.2b), boundary layers appear in a neighborhood of the set S^1. The first space derivative of the solution tends to infinity in a neighborhood of the set S^1, however, the function $P(x, t)$ remains bounded uniformly with respect to the parameter ε.

Besides the one-dimensional problem (3.2), (3.1), we consider a two-dimensional problem on a rectangle. On the set

$$G = D \times (0, T] \tag{3.3}$$

where $D = D_{(2.13b)}$ is a rectangular domain, we consider the following singularly perturbed equation of parabolic type [see Eq. (2.12a)]

$$L_{(3.4)} u(x, t) \equiv \left\{ \varepsilon^2 \sum_{s=1,2} a_s(x, t) \frac{\partial^2}{\partial x^2} - p(x, t) \frac{\partial}{\partial t} - c(x, t) \right\} u(x, t) = f(x, t)$$

$$(x, t) \in G \tag{3.4a}$$

On the lateral boundary $S^1 = \Gamma \times (0, T]$, $\Gamma = \bar{D} \backslash D$ the following condition is given:

$$l_{(3.4)} u(x, t) \equiv \varepsilon \frac{\partial}{\partial n} u(x, t) = \psi(x, t) \qquad (x, t) \in S^1 \tag{3.4b}$$

where

$$\frac{\partial}{\partial n} u(x, t) = -\frac{\partial}{\partial x_s} u(x, t) \qquad \text{for} \qquad x_s = d_{0s} \qquad x_{3-s} \in (d_{0,3-s}, d_{1,3-s})$$

$$\frac{\partial}{\partial n} u(x, t) = \frac{\partial}{\partial x_s} u(x, t) \qquad \text{for} \qquad x_s = d_{1s}$$

$$x_{3-s} \in (d_{0,3-s}, d_{1,3-s}) \qquad s = 1, 2$$

On the set $S^0 = \bar{D} \times \{t=0\}$ (the lower base of G) the function $u(x, t)$ takes the given values

$$u(x, t) = \varphi(x) \qquad (x, t) \in S^0 \tag{3.4c}$$

The coefficients and the right-hand side of the equation and also the initial function are assumed to be sufficiently smooth, $a_s(x, t) \geq a_0 > 0$, $c(x, t) \geq 0$, $p(x, t) \geq p_0 > 0$, $S = S^1 \cup S^0$, $S = \bar{G} \backslash G$. The boundary function $\psi(x, t)$ is sufficiently smooth on each side of the set S^1. However, this function is discontinuous on the set $S^c = \Gamma^c \times (0, T]$, where Γ^c is a set of corner points of the rectangle D.

We want to find the solution of problem (3.4), (3.3).

In the case of problem (3.4), we can find the derivatives $(\partial/\partial x_1)u(x, t)$ and $(\partial/\partial x_2)u(x, t)$. The functions

$$P_s(x, t) = \varepsilon \frac{\partial}{\partial x_s} u(x, t) \qquad s = 1, 2$$

are called the normalized diffusion fluxes with respect to the variables x_s.

Generally speaking, the functions $P_s(x, t)$ can have discontinuities on the set $S_0^1 = \Gamma \times \{t=0\}$. Therefore, we consider these functions on the set $\bar{G}_0 = \bar{G} \backslash S_0^1$. The functions $P_s(x, t)$ are continuous on $\bar{G}$ if, on the set S_0^1, the data of problem (3.4) [namely, the functions $f(x, t)$, $\varphi(x)$, $\psi(x, t)$] satisfy special compatibility conditions that ensure sufficient smoothness of the problem solution on $\bar{G}$.

When the parameter ε tends to zero in problem (3.4), a boundary layer appears in a neighborhood of the set S^1.

Many applied problems often require that one should find the distribution of diffusion fluxes in a material at a certain time. Therefore, together with the principal problem

$$\begin{cases} \textit{find the function } u(x, t),\ (x, t) \in \bar{G}, \textit{ that is,} \\ \textit{the solution of the boundary value problem,} \end{cases} \tag{3.5}$$

we also consider the following problem:

$$\begin{cases} \textit{find the function } u(x, t),\ (x, t) \in \bar{G}, \textit{ that is,} \\ \textit{the solution of the boundary value problem,} \\ \textit{and also the normalized diffusion fluxes.} \end{cases} \tag{3.6}$$

We say that the finite difference scheme solves the Neumann problem (3.5), if the grid solution converges to the solution of the boundary value problem, and for problem (3.6), if, in addition, the grid solution allows us to construct approximations that are convergent to the normalized diffusion fluxes.

B. Numerical Experiments with the Classical Difference Scheme

In Section II.B we saw that for any small step size of the grid a value of the parameter ε could be found such that the error in the approximate solution became comparable to the exact solution. In principle, we surmise that computational problems can arise also in the case of singularly perturbed equations, when the flux is given on the domain boundary.

We state a problem and thereafter study classical and new special finite difference schemes to solve it.

Suppose that we want to find the solution of the Neumann problem for the singularly perturbed diffusion equation

$$L_{(3.7)}u(x,t) \equiv \varepsilon^2 \frac{\partial^2}{\partial x^2} u(x,t) - \frac{\partial}{\partial t} u(x,t) = f(x,t) \qquad (x,t) \in G \quad (3.7\text{a})$$

$$l_{(3.7)}u(x,t) \equiv \varepsilon \frac{\partial}{\partial n} u(x,t) = \psi(x,t) \qquad (x,t) \in S^1 \quad (3.7\text{b})$$

$$u(x,t) = \varphi(x) \qquad (x,t) \in S^0 \quad (3.7\text{c})$$

Here the domain of definition for Eq. (3.7a) is established by the relation

$$G = D \times (0, T] \quad (3.8)$$

where $D = (0, 1)$, S is the boundary of the domain G, $S = \bar{G} \backslash G$, S^1 and S^0 are the lateral and lower boundaries of the set G, $f(x,t)$, $(x,t) \in \bar{G}$, $\psi(x,t)$, $(x,t) \in S^1$, $\varphi(x)$, $x \in \bar{D}$ are given functions. We define these functions as follows:

$$f(x,t) = -2t \qquad (x,t) \in \bar{G}$$

$$\psi(0,t) = \frac{8}{3\sqrt{\pi}} t^{3/2} \qquad \psi(1,t) = 0 \qquad t \in [0, T] \quad (3.9)$$

$$\varphi(x) = \cos(\pi x) \qquad x \in \bar{D} \qquad T = 1$$

Note that the function $P(x,t)$ is continuous on the set $\bar{G}$ in the case of problem (3.7), (3.9), (3.8).

According to the theory, the solution of the above problem can be represented as a sum of two functions

$$u(x,t) = U(x,t) + V(x,t) \qquad (x,t) \in \bar{G}$$

One of them, $U(x, t)$, is the regular part of the solution and the other, $V(x, t)$, is the singular part. The functions $U(x, t)$ and $V(x, t)$ are the solutions of the following problems:

$$
\begin{aligned}
L_{(3.7)}U(x, t) &= -2t \qquad (x, t) \in G \\
l_{(3.7)}U(x, t) &= 0 \qquad (x, t) \in S^1 \\
U(x, t) &= \cos(\pi x) \qquad (x, t) \in S^0
\end{aligned} \tag{3.10}
$$

$$
\begin{aligned}
L_{(3.7)}V(x, t) &= 0 \qquad (x, t) \in G \\
l_{(3.7)}V(x, t) &= \begin{Bmatrix} \dfrac{8}{3\sqrt{\pi}} t^{3/2} & x = 0 \\ 0 & x = 1 \end{Bmatrix} \qquad (x, t) \in S^1 \\
V(x, t) &= 0 \qquad (x, t) \in S^0
\end{aligned} \tag{3.11}
$$

The solution of problem (3.10) is as follows:

$$
U(x, t) = t^2 + \cos(\pi x)\exp(-\varepsilon^2\pi^2 t) \qquad (x, t) \in \bar{G}
$$

Its graph is shown in Fig. 15 for $\varepsilon = 1.0$, 0.1, and 0.01. The graphs of the solution of problem (3.11) for $\varepsilon = 1.0$, 0.1, and 0.01 are given in Fig. 16.

It is convenient to represent the function $V(x, t)$ as a sum of two functions

$$
V(x, t) = W(x, t) + v(x, t) \qquad (x, t) \in \bar{G}
$$

where

$$
\begin{aligned}
W(x, t) = {} & \left(\frac{x^4}{12\varepsilon^4} + \frac{x^2}{\varepsilon^2} t + t^2 \right) \operatorname{erfc}\left(\frac{x}{2\varepsilon\sqrt{t}} \right) \\
& - \frac{1}{\sqrt{\pi}} \left(\frac{x^3}{6\varepsilon^3} t^{1/2} + \frac{5x}{3\varepsilon} t^{3/2} \right) \exp\left(- \frac{x^2}{4\varepsilon^2 t} \right) \qquad 0 \le x < \infty \qquad t \ge 0
\end{aligned} \tag{3.12}
$$

For the function $v(x, t)$ (i.e., the remainder term) and its derivative the following estimates are fulfilled

$$
\max_{\bar{G}} |v(x, t)|, \ \max_{\bar{G}} \left| \frac{\partial}{\partial x} v(x, t) \right| \le M\varepsilon^n
$$

where the integer n can be chosen arbitrarily large and the constant M is

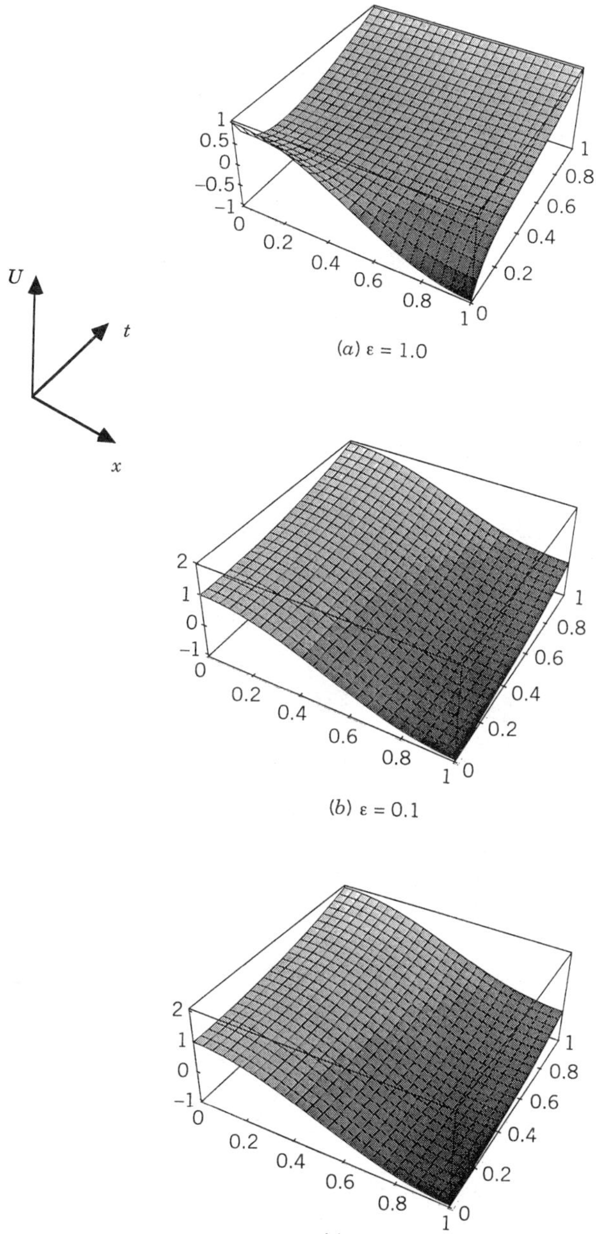

Figure 15. Regular part of the solution $U(x, t)$ of problem (3.7), (3.9) for $\varepsilon = 1.0$, 0.1, and 0.01.

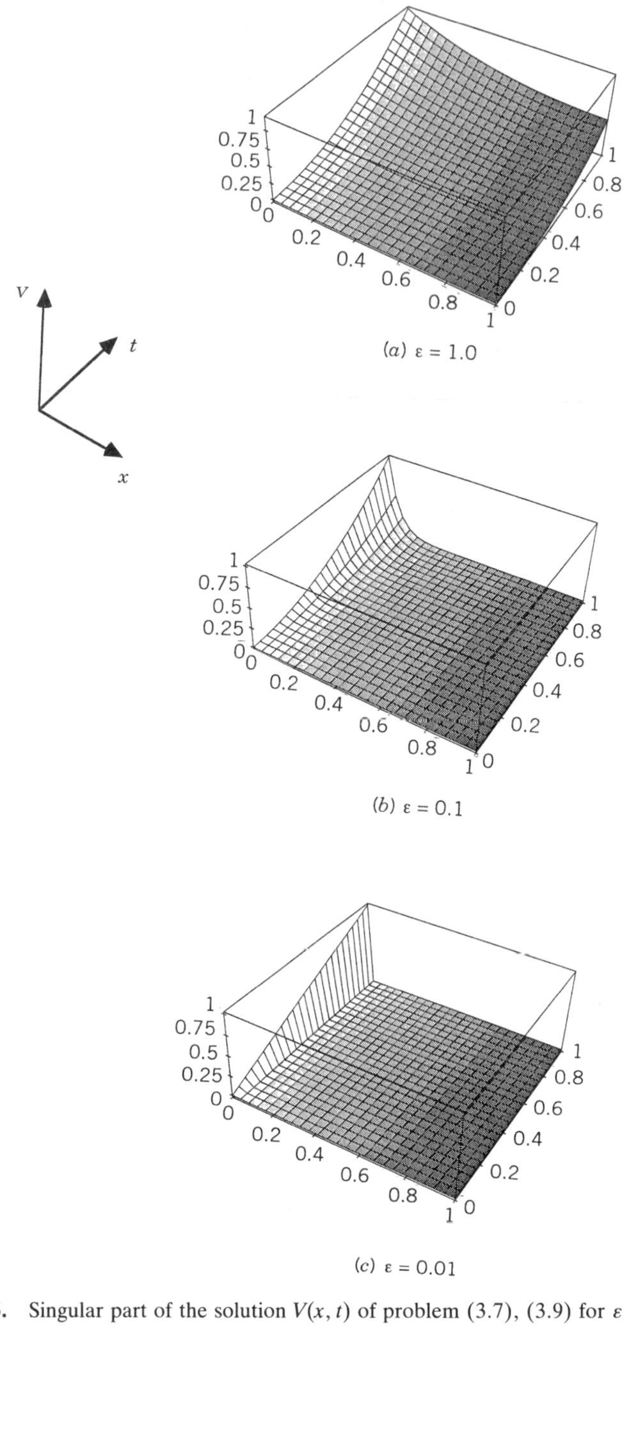

Figure 16. Singular part of the solution $V(x, t)$ of problem (3.7), (3.9) for $\varepsilon = 1.0$, 0.1, and 0.01.

independent of x, t, and ε. Thus, the function $V(x, t)$ approximates the function $W(x, t)$ with an accuracy bounded by the value $M\varepsilon^n$. The function $W(x, t)$ is the solution of the half-line problem

$$L_{(3.7)}W(x,t) = 0 \qquad 0 < x < \infty \qquad t > 0$$

$$\varepsilon \frac{\partial}{\partial x} W(0,t) = -\frac{8}{3\sqrt{\pi}} t^{3/2} \qquad t \geq 0$$

$$W(x,0) = 0 \qquad 0 \leq x < \infty$$

This function satisfies the relations

$$\max_{\substack{0 \leq x < \infty, \\ 0 \leq \vartheta \leq t}} W(x, \vartheta) = W(0,t) = t^2$$

$$\frac{\partial}{\partial x} W(x,t) \leq 0 \qquad 0 \leq x < \infty \qquad t \geq 0$$

$$\max_{0 \leq x \leq \infty} \varepsilon \left| \frac{\partial}{\partial x} W(x,t) \right| = \varepsilon \left| \frac{\varepsilon}{\partial x} W(0,t) \right| = \frac{8}{3\sqrt{\pi}} t^{3/2} \qquad t \geq 0$$

As ε decreases, the function $W(x, t)$ and the product $\varepsilon(\partial/\partial x)W(x, t)$ vanish, for $x \geq x_0 \geq m$, $0 \leq t \leq 1$, more rapidly than any power of the parameter ε

$$|W(x,t)|, \ \varepsilon \left| \frac{\partial}{\partial x} W(x,t) \right| \leq M\varepsilon^n$$

Thus, we get the representation

$$u(x,t) = U(x,t) + W(x,t) + v(x,t) \qquad (x,t) \in \bar{G} \tag{3.13}$$

for the function $u_{(3.7)}(x, t)$. There are explicit analytical representations for the functions $U(x, t)$ and $W(x, t)$; the function $v(x, t)$, together with its derivatives, is small for small values of the parameter. The function $U(x, t)$ is the solution of problem (3.10), while the function $W(x, t)$ is the solution of the problem

$$L_{(3.7)}W(x,t) = 0 \qquad (x,t) \in G$$

$$l_{(3.7)}W(x,t) = \left\{ \begin{array}{ll} \dfrac{8}{3\sqrt{\pi}} t^{3/2} & x = 0 \\ \varepsilon \dfrac{\partial}{\partial x} W(1,t) & x = 1 \end{array} \right\} \qquad (x,t) \in S^1 \tag{3.14}$$

$$W(x,t) = 0 \qquad (x,t) \in S^0$$

and the function $v(x, t)$ is the solution for the problem

$$L_{(3.7)}v(x,t)=0 \qquad (x,t)\in G$$
$$l_{(3.7)}v(x,t)=\left\{\begin{matrix} 0 & x=0 \\ -\varepsilon \frac{\partial}{\partial x} W(1,t) & x=1 \end{matrix}\right\} \qquad (x,t)\in S^1 \tag{3.15}$$
$$v(x,t)=0 \qquad (x,t)\in S^0$$

Here the product $\varepsilon(\partial/\partial x)W(x, t)$ can be defined according to formula (3.12)

$$\varepsilon \frac{\partial}{\partial x} W(x,t)=\left(\frac{x^3}{3\varepsilon^3}+\frac{2x}{\varepsilon}t\right)\operatorname{erfc}\left(\frac{x}{2\varepsilon\sqrt{t}}\right)$$
$$-\frac{1}{\sqrt{\pi}}\left(\frac{2x^2}{3\varepsilon^2}t^{1/2}+\frac{8}{3}t^{3/2}\right)\exp\left(-\frac{x^2}{4\varepsilon^2 t}\right) \qquad 0\le x<\infty \qquad t\ge 0$$

Now we study the behavior of the error in the approximate solution for various values of the parameter ε. To solve the boundary value problem (3.7), (3.9) numerically, we apply a classical finite difference scheme. On the set $\bar{G}_{(3.8)}$ we introduce the uniform rectangular grid

$$\bar{G}_h=\bar{\omega}_1\times\bar{\omega}_0 \tag{3.16}$$

where $\bar{\omega}_1$ and $\bar{\omega}_0$ are uniform grids with the step sizes $h=N^{-1}$ and $h_0=N_0^{-1}$, respectively, $N+1$ and N_0+1 are the number of nodes in the grids $\bar{\omega}_1$ and $\bar{\omega}_0$. On the grid $\bar{G}_h$ the problem is approximated by the finite difference scheme

$$\Lambda_{(3.17)}z(x,t)\equiv \varepsilon^2\delta_{x\bar{x}}z(x,t)-\delta_{\bar{t}}z(x,t)=f(x,t) \qquad (x,t)\in G_h$$
$$\lambda_{(3.17)}z(x,t)=\psi(x,t) \qquad (x,t)\in S_h^1 \tag{3.17}$$
$$z(x,t)=\varphi(x) \qquad (x,t)\in S_h^0$$

Here

$$G_h=G\cap\bar{G}_h \qquad S_h^1=S^1\cap\bar{G}_h \qquad S_h^0=S^0\cap\bar{G}_h$$
$$\lambda_{(3.17)}z(x,t)=\begin{cases} -\varepsilon\delta_x z(x,t) & x=0 \\ \varepsilon\delta_{\bar{x}}z(x,t) & x=1 \end{cases}$$

$\delta_x z(x,t)$, $\delta_{\bar{x}} z(x,t)$ are the first forward and backward difference derivatives.

To approximate the function $P(x,t)$, we use the grid function

$$P^{h+}(x,t) \equiv \varepsilon \delta_x z(x,t) \qquad (x,t) \in \bar{G}_h^- \tag{3.18}$$

where $\bar{G}_h^- = \bar{\omega}_1^- \times \bar{\omega}_0$ is a grid on $\bar{G}$, $\bar{\omega}_1^- = \omega_1 \cup \{x = 0\}$.

The finite difference scheme (3.17), (3.16) allows us to find the function $z(x,t)$, $(x,t) \in \bar{G}_h$. Therefore, from formula (3.18), we can compute an approximation of the normalized flux $P(x,t)$.

From the theory of finite difference schemes it is known that, for a fixed value of the parameter, the solution of the difference scheme (3.17), (3.16) converges to the solution of the boundary value problem (3.7) when $N, N_0 \to \infty$. For sufficiently smooth data in problem (3.7), for example, under condition (3.9), the following estimate is valid

$$\max_{\bar{G}_h} |u(x,t) - z(x,t)| \le Q(\varepsilon)[N^{-1} + N_0^{-1}]$$

We are interested in problem (3.5) for the boundary value problem (3.7), more precisely, we want to know the behavior of the quantities $E(\varepsilon, N)$ and $\bar{E}(N)$, where

$$E(\varepsilon, N) = E(\varepsilon, N; u(\cdot)) \equiv E(\varepsilon, N, N)$$

$$\bar{E}(N) = \bar{E}(N; u(\cdot)) \equiv \bar{E}(N, N)$$

$$E(\varepsilon, N, N_0) = E(\varepsilon, N, N_0; u(\cdot)) = \max_{\bar{G}_h} |u(x,t;\varepsilon) - z(x,t;\varepsilon, N, N_0)|$$

$$\bar{E}(N, N_0) = \bar{E}(N, N_0; u(\cdot)) = \max_{\varepsilon} E(\varepsilon, N, N_0)$$

$u(x,t) = u(x,t;\varepsilon)$ and $z(x,t) = z(x,t;\varepsilon, N, N_0)$ are the solutions of problems (3.7), (3.9), and (3.17), (3.16), respectively. We are also interested in the behavior of the errors $E(\varepsilon, N)$ and $\bar{E}(N)$ for each component in the representation (3.13) of the solution of the boundary value problem (3.7), (3.9). We want to analyze the values $E(\varepsilon, N; U(\cdot))$, $\bar{E}(N; U(\cdot))$, $E(\varepsilon, N; W(\cdot))$, $\bar{E}(N; W(\cdot))$, $E(\varepsilon, N; v(\cdot))$, $\bar{E}(N; v(\cdot))$. Note that we construct the difference schemes

$$\Lambda_{(3.17)} z(x,t) = f(x,t) \qquad (x,t) \in G_h$$

$$\lambda_{(3.17)} z(x,t) = 0 \qquad (x,t) \in S_h^1 \tag{3.19}$$

$$z(x,t) = \varphi(x) \qquad (x,t) \in S_h^0$$

$$\begin{aligned} &\Lambda_{(3.17)} z(x,t) = 0 \qquad (x,t) \in G_h \\ &\lambda_{(3.17)} z(x,t) = l_{(3.7)} W(x,t) \qquad (x,t) \in S_h^1 \\ &z(x,t) = 0 \qquad (x,t) \in S_h^0 \end{aligned} \tag{3.20}$$

$$\begin{aligned} &\Lambda_{(3.17)} z(x,t) = 0 \qquad (x,t) \in G_h \\ &\lambda_{(3.17)} z(x,t) = l_{(3.7)} v(x,t) \qquad (x,t) \in S_h^1 \\ &z(x,t) = 0 \qquad (x,t) \in S_h^0 \end{aligned} \tag{3.21}$$

to problems (3.10), (3.14), and (3.15), respectively. One can easily see that we obtain

$$z_{(3.17)}(x,t) = z_{(3.19)}(x,t) + z_{(3.20)}(x,t) + z_{(3.21)}(x,t) \qquad (x,t) \in \bar{G}_h$$

In Tables XVII and XVIII we give the values of $E(\varepsilon, N)$ and $\bar{E}(N)$ computed for problems (3.10) and (3.14) for different values of ε and N, and $N_0 = N$.

From Table XVII it is obvious that, as N increases, the error $E(\varepsilon, N)$ decreases for every fixed value of the parameter ε, and the value of $\bar{E}(N)$ also decreases. This type of behavior means that, for $N_0 = N$, as N increases, the approximate solution $z_{(3.19)}(x,t)$ converges to the function $U(x,t)$ both for a fixed value of the parameter ε and ε-uniformly.

In Table XIX the values of $E(\varepsilon, N, N_0; U(\cdot))$ are given for $\varepsilon = 2^{-4}$ and for various N and N_0. The errors $E(\varepsilon, N, N_0)$ decrease when both N and N_0 increase. Similar behavior of $E(\varepsilon, N, N_0)$ can be observed for other values of the parameters ε. Thus, as N and N_0 increase, the solution

TABLE XVII

Table of Errors $E(\varepsilon, N)$ for the Classical Scheme (3.19), (3.16) in the Case of Problem (3.10)

$\varepsilon \backslash N$	4	16	64	256	1024
1	2.511e − 1	6.265e − 2	1.636e − 2	4.259e − 3	1.076e − 3
2^{-2}	5.000e − 1	1.175e − 1	2.882e − 2	7.168e − 3	1.790e − 3
2^{-4}	5.476e − 1	9.389e − 2	2.149e − 2	5.264e − 3	1.309e − 3
2^{-6}	5.432e − 1	8.282e − 2	1.767e − 2	4.284e − 3	1.064e − 3
2^{-8}	5.429e − 1	8.179e − 2	1.690e − 2	4.034e − 3	1.000e − 3
2^{-10}	5.429e − 1	8.172e − 2	1.683e − 2	3.986e − 3	9.846e − 4
2^{-12}	5.429e − 1	8.171e − 2	1.683e − 2	3.982e − 3	9.816e − 4
$\bar{E}(N)$	5.476e − 1	1.175e − 1	2.882e − 2	7.168e − 3	1.790e − 3

TABLE XVIII
Table of Errors $E(\varepsilon, N)$ for the Classical Scheme (3.20), (3.16) in the Case of Problem (3.14)

$\varepsilon \backslash N$	4	16	64	256	1024
1	4.736e − 1	9.723e − 2	2.332e − 2	5.771e − 3	1.439e − 3
2^{-2}	1.098e + 0	2.347e − 1	5.581e − 2	1.377e − 2	3.431e − 3
2^{-4}	5.213e + 0	1.014e + 0	2.114e − 1	4.994e − 2	1.230e − 2
2^{-6}	2.312e + 1	5.177e + 0	9.935e − 1	2.056e − 1	4.848e − 2
2^{-8}	9.530e + 1	2.311e + 1	5.169e + 0	9.884e − 1	2.041e − 1
2^{-10}	3.842e + 2	9.530e + 1	2.311e + 1	5.167e + 0	9.871e − 1
2^{-12}	1.540e + 3	3.842e + 2	9.530e + 1	2.311e + 1	5.166e + 0
$\bar{E}(N)$	1.540e + 3	3.842e + 2	9.530e + 1	2.311e + 1	5.166e + 0

$z_{(3.19)}(x, t)$ converges to the function $U(x, t)$ both for a fixed value of the parameter ε and ε-uniformly.

From Table XVIII we can see that, as N increases, the error $E(\varepsilon, N)$ decreases for every fixed value of the parameter ε. However, this error grows for a fixed value of N when ε decreases. The error $E(\varepsilon, N)$ grows when the product εN decreases, moreover, the error exceeds the value $\max_{\bar{G}_h} W(x, t) = 1$ for $\varepsilon N < 1$. Thus, for $N_0 = N$, the approximate solution $z_{(3.20)}(x, t)$ does not converge to the function $W(x, t)$ ε-uniformly when N increases. Consequently, as N and N_0 increase, the function $z_{(3.20)}(x, t)$ does not converge to the function $W(x, t)$ ε-uniformly.

From Table XVIII we see that the error $E(\varepsilon, N)$ grows without bound when the product εN tends to zero, for example, $E(\varepsilon, N) \geq 1000$ [i.e., the error is more than 1000 times the maximum value of the solution $W(x, t)$ itself] for $\varepsilon N \leq 2^{-10} = 1/1024$.

Note that we cannot compute the values of $E(\varepsilon, N, N_0; v(\cdot))$ and $\bar{E}(N, N_0; v(\cdot))$ because we have no analytical representation of the function $v(x, t)$. As the derivatives of the function $v(x, t)$ vanish rapidly when $\varepsilon \to 0$, the errors $E(\varepsilon, N, N_0; v(\cdot))$ and $\bar{E}(N, N_0; v(\cdot))$ are well

TABLE XIX
Table of Errors $E(\varepsilon, N, N_0)$ for the Classical Scheme (3.19), (3.16) in the Case of Problem (3.10) with $\varepsilon = 2^{-4}$

$N_0 \backslash N$	4	16	64	256	1024
4	5.476e − 1	2.809e − 1	2.555e − 1	2.511e − 1	2.501e − 1
16	3.603e − 1	9.389e − 2	6.830e − 2	6.381e − 2	6.279e − 2
64	3.135e − 1	4.715e − 2	2.149e − 2	1.697e − 2	1.595e − 2
256	3.018e − 1	3.547e − 2	9.792e − 3	5.264e − 3	4.237e − 3
1024	2.989e − 1	3.255e − 2	6.866e − 3	2.337e − 3	1.309e − 3

approximated by the values of $E_{N^*,N_0^*}(\varepsilon, N, N_0)$ and $E_{N^*,N_0^*}(N, N_0)$, where

$$\begin{aligned} E_{N^*,N_0^*}(\varepsilon, N, N_0) &= E_{N^*,N_0^*}(\varepsilon, N, N_0; v_{N^*,N_0^*}(\cdot)) \\ &= \max_{\bar{G}_h} |v_{N^*,N_0^*}(x, t; \varepsilon) - z(x, t; \varepsilon, N, N_0)| \\ \bar{E}_{N^*,N_0^*}(N, N_0) &= \bar{E}_{N^*,N_0^*}(N, N_0; v_{N^*,N_0^*}(\cdot)) \\ &= \max_{\varepsilon} E_{N^*,N_0^*}(\varepsilon, N, N_0; v_{N^*,N_0^*}(\cdot)) \end{aligned}$$

Here

$$v_{N^*,N_0^*}(x, t) \qquad (x, t) \in \bar{G} \tag{3.22}$$

is a continuous function constructed by linear interpolation with respect to x and t from the values of the function $z(x, t; \varepsilon, N^*, N_0^*)$, that is, the solution of problem (3.21), (3.16) for $N = N^*$, $N_0 = N_0^*$; $z(x, t) = z(x, t; \varepsilon, N, N_0)$ is the solution of problem (3.21), (3.16); the values of N^*, N_0^* exceed N, N_0 substantially. We define

$$E_{N^*,N_0^*}(\varepsilon, N) = E_{N^*,N_0^*}(\varepsilon, N, N) \qquad \bar{E}_{N^*,N_0^*}(N) = \bar{E}_{N^*,N_0^*}(N, N)$$

In Table XX we give the computed values $E_{1024,1024}(\varepsilon, N)$ and $\bar{E}_{1024,1024}(N)$ for different values of the parameter ε and the number of nodes N. It follows from Table XX that, as N increases, the error $E(\varepsilon, N)$ decreases for every fixed value of the parameter ε, and the error $\bar{E}(N)$ decreases too. That is, for $N_0 = N$, when N increases, the approximate solution $z_{(3.21)}(x, t)$ converges to the function $v(x, t)$ both for a fixed value of the parameter ε and ε-uniformly. As N and N_0 increase, the solution $z_{(3.21)}(x, t)$ also converges to the function $v(x, t)$ both for a fixed value of the parameter ε and ε-uniformly.

TABLE XX

Table of Errors $E_{1024,1024}(\varepsilon, N)$ for the Classical Scheme (3.21), (3.16) in the Case of Problem (3.15)

$\varepsilon \backslash N$	4	16	64	256
1	1.260e 01	2.511e − 02	5.713e − 03	1.130e − 03
2^{-2}	4.944e − 04	1.043e − 04	2.295e − 05	4.485e − 06
2^{-4}	1.565e − 31	3.698e − 32	7.664e − 33	1.315e − 33
2^{-6}	0.000e − 00	0.000e − 00	0.000e − 00	0.000e − 00
$\bar{E}(N)$	1.260e − 01	2.511e − 02	5.713e − 03	1.130e − 03

It follows from the analysis of the tables that the largest part of the error in the approximate solution $z_{(3.17)}(x, t)$, that is, the solution of problem (3.17), (3.16), is the error in the singular part of the solution. The function $z_{(3.17)}(x, t)$ converges to the solution of problem (3.7), (3.9), as N and N_0 grow, for fixed values of the parameter ε. However, the function $z_{(3.17)}(x, t)$ does not converge ε-uniformly. Moreover, the error in the approximate solution can be many times the maximal value of the exact solution. The error in the approximate solution increases unboundedly when the product εN (for $N_0 = N$) tends to zero. Thus, in the case of classical difference schemes, the approximate solution does not approximate the unknown solution ε-uniformly even qualitatively.

In the case of problem (3.6), we are interested in the approximation of the normalized diffusion flux $P(x, t)$. It is known from the theory of finite difference schemes that, for a fixed value of the parameter ε and $N, N_0 \to \infty$, the discrete normalized diffusion flux $P^{h+}_{(3.18)}(x, t)$ converges to the function $P(x, t)$ [for a sufficiently smooth solution of problem (3.7)]. Under condition (3.9) the following estimate is valid

$$\max_{\bar{G}_h^-} |P(x, t) - P^{h+}(x, t)| \leq Q(\varepsilon)[N^{-1} + N_0^{-1}]$$

According to the explicit form of the functions $U(x, t)$ and $W(x, t)$ and the estimates for the function $v(x, t)$ and its derivatives, the following estimates and relations hold

$$\max_{\bar{G},\varepsilon} |P(x, t; U(\cdot))|, \ \max_{\bar{G},\varepsilon} |P(x, t; W(\cdot))|, \ \max_{\bar{G},\varepsilon} |P(x, t; v(\cdot))| \leq M$$

$$\lim_{\varepsilon \to 0} |P(x, t; U(\cdot))|, \ \lim_{\varepsilon \to 0} |P(x, t; v(\cdot))| = 0 \qquad (x, t) \in \bar{G}$$

$$\max_{x} |P(x, t; W(\cdot))| = |P(x = 0, t; W(\cdot))| = \frac{8}{3\sqrt{\pi}} t^{3/2} \qquad 0 \leq t \leq T$$

Taking into consideration the above relations, we obtain the following:

$$\max_{\bar{G},\varepsilon} |P(x, t; u_{(3.7)}(\cdot))| \leq M$$

$$\lim_{\varepsilon \to 0} \max_{x} |P(x, t; u_{(3.7)}(\cdot))| = \frac{8}{3\sqrt{\pi}} t^{3/2} \qquad 0 \leq t \leq T$$

Graphs of the functions $P(x, t; U(\cdot))$ and $P(x, t; V(\cdot))$ are shown in Figs. 17 and 18.

To estimate the closeness of the functions $P(x, t)$ and $P^{h+}(x, t)$, we use

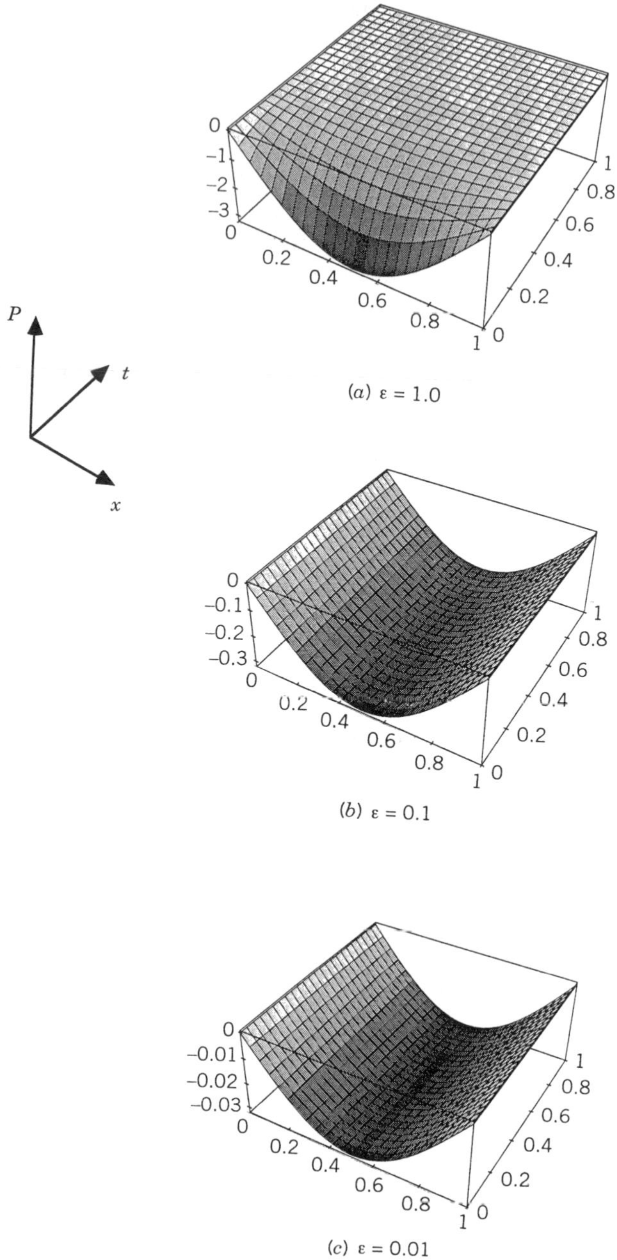

Figure 17. Normalized flux $P(x, t; U(\cdot))$ for problem (3.7), (3.9) for $\varepsilon = 1.0$, 0.1, and 0.01.

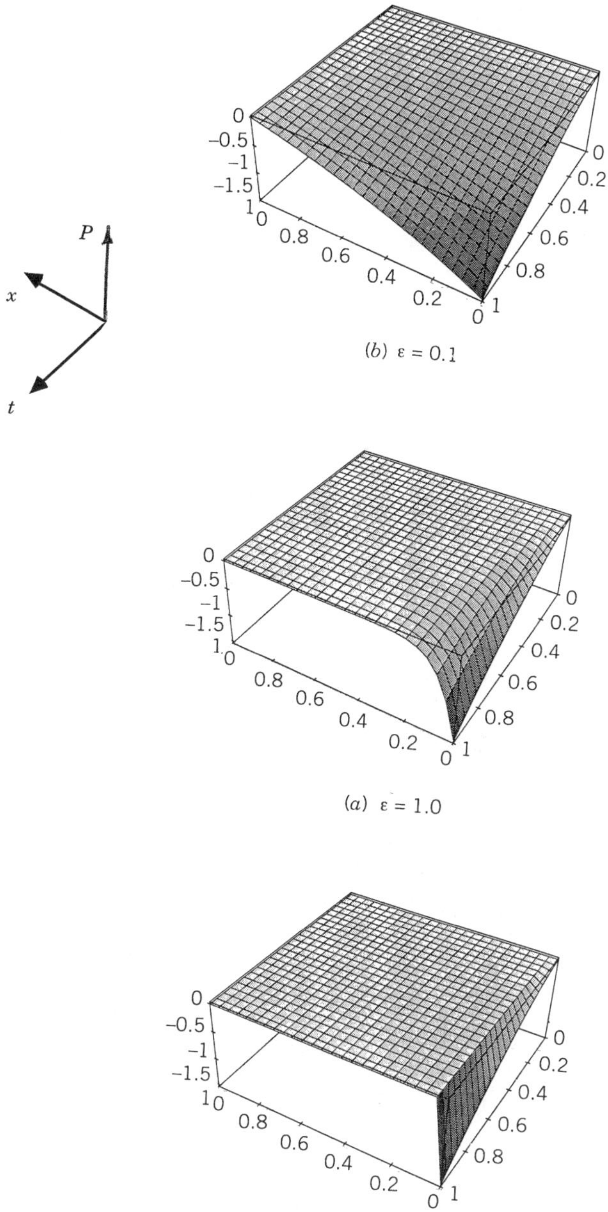

Figure 18. Normalized flux $P(x, t; V(\cdot))$ for problem (3.7), (4.9) for $\varepsilon = 1.0$, 0.1, and 0.01.

TABLE XXI
Table of Errors of the Normalized Flux $Q(\varepsilon, N)$ for the Classical Scheme (3.19), (3.16) in the Case of Problem (3.10)

$\varepsilon \backslash N$	4	16	64	256	1024
1	3.773e − 1	3.307e − 1	1.321e − 1	3.709e − 2	9.546e − 3
2^{-2}	4.760e − 1	1.474e − 1	3.817e − 2	9.615e − 3	2.408e − 3
2^{-4}	1.375e − 1	3.821e − 2	9.629e − 3	2.409e − 3	6.024e − 4
2^{-6}	3.469e − 2	9.575e − 3	2.409e − 3	6.024e − 4	1.506e − 4
2^{-8}	8.677e − 3	2.394e − 3	6.021e − 4	1.506e − 4	3.765e − 5
2^{-10}	2.169e − 3	5.985e − 4	1.505e − 4	3.765e − 5	9.412e − 6
2^{-12}	5.423e − 4	1.496e − 4	3.763e − 5	9.412e − 6	2.353e − 7
$\bar{Q}(N)$	4.760e − 1	3.307e − 1	1.321e − 1	3.709e − 2	9.546e − 3

the quantities $Q(\varepsilon, N)$ and $\bar{Q}(N)$, where

$$Q(\varepsilon, N) = Q(\varepsilon, N; u(\cdot)) \equiv Q(\varepsilon, N, N) \qquad \bar{Q}(N) = \bar{Q}(N; u(\cdot)) \equiv \bar{Q}(N, N)$$

$$Q(\varepsilon, N, N_0) = Q(\varepsilon, N, N_0; u(\cdot)) \equiv \max_{\bar{G}_h^-} |P(x, t; \varepsilon) - P^{h+}(x, t; \varepsilon, N, N_0)|$$

$$\bar{Q}(N, N_0) = \bar{Q}(N, N_0; u(\cdot)) \equiv \max_{\varepsilon} Q(\varepsilon, N, N_0)$$

Here the functions $P(x, t) = P(x, t; \varepsilon)$ and $P^{h+}(x, t) = P^{h+}(x, t; \varepsilon, N, N_0)$ are found from the solutions of problems (3.7), (3.9) and (3.17), (3.16).

The results of computing the values of $Q(\varepsilon, N)$ and $\bar{Q}(N)$ for problems (3.10) and (3.14) are given in Tables XXI and XXII.

It is clear from Table XXI that, with $N_0 = N$, as N increases, the computed diffusion flux $P^{h+}(x, t; U(\cdot))$ converges to the real flux $P(x, t; U(\cdot))$ for fixed values of the parameter ε and ε-uniformly.

We see from Table XXII that, with $N_0 = N$, when N increases, the diffusion flux $P^{h+}(x, t; W(\cdot))$ converges to the real flux $P(x, t; W(\cdot))$ for

TABLE XXII
Table of Errors of the Normalized Flux $Q(\varepsilon, N)$ for the Classical Scheme (3.20), (3.16) in the Case of Problem (3.14)

$\varepsilon \backslash N$	4	16	64	256	1024
1	1.675e − 1	3.658e − 2	8.844e − 3	2.193e − 3	5.470e − 4
2^{-2}	8.878e − 2	1.728e − 2	3.903e − 3	9.482e − 4	2.353e − 4
2^{-4}	4.679e − 2	5.094e − 2	6.211e − 3	1.095e − 3	2.444e − 4
2^{-6}	3.116e − 3	3.844e − 2	4.071e − 2	3.421e − 3	3.910e − 4
2^{-8}	1.954e − 4	2.531e − 2	3.642e − 2	3.813e − 2	2.722e − 3
2^{-10}	1.221e − 5	1.585e − 4	2.391e − 3	3.591e − 2	3.749e − 2
2^{-12}	7.633e − 7	9.911e − 6	1.498e − 4	2.357e − 3	3.579e − 2
$\bar{Q}(N)$	1.675e − 1	5.094e − 2	4.071e − 2	3.813e − 2	3.749e − 2

fixed values of the parameter. However, the function $P^{h+}(x, t; W(\cdot))$ does not converge to $P(x, t; W(\cdot))$ ε-uniformly. The function $P^{h+}(x, t; W(\cdot))$ reflects qualitatively and ε-uniformly the behavior of the function $P(x, t; W(\cdot))$.

The analytical representation of the function $v(x, t)$ is unknown to us, therefore, in order to analyze the convergence of the fluxes, we use the quantities $Q_{N^*,N_0^*}(\varepsilon, N, N_0)$ and $\bar{Q}_{N^*,N_0^*}(N, N_0)$, where

$$\begin{aligned} Q_{N^*,N_0^*}(\varepsilon, N, N_0) &= Q_{N^*,N_0^*}(\varepsilon, N, N_0; v_{N^*,N_0^*}(\cdot)) \\ &= \max_{\bar{G}_h^-} |P(x, t; \varepsilon, v_{N^*,N_0^*}(\cdot)) - P^{h+}(x, t; \varepsilon, N, N_0)| \end{aligned}$$

$$\begin{aligned} \bar{Q}_{N^*,N_0^*}(N, N_0) &= \bar{Q}_{N^*,N_0^*}(N, N_0; v_{N^*,N_0^*}(\cdot)) \\ &= \max_{\varepsilon} Q_{N^*,N_0^*}(\varepsilon, N, N_0; v_{N^*,N_0^*}(\cdot)) \end{aligned}$$

Here $v_{N^*,N_0^*}(x, t) = v_{N^*,N_0^*(3.22)}(x, t)$ and the function $P(x, t; \varepsilon, v_{N^*,N_0^*}(\cdot))$, $(x, t) \in \bar{G}$ are defined by the relations

$$P(x, t; \varepsilon, v_{N^*,N_0^*}(\cdot)) = \begin{cases} (h^*)^{-1}[v_{N^*,N_0^*}(h^*, t) - v_{N^*,N_0^*}(0, t)] & 0 \le x < 2^{-1}h^* \\ (h^*)^{-1}[v_{N^*,N_0^*}(x + 2^{-1}h^*, t) - v_{N^*,N_0^*}(x - 2^{-1}h^*, t)] & \\ \qquad 2^{-1}h^* \le x < 1 - 2^{-1}h^* & \\ (h^*)^{-1}[v_{N^*,N_0^*}(1, t) - v_{N^*,N_0^*}(1 - h^*, t)] & 1 - 2^{-1}h^* \le x \le 1 \end{cases}$$

$$(x, t) \in \bar{G}$$

where h^* is the step size of the grid $\bar{\omega}_1$ defining the grid $\bar{G}_{h(3.16)}(N^*, N_0^*)$. We define

$$Q_{N^*,N_0^*}(\varepsilon, N) = Q_{N^*,N_0^*}(\varepsilon, N; v_{N^*,N_0^*}(\cdot)) \equiv Q_{N^*,N_0^*}(\varepsilon, N, N)$$

$$\bar{Q}_{N^*,N_0^*}(N) = \bar{Q}_{N^*,N_0^*}(N; v_{N^*,N_0^*}(\cdot)) \equiv \bar{Q}_{N^*,N_0^*}(N, N)$$

In Table XXIII we give the errors in the normalized diffusion fluxes $Q_{1024,1024}(\varepsilon, N)$ and $\bar{Q}_{1024,1024}(N)$ for problem (3.17) computed with the finite difference scheme (3.21), (3.16) for various values of the parameter ε and the number of nodes N with $N_0 = N$. One can see from the table that, as N increases, the error $Q_{1024,1024}(\varepsilon, N)$ tends to zero (while the computed flux $P^{h+}(x, t; v_{N^*,N_0^*}(\cdot))$ converges to the function $P(x, t; v_{N^*,N_0^*}(\cdot)))$ for a fixed value of the parameter ε and ε-uniformly. When N and N_0 increase, the computed flux $P^{h+}(x, t; v_{N^*,N_0^*}(\cdot))$ also

TABLE XXIII
Table of Errors of the Normalized Flux $Q_{1024,1024}(\varepsilon, N)$ for the Classical Scheme (3.21), (3.16) in the Case of Problem (3.15)

$\varepsilon \backslash N$	4	16	64	256
1	1.258e − 01	3.549e − 02	9.153e − 03	2.306e − 03
2^{-2}	5.353e − 04	2.816e − 04	8.831e − 05	2.342e − 05
2^{-4}	3.980e − 32	3.979e − 32	3.464e − 32	1.585e − 32
2^{-6}	0.000e − 00	0.000e − 00	0.000e − 00	0.000e − 00
$\bar{Q}(N)$	1.258e − 01	3.549e − 02	9.153e − 03	2.306e − 03

converges to the function $P(x, t; v_{N^*,N_0^*}(\cdot))$ for a fixed value of the parameter and ε-uniformly.

It follows from the results given in Tables XXI–XXIII that the largest error in the computed flux arises from the approximation error for the singular part of the problem solution. We see that, as N and N_0 increase, the computed diffusion flux $P^{h+}(x, t; z_{(3.17)}(\cdot))$ approximates the real flux $P(x, t; u_{(3.7)}(\cdot))$ for a fixed value of the parameter. However, the computed flux does not converge ε-uniformly. Nevertheless, the computed flux qualitatively well approximates the actual flux ε-uniformly.

The foregoing analysis shows that, using the classical finite difference scheme (3.17), (3.16) for the solution of the Newmann boundary value problem (3.7), (3.9), the approximate solution $z_{(3.17)}(x, t)$ converges to the solution $u_{(3.7)}(x, t)$ for a fixed value of the parameter ε and $N, N_0 \to \infty$ (as was predicted by the theory). The function $z_{(3.17)}(x, t)$ does not converge to the function $u_{(3.7)}(x, t)$ ε-uniformly. The error in the approximate solution can be many times the magnitude of the actual solution. The error in the approximate solution increases without bound under the conditions $\varepsilon \leq N^{-1}$, $\varepsilon N \to 0$ for $N, N_0 \to \infty$. For small values of the parameter the approximate solution docs not approximate the actual solution qualitatively, even roughly. Consequently, the approximate solution does not qualitatively approach the actual solution ε-uniformly even roughly.

The approximate solutions obtained by the classical difference schemes (3.17), (3.16) allow us to approach the normalized diffusion flux for a fixed value of the parameter, as was predicted by the theory. The computed flux $P^{h+}(x, t; z_{(3.17)}(\cdot))$ does not approximate the actual flux $P(x, t; u_{(3.7)}(\cdot))$ ε-uniformly. The functions $P^{h+}(x, t; z_{(3.17)}(\cdot))$ qualitatively well approximate the function $P(x, t; u_{(3.7)}(\cdot))$ ε-uniformly.

Thus, classical finite difference schemes are not appropriate for the solution of problems (3.5) and (3.6) in the case of the singularly perturbed Neumann problems (3.2), (3.1) and (3.4), (3.3).

C. Principles of Constructing Special Finite Difference Schemes for the Neumann Problem

Using the example of a boundary value problem for a singularly perturbed ordinary differential equation with Neumann boundary conditions, we discuss some principles of constructing special finite difference schemes. These principles will be used in Section III.D to construct special finite difference schemes for singularly perturbed equations of the parabolic type.

In the case of Dirichlet problem (2.37) considered in Section 2.3, the unknown solution and the solution of the finite difference scheme take the same given values on the boundary. Thus, in the Dirichlet problem no error appears on the boundary. The error is generated only by an error in the approximation of the differential equation by the difference equation.

We consider the Neumann problem for the ordinary differential equation (1.13a) on the unit segment $\bar{D}$, $\bar{D} = [0, 1]$:

$$L_{(3.23)}u(x) \equiv \varepsilon^2 \frac{d^2}{dx^2} u(x) - c(x)u(x) = f(x) \qquad x \in D \tag{3.23a}$$

$$l_{(3.23)}u(x) \equiv \varepsilon \frac{\partial}{\partial n} u(x) = \begin{Bmatrix} -\varepsilon \dfrac{d}{dx} u(x) & x = 0 \\ \varepsilon \dfrac{d}{dx} u(x) & x = 1 \end{Bmatrix} = \psi(x) \qquad x \in \Gamma \tag{3.23b}$$

For this problem, we use a finite difference scheme on a grid with an arbitrary distribution of $N+1$ nodes and with fixed N. We show that the nodes of this grid can be distributed in such a way that the error in the approximate solution tends to zero ε-uniformly when N increases.

We obtain estimates of the solution and its derivatives for problem (3.23). Similarly to (2.40) we arrive at the estimate

$$\max_{\bar{D}} |u(x)| \leq M_{(3.24)} \equiv (c_0)^{-1/2} \max_{\Gamma} |\psi(\xi)| + (c_0)^{-1} \max_{\bar{D}} |f(\xi)| \tag{3.24}$$

According to the theory of differential equations, the solution of a boundary value problem can be represented as a sum of two functions, that is,

$$u(x) = U(x) + V(x) \qquad x \in \bar{D} \tag{3.25}$$

Here $U(x)$ and $V(x)$ are the regular and singular parts of the problem solution. The following estimates are fulfilled for the functions $U(x)$ and

$V(x)$ and their derivatives:

$$\left|\frac{d^k}{dx^k}U(x)\right| \leq M$$

$$\left|\frac{d^k}{dx^k}V(x)\right| \leq M\varepsilon^{-k}\exp(-m_0\varepsilon^{-1}r(x)) \qquad x\in\bar{D} \qquad 0\leq k\leq 4 \tag{3.26}$$

where $r(x)$ is the distance from the point x to the ends of the segment $\bar{D}$, $m_0 < m_{(3.26)}$, $m_{(3.26)} = (c_0)^{1/2}$. Note that, for the Neumann problem (3.23), the estimates of the solution and the components $U(x)$, $V(x)$ from representation (3.25) are like those for the Dirichlet problem [see estimates (2.40), (2.42), and (2.43)].

For problem (3.23), we use a classical finite difference approximation on a grid with an arbitrary distribution of nodes. On the set $\bar{D}$ we introduce the grid

$$\bar{D} = \bar{\omega}_1 \tag{3.27}$$

where $\bar{\omega}_1$ is the grid $\bar{\omega}_{1(2.44)}$ with an arbitrary distribution of nodes. On the grid $\bar{D}_h$ we consider the finite difference scheme

$$\Lambda_{(3.28)}z(x) \equiv \varepsilon^2\delta_{\bar{x}\hat{x}}z(x) - c(x)z(x) = f(x) \qquad x\in D_h \tag{3.28a}$$

$$\lambda_{(3.28)}z(x) \equiv \left\{\begin{matrix} -\varepsilon\delta_x z(x) & x=0 \\ \varepsilon\delta_{\bar{x}}z(x) & x=1 \end{matrix}\right\} = \psi(x) \qquad x\in\Gamma_h \tag{3.28b}$$

Here $\Lambda_{(3.28)} = \Lambda_{(2.46)}$.

We can obtain an estimate for the solution of the discrete Neumann problem (3.28), (3.27) that is similar to the estimate (2.48) for the solution of the discrete Dirichlet problem (2.46), (2.44). For the nodes of the grid (3.27), let us assume that the condition

$$x^1 - x^0, x^N - x^{N-1} \leq M\varepsilon \qquad x^0 = 0 \qquad x^N = 1 \qquad x^1, x^{N-1} \in D_h \tag{3.29}$$

is fulfilled. Then the following estimate is valid for the solution of problem (3.28), (3.27), (3.29):

$$\max_{\bar{D}_h}|z(x)| \leq M_{(3.30)} \qquad M_{(3.30)} \equiv M^1[\max_{\Gamma_h}|\psi(\xi)| + \max_{\bar{D}_h}|f(\xi)|] \tag{3.30}$$

where the constant M^1 is independent of ε and the functions $\psi(x)$, $f(x)$, but it can be defined only by the value of $M_{(3.29)}$, $M^1 = M^1(M_{(3.29)})$.

We are interested in the function

$$w(x) = z(x) - u(x) \qquad x \in \bar{D}_h$$

which is the error in the solution of the finite difference scheme (3.28), (3.27). In the case of the boundary value problem (3.23), according to the definition, the solution of the finite difference scheme (3.28), (3.27) converges ε-uniformly as $N \to \infty$ if the following estimate is fulfilled

$$\max_{\bar{D}_h} |z(x) - u(x)| \le \lambda(N)$$

where the function $\lambda(N)$ is independent of the parameter ε and tends to zero when $N \to \infty$.

The function $w(x)$ is the solution of the discrete problem

$$\begin{aligned} \Lambda_{(3.28)} w(x) &= -\varepsilon^2 \eta(x) \qquad x \in D_h \\ \lambda_{(3.28)} w(x) &= \varepsilon \nu(x) \qquad x \in \Gamma_h \end{aligned} \tag{3.31}$$

Here $\eta(x)$ is the error in the approximation of the second-order derivative by the difference derivative inside the domain (see Section II.C), and $\nu(x)$ is the error in the approximation of the first-order derivative by the difference derivative on the boundary

$$\nu(x) = \nu(x; u(\cdot)) \qquad \nu(x) \equiv \begin{cases} \delta_x u(x) - \dfrac{d}{dx} u(x) & x = 0 \\ -\delta_{\bar{x}} u(x) + \dfrac{d}{dx} u(x) & x = 1 \end{cases}$$

Thus, the function $w(x)$ is the solution of the discrete problem similar to problem (3.28), (3.27).

The estimate of the function $w(x)$ [the solution of problem (3.31), (3.27)] follows from the estimate (3.30)

$$\max_{\bar{D}_h} |w(x)| \le M[\varepsilon \max_{\Gamma_h} |\nu(\xi)| + \varepsilon^2 \max_{D_h} |\eta(\xi)|] \tag{3.32}$$

In this way, for the solution of the finite difference scheme (3.28), (3.27) to converge ε-uniformly, it is sufficient that the functions $\nu(x)$ and $\eta(x)$ satisfy the inequality

$$\varepsilon \max_{\Gamma_h} |\nu(x)|, \ \varepsilon^2 \max_{D_h} |\eta(x)| \le \lambda(N)$$

In Section II.C the special grid $\bar{\omega}_{*(2.58)}$ was constructed. On this grid the estimate (2.60) is valid for the function $\eta(x)$ in the case of the Dirichlet problem. On the same grid $\bar{\omega}_{*(2.58)}$, in the case of the Neumann problem, a similar estimate is true for the function $\eta(x)$

$$\varepsilon^2 \max_{D_h} |\eta(x; u_{(3.23)}(\cdot))| \le MN^{-1}\ln N \qquad \bar{D} = \bar{D}_h^{sp} \tag{3.33}$$

where

$$\bar{D}_h^{sp} = \bar{D}_{h(2.59)}^{sp} \tag{3.34}$$

On the grid $\bar{D}_h^{sp}$ the following estimate is also valid:

$$\varepsilon \max_{\Gamma_h} |\nu(x; u_{(3.23)}(\cdot))| \le MN^{-1}\ln N \qquad \bar{D} = \bar{D}_h^{sp} \tag{3.35}$$

The ε-uniform convergence of the solution for the finite difference scheme (3.28), (3.34) is established from the estimates (3.32), (3.33) and (3.35)

$$\max_{\bar{D}_h} |u(x) - z(x)| \le MN^{-1}\ln N \tag{3.36}$$

where $\bar{D}_h = \bar{D}_{h(3.34)}^{sp}$, $z(x)$ is the solution of the finite difference scheme (3.28), (3.34). Thus, in the case of the Neumann boundary value problem (3.23), we have constructed a special finite difference scheme to solve problem (3.5) ε-uniformly.

Now we turn to problem (3.6). Taking into account Eqs. (3.23a), (3.28a) and the ε-uniform convergence of the solution of the finite difference scheme (3.28), (3.34), we find that the product $\varepsilon^2\delta_{\bar{x}\hat{x}}z(x)$ converges to the product $\varepsilon^2(d^2/dx^2)u(x)$ ε-uniformly

$$\varepsilon^2 \max_{D_h} \left| \frac{d^2}{dx^2} u(x) - \delta_{\bar{x}\hat{x}} z(x) \right| \le MN^{-1}\ln N \qquad \bar{D} = \bar{D}_h^{sp} \tag{3.37}$$

Using estimates (3.36), (3.37), we establish that the function $P^{h+}(x) = P^{h+}(x; z(\cdot))$, where $z(x)$ is the solution of problem (3.28), (3.34), approximates the function $P(x) = P(x; u_{(3.23)}(\cdot))$ ε-uniformly, that is, the computed diffusion fluxes approximate the actual diffusion fluxes ε-uniformly

$$\max_{\bar{D}_{\bar{h}}} |P(x) - P^{h+}(x)| \le MN^{-1}\ln N \qquad \bar{D} = \bar{D}_h^{sp}$$

The estimate

$$\max_{x^i \le x \le x^{i+1}} \max_{x^i \in \bar{D}_h^{sp-}} |P(x) - P^{h+}(x^i)| \le MN^{-1} \ln N$$

is also fulfilled.

Thus, for the Neumann boundary value problem (3.23), a special finite difference scheme has been constructed. Its solution $z(x)$, $x \in \bar{D}_h$, and the function $P^{h+}(x)$, $x \in \bar{D}_h^-$ allow us to approximate the solution of the boundary value problem and the normalized diffusion flux ε-uniformly. In other words, they solve problem (3.6) ε-uniformly.

D. Special Finite Difference Schemes for Problems (3.2), (3.1) and (3.4), (3.3): Numerical Experiments with the Special Finite Difference Scheme

We construct a special finite difference scheme for the one-dimensional problem (3.2), (3.1). On the set $\bar{G}_{(3.1)}$ we introduce the rectangular grid

$$\bar{G}_h \equiv \bar{G}_{h(3.38)} = \bar{G}_{h(2.66)} = \bar{\omega}_{1(2.66)} \times \bar{\omega}_{0(2.66)} \tag{3.38}$$

where $\bar{\omega}_1$ is a grid with an arbitrary distribution of nodes. On this grid we consider the set of difference equations corresponding to problem (3.2)

$$\Lambda_{(3.39)} z(x,t) \equiv \{\varepsilon^2 a(x,t)\delta_{\bar{x}\hat{x}} - p(x,t)\delta_{\bar{t}} - c(x,t)\} z(x,t) = f(x,t) \tag{3.39a}$$
$$(x,t) \in G_h$$

$$\lambda_{(3.39)} z(x,t) \equiv \begin{Bmatrix} -\varepsilon\delta_x z(x,t) & x = d_0 \\ \varepsilon\delta_{\bar{x}} z(x,t) & x = d_1 \end{Bmatrix} = \psi(x,t) \qquad (x,t) \in S_h^1 \tag{3.39b}$$

$$z(x,t) = \varphi(x) \qquad (x,t) \in S_h^0 \tag{3.39c}$$

The difference scheme (3.39), (3.38) allows us to find the function $z(x,t)$, $(x,t) \in \bar{G}_h$. The approximation of the normalized diffusion flux $P(x,t)$, when the function $P(x,t)$ is continuous on $\bar{G}$, can be computed, for example, by the formula

$$P^{h+}(x,t) \equiv \varepsilon\delta_x z(x,t) \qquad (x,t) \in \bar{G}_h^-$$

where $\bar{G}_h^-$ is a set of nodes from $\bar{G}_{0h} = \bar{G}_0 \cap \bar{G}_h$ for which the operator δ_x is defined, that is, the node $(x,t) = (x^i, t)$ belongs to $\bar{G}_{0h}^-$ if the nodes (x^i, t), (x^{i+1}, t) belong to $\bar{G}_{0h}$.

As is known from the theory of finite difference schemes, the solution of the finite difference scheme (3.39), (3.38) for N, $N_0 \to \infty$ converges for a fixed value of the parameter ε as follows:

$$\max_{\bar{G}_h} |u(x,t) - z(x,t)| \le Q(\varepsilon)[N^{-1} + N_0^{-1}] \qquad (x,t) \in \bar{G}_h$$

If the grid

$$\bar{G}_h = \bar{\omega}_1 \times \bar{\omega}_0 \tag{3.40}$$

is uniform with respect to all its variables, then, when the function $P(x, t)$ is sufficiently smooth on $\bar{G}$, we obtain

$$\max_{\bar{G}_h^-} |P(x,t) - P^{h+}(x,t)| \le Q(\varepsilon)[N^{-1} + N_0^{-1}]$$

that is, the computed diffusion flux converges to the actual flux for a fixed value of the parameter ε.

The function $w(x,t) = z(x,t) - u(x,t)$, $(x,t) \in \bar{G}_h$ (the error in the approximate solution) is the solution of the grid problem

$$\Lambda_{(3.39)} w(x,t) = \beta(x,t) \qquad (x,t) \in G_h$$

$$\lambda_{(3.39)} w(x,t) = \gamma(x,t) \qquad (x,t) \in S_h^1$$

$$w(x,t) = 0 \qquad (x,t) \in S_h^0$$

Here $\beta(x,t)$, $\gamma(x,t)$ are truncation errors for Eq. (3.39a) and the boundary condition (3.39b)

$$\begin{aligned}\beta(x,t) &\equiv f(x,t) - \Lambda_{(3.39)} u(x,t) = L_{(3.2)} u(x,t) - \Lambda_{(3.39)} u(x,t)\\ &= \varepsilon^2 a(x,t)\left[\frac{\partial^2}{\partial x^2} u(x,t) - \delta_{\bar{x}\hat{x}} u(x,t)\right] + p(x,t)\left[\delta_{\bar{t}} u(x,t) - \frac{\partial}{\partial t} u(x,t)\right]\end{aligned}$$

$$\gamma(x,t) \equiv \psi(x,t) - \lambda_{(3.39)} u(x,t)$$

$$\equiv \left\{\begin{array}{ll} \varepsilon\left(\delta_x u(x,t) - \dfrac{\partial}{\partial x} u(x,t)\right) & x = d_0 \\ \varepsilon\left(-\delta_{\bar{x}} u(x,t) + \dfrac{\partial}{\partial x} u(x,t)\right) & x = d_1 \end{array}\right\}$$

When the condition

$$x^1 - x^0, x^N - x^{N-1} \le M\varepsilon \qquad x^0, x^1, x^{N-1}, x^N \in \bar{\omega}_{1(3.38)}$$

is fulfilled for the grid $\bar{G}_{(3.38)}$, the following estimate is valid for the function $w(x, t)$

$$\max_{\bar{G}_h} |w(x, t)| \leq M[\max_{S_h^1} |\gamma(\xi, \tau)| + \max_{G_h} |\beta(\xi, \tau)|] \tag{3.41}$$

Thus, for the ε-uniform convergence of the solution of the finite difference scheme (3.39), (3.38) it suffices to construct the grid $\bar{G}_h$ so that the inequalities

$$\max_{S_h^1} |\gamma(\xi, \tau)|, \ \max_{G_h} |\beta(\xi, \tau)| \leq \lambda(N, N_0)$$

are fulfilled, where the function $\lambda(N, N_0)$ does not depend on the parameter ε and tends to zero when $N, N_0 \to \infty$.

It is known from the theory of differential equations that a sufficiently smooth solution of problem (3.2) can be expressed as a sum of the two functions $U(x, t)$ and $V(x, t)$, which are the regular and singular parts of the solution

$$u(x, t) = U(x, t) + V(x, t) \qquad (x, t) \in \bar{G}$$

For the components $U(x, t)$, $V(x, t)$ a priori estimates similar to estimates (2.70b), (2.70c) are valid.

Let us construct the special grid

$$\bar{G}_h^{sp} \equiv \bar{G}_{h(3.42)}^{sp} = \bar{G}_{h(2.72)}^{sp} \tag{3.42}$$

For the Neumann problem, we obtain estimates of the functions $\beta(\xi, \tau)$ and $\gamma(\xi, \tau)$ that are similar to the function $\beta(\xi, \tau)$ in the case of the Dirichlet problem (2.14) (see Section II.D)

$$\max_{S_h^1} |\gamma(\xi, \tau)|, \ \max_{G_h} |\beta(\xi, \tau) \leq M[N^{-1} \ln N + N_0^{-1}]$$

where $\bar{G}_h = \bar{G}_{h(3.42)}^{sp}$. From these, by virtue of estimate (3.41), we can establish the ε-uniform convergence of the solution for finite difference scheme (3.39), (3.42)

$$\max_{\bar{G}_h^{sp}} |u(x, t) - z(x, t)| \leq M[N^{-1} \ln N + N_0^{-1}]$$

The computed normalized diffusion flux $P^{h+}(x, t)$, $(x, t) \in \bar{G}_h^-$ approximates the actual flux $P(x, t)$, $(x, t) \in \bar{G}$ ε-uniformly

$$\max_{\bar{G}_h^{sp-}} |P(x, t) - P^{h+}(x, t)| \leq M[N^{-1} \ln N + N_0^{-1}]$$

Now we construct a finite difference scheme for the boundary value problem (3.4), (3.3). On the set $\bar{G}_{(3.3)}$ we introduce the rectangular grid

$$\bar{G}_h \equiv \bar{G}_{h(3.43)} = \bar{G}_{h(2.73)} = \bar{D}_h \times \bar{\omega}_0 = \bar{\omega}_1 \times \bar{\omega}_2 \times \bar{\omega}_0 \qquad (3.43)$$

where $\bar{\omega}_1, \bar{\omega}_2$ are grids with an arbitrary distribution of nodes. To approximate problem (3.4) on the grid $\bar{G}_{h(3.43)}$, we use the difference scheme

$$\begin{aligned} \Lambda_{(3.44)} z(x,t) &= f(x,t) \qquad (x,t) \in G_h \\ \lambda_{(3.44)} z(x,t) &= \psi(x,t) \qquad (x,t) \in S_h^1 \setminus S_h^c \\ z(x,t) &= \varphi(x) \qquad (x,t) \in S_h^0 \end{aligned} \qquad (3.44a)$$

Here, $S_h^c = S^c \cap \bar{G}_h$,

$$\Lambda_{(3.44)} z(x,t) = \Lambda_{(2.74)} z(x,t)$$
$$\equiv \Big\{ \varepsilon^2 \sum_{s=1,2} a_s(x,t) \delta_{\bar{x}s\hat{x}s} - p(x,t)\delta_{\bar{t}} - c(x,t) \Big\} z(x,t)$$

$$\lambda_{(3.44)} z(x,t) \equiv \begin{cases} -\varepsilon \delta_{xs} z(x,t) & \text{for} \quad x_s = d_{0s}, x_{3-s} \in \omega_{3-s} \\ \varepsilon \delta_{\bar{x}s} z(x,t) & \text{for} \quad x_s = d_{1s}, x_{3-s} \in \omega_{3-s}, s = 1,2 \end{cases}$$

The set of finite difference equations (3.44a), (3.43) allows us to find the function $z(x,t)$ for $(x,t) \in \bar{G}_h \setminus S_h^c$. We write down a relation that allows us to determine the function $z(x,t)$ for $(x,t) \in S_h^c$. Let the node $(x^c, t) = (x_1^c, x_2^c, t) \in S_h^c$ be one of the following four nodes: (x_1^i, x_2^k, t), (x_1^{i+1}, x_2^k, t), (x_1^i, x_2^{k+1}, t), $(x_1^{i+1}, x_2^{k+1}, t)$. Suppose that

$$z(x_1^i, x_2^k, t) + z(x_1^{i+1}, x_2^{k+1}, t) = z(x_1^{i+1}, x_2^k, t) + z(x_1^i, x_2^{k+1}, t)$$
$$(x^c, t) \in S_h^c \qquad (3.44b)$$
$$x^c \in \{(x_1^i, x_2^k), (x_1^{i+1}, x_2^k), (x_1^i, x_2^{k+1}), (x_1^{i+1}, x_2^{k+1})\}$$

that is, the values of the function $z(x,t)$ at these nodes lie on a plane. The finite difference scheme (3.44), (3.43), that is, the set of finite difference equations (3.44), (3.43), allows us to find the function $z(x,t)$ for $(x,t) \in \bar{G}_h$.

To approximate the normalized diffusion fluxes $P_s(x,t)$, we use the formulas

$$P_s^{h+}(x,t) \equiv \varepsilon \delta_{xs} z(x,t) \qquad (x,t) \in \bar{G}_h^{s-} \qquad s = 1,2$$

if the functions $P_s(x,t)$, $s=1,2$ are continuous on $\bar{G}$. Here

$$\bar{G}_h^{s-} = \bar{D}_h^{s-} \times \bar{\omega}_0 \qquad s=1,2$$

$$\bar{D}_h^{1-} = \bar{\omega}_1^- \times \bar{\omega}_2 \qquad \bar{D}_h^{2-} = \bar{\omega}_1 \times \bar{\omega}_2^- \qquad \bar{\omega}_s^- = \omega_s \cup \{x_s = d_{0s}\}$$

If the functions $P_s(x,t)$, $s=1,2$ are continuous only on $\bar{G}_0$, then we suppose that

$$P_s^{h+}(x,t) \equiv \varepsilon \delta_{xs} z(x,t) \qquad (x,t) \in \bar{G}_{0h}^{s-} \qquad s=1,2$$

where $\bar{G}_{0h}^{s-}$ is a set of nodes from $\bar{G}_{0h} = \bar{G} \cap \bar{G}_h$ for which the operator δ_{xs} is defined, that is, the nodes (x_1^i, x_2, t) and (x_1, x_2^k, t) belong to the sets $\bar{G}_{0h}^{1-}$ and $\bar{G}_{0h}^{2-}$, respectively, if the nodes (x_1^i, x_2, t), (x_1^{i+1}, x_2, t) and (x_1, x_2^k, t), (x_1, x_2^{k+1}, t) belong to the set $\bar{G}_{0h}$.

The investigation of convergence for the finite difference scheme (3.44), (3.43) is similar to the investigation for the finite difference scheme (3.39), (3.38) for problem (3.2), (3.1). We know from the theory of finite difference schemes that, if the solution of the boundary value problem (3.4), (3.3) is sufficiently smooth (e.g., if it has continuous derivatives up to fourth order in x_1 and x_2 and up to second order in t on the set $\bar{G}$) for each fixed value of the parameter, then the solution of the finite difference scheme (3.44), (3.43) converges for $N, N_0 \to \infty$ (where $N = \min[N_1, N_2]$) to the solution of the boundary value problem (3.4), (3.3) for a fixed value of the parameter, and the estimate

$$\max_{\bar{G}_h} |u(x,t) - z(x,t)| \le Q(\varepsilon)[N^{-1} + N_0^{-1}]$$

is valid. If the grid

$$\bar{G}_h = \bar{\omega}_1 \times \bar{\omega}_2 \times \bar{\omega}_0 \tag{3.45}$$

is uniform in all of its variables, and the functions $P_s(x,t)$, $s=1,2$ are sufficiently smooth on $\bar{G}$, the following estimate holds

$$\max_{\bar{G}_h^{s-}} |P_s(x,t) - P_s^{h+}(x,t)| \le Q(\varepsilon)[N^{-1} + N_0^{-1}] \qquad s=1,2$$

where $\bar{G}_h = \bar{G}_{h(3.45)}$.

On the set $\bar{G}$ we introduce the special grid

$$\bar{G}_h^{sp} \equiv \bar{G}_{h(3.46)}^{sp} = \bar{G}_{h(2.76)}^{sp} \tag{3.46}$$

Taking into account the estimates of the solution and its derivatives for problem (3.4), (3.3), we establish the ε-uniform convergence of the solution of the finite difference scheme (3.44), (3.46)

$$\max_{\bar{G}_h^{sp}} |u(x, t) - z(x, t)| \leq M[N^{-1}\ln N + N_0^{-1}]$$

as well as the ε-uniform convergence of the discrete diffusion fluxes

$$\max_{\bar{G}_h^{sps-}} |P_s(x, t) - P_s^{h+}(x, t)| \leq M[N^{-1}\ln N + N_0^{-1}] \qquad s = 1, 2$$

Thus, the finite difference schemes (3.39), (3.42) and (3.44), (3.46) and, respectively, the grid fluxes $P^{h+}(x, t)$ and $P_s^{h+}(x, t)$ allow us to approximate ε-uniformly both the solutions of the boundary value problems (3.2), (3.1) and (3.4), (3.3) and the corresponding diffusion fluxes, that is, these schemes and grid fluxes solve problem (3.6) ε-uniformly.

Now we return to the study of the normalized diffusion fluxes for boundary value problems with Dirichlet boundary condition. In Section II.D the ε-uniformly convergent finite difference schemes (2.74), (2.76) and (2.67), (2.72) were constructed for the Dirichlet problems (2.12), (2.13) and (2.14), (2.15), respectively. For these problems, we now construct and analyze the approximations of the normalized diffusion fluxes. We consider the normalized diffusion fluxes for problem (2.14), (2.15) in the form

$$P(x, t) \equiv \varepsilon \frac{\partial}{\partial x} u(x, t) \qquad (x, t) \in \bar{G}_{(2.15)}$$

and for problem (2.12), (2.13) in the form

$$P_s(x, t) \equiv \varepsilon \frac{\partial}{\partial x_s} u(x, t) \qquad (x, t) \in \bar{G}_{(2.13)} \qquad s = 1, 2$$

that is, fluxes with respect to the variables x_1 and x_2. To approximate these fluxes, we use the discrete fluxes

$$P^{h+}(x, t) \equiv \varepsilon\delta_x z(x, t) \qquad (x, t) \in \bar{G}_{h(2.72)}^{sp-}$$

where $z(x, t)$ is the solution of the difference scheme (2.67), (2.72), and

$$P_s^{h+}(x, t) \equiv \varepsilon\delta_{xs} z(x, t) \qquad (x, t) \in \bar{G}_{h(2.76)}^{sps-} \qquad s = 1, 2$$

where $z(x, t)$ is the solution of the difference scheme (2.74), (2.76). For the discrete diffusion fluxes (on the special grids) the following estimates are satisfied in the case of problem (2.14), (2.15):

$$\max_{\bar{G}_h^{sp-}} |P(x,t) - P^{h+}(x,t)| \le M[N^{-1}\ln N + N_0^{-1}] \qquad \bar{G}_h^{sp-} = \bar{G}_{h(2.72)}^{sp-}$$

and in the case of problem (2.12), (2.13)

$$\max_{\bar{G}_h^{sps-}} |P_s(x,t) - P_s^{h+}(x,t)| \le M[N^{-1}\ln N + N_0^{-1}]$$

$$s = 1, 2 \qquad \bar{G}_h^{sps-} = \bar{G}_{h(2.76)}^{sps-}$$

That is, for the Dirichlet boundary value problems (2.12), (2.13) and (2.14), (2.15), respectively, finite difference schemes (2.74), (2.76) and (2.67), (2.72), and the grid fluxes $P^{h+}(x,t)$ and $P_s^{h+}(x,t)$ solve problem (3.6) ε-uniformly.

Now we study the efficiency of the special finite difference scheme in the case of the Neumann boundary value problem (3.7), (3.9). On the set $\bar{G}_{(3.8)}$ we introduce the special grid

$$\bar{G}_h^{sp} \equiv \bar{G}_{h(3.47)}^{sp} = \bar{G}_{h(2.77)}^{sp} \tag{3.47}$$

To solve the problem, we use the finite difference scheme

$$\begin{aligned} \Lambda_{(3.48)} z(x,t) &= f(x,t) \qquad (x,t) \in G_h^{sp} \\ \lambda_{(3.48)} z(x,t) &= \psi(x,t) \qquad (x,t) \in S_h^{sp1} \\ z(x,t) &= \varphi(x) \qquad (x,t) \in S_h^{sp0} \end{aligned} \tag{3.48}$$

Here

$$f(x,t) = f_{(3.9)}(x,t) \qquad \psi(x,t) = \psi_{(3.9)}(x,t) \qquad \varphi(x) = \varphi_{(3.9)}(x)$$

$$\Lambda_{(3.48)} = \Lambda_{(2.78)} \qquad \lambda_{(3.48)} = \lambda_{(3.17)}$$

In order to analyze the error in the approximate solution, we find the errors in each component of the representation (3.13) of the boundary value problem solution. The boundary value problems (3.10), (3.14), and

TABLE XXIV
Table of Errors $E(\varepsilon, N)$ for the Special Scheme (3.49), (3.47) in the Case of Problem (3.10)

$\varepsilon \backslash N$	4	16	64	256	1024
1	2.511e − 1	6.265e − 2	1.636e − 2	4.259e − 3	1.076e − 3
2^{-2}	5.000e − 1	1.175e − 1	2.882e − 2	7.168e − 3	1.790e − 3
2^{-4}	3.965e − 1	9.389e − 2	2.149e − 2	5.264e − 3	1.309e − 3
2^{-6}	2.590e − 1	6.577e − 2	1.650e − 2	4.160e − 3	1.052e − 3
2^{-8}	2.506e − 1	6.270e − 2	1.568e − 2	3.922e − 3	9.813e − 4
2^{-10}	2.500e − 1	6.251e − 2	1.563e − 2	3.907e − 3	9.769e − 4
2^{-12}	2.500e − 1	6.250e − 2	1.563e − 2	3.906e − 3	9.766e − 4
$\bar{E}(N)$	5.000e − 1	1.175e − 1	2.882e − 2	7.169e − 3	1.790e − 3

(3.15) are approximated by the finite difference schemes

$$\begin{aligned} \Lambda_{(3.48)} z(x,t) &= f(x,t) \qquad (x,t) \in G_h^{sp} \\ \lambda_{(3.48)} z(x,t) &= 0 \qquad (x,t) \in S_h^{sp1} \\ z(x,t) &= \varphi(x) \qquad (x,t) \in S_h^{sp0} \end{aligned} \tag{3.49}$$

$$\begin{aligned} \Lambda_{(3.48)} z(x,t) &= 0 \qquad (x,t) \in G_h^{sp} \\ \lambda_{(3.48)} z(x,t) &= l_{(3.7)} W(x,t) \qquad (x,t) \in S_h^{sp1} \\ z(x,t) &= 0 \qquad (x,t) \in S_h^{sp0} \end{aligned} \tag{3.50}$$

$$\begin{aligned} \Lambda_{(3.48)} z(x,t) &= 0 \qquad (x,t) \in G_h^{sp} \\ \lambda_{(3.48)} z(x,t) &= l_{(3.7)} v(x,t) \qquad (x,t) \in S_h^{sp1} \\ z(x,t) &= 0 \qquad (x,t) \in S_h^{sp0} \end{aligned} \tag{3.51}$$

In Tables XXIV and XXV we give the values of the errors $E(\varepsilon, N)$ and

TABLE XXV
Table of Errors $E(\varepsilon, N)$ for the Special Scheme (3.50), (3.47) in the Case of Problem (3.14)

$\varepsilon \backslash N$	4	16	64	256	1024
1	4.736e − 1	9.723e − 2	2.332e − 2	5.771e − 3	1.439e − 3
2^{-2}	1.098e + 0	2.347e − 1	5.581e − 2	1.377e − 2	3.431e − 3
2^{-4}	3.367e + 0	1.014e + 0	2.114e − 1	4.994e − 2	1.230e − 2
2^{-6}	3.221e + 0	1.492e + 0	4.650e − 1	1.398e − 1	4.196e − 2
2^{-8}	3.184e + 0	1.492e + 0	4.650e − 1	1.398e − 1	4.196e − 2
2^{-10}	3.175e + 0	1.492e + 0	4.650e − 1	1.398e − 1	4.196e − 2
2^{-12}	3.172e + 0	1.492e + 0	4.650e − 1	1.398e − 1	4.196e − 2
$\bar{E}(N)$	3.367e + 0	1.492e + 0	4.650e − 1	1.398e − 1	4.196e − 2

TABLE XXVI
Table of Errors $E_{1024,1024}(\varepsilon, N)$ for the Special Scheme (3.51), (3.47) in the Case of Problem (3.15)

$\varepsilon \backslash N$	4	16	64	256
1	1.260e − 01	2.511e − 02	5.713e − 03	1.130e − 03
2^{-2}	4.944e − 04	1.043e − 04	2.295e − 05	4.485e − 06
2^{-4}	1.076e − 31	3.698e − 32	7.664e − 33	1.315e − 33
2^{-6}	0.000e − 00	0.000e − 00	0.000e − 00	0.000e − 00
$E(N)$	1.260e − 01	2.511e − 02	5.713e − 03	1.130e − 03

$\bar{E}(N)$ computed for problems (3.10) and (3.14) with various values of ε and N and $N_0 = N$. It is seen from the tables that, as N increases, the functions $z_{(3.49)}(x, t)$ and $z_{(3.50)}(x, t)$ converge ε-uniformly to the functions $U(x, t)$ and $W(x, t)$, respectively.

The values of $E_{1024,1024}(\varepsilon, N)$ and $\bar{E}_{1024,1024}(N)$ for problem (3.15) are given in Table XXVI. Note that the special grid is uniform for the particular values of ε and N given in the table (except for $\varepsilon = 2^{-4}$, $N = 4$). Therefore, this table and Table XX differ only in the result for $\varepsilon = 2^{-4}$, $N = 4$. We see from the results given in Table XX that the error $E_{1024,1024}(\varepsilon, N)$ tends to zero ε-uniformly and, consequently, the function $z_{(3.51)}(x, t)$ converges to the function $v(x, t)$ ε-uniformly.

The results given in Tables XXIV–XXVI lead to the conclusion that the solution of the finite difference scheme (3.48), (3.47) converges to the solution of the boundary value problem ε-uniformly.

For the discrete problems (3.49), (3.47); (3.50), (3.47) and (3.51), (3.47) the discrete diffusion fluxes were computed and their convergence was studied. The results of computing the values of $Q(\varepsilon, N)$ and $\bar{Q}(N)$ for problems (3.10) and (3.14) are given in Tables XXVII and XXVIII. It is

TABLE XXVII
Table of Errors of the Normalized Flux $Q(\varepsilon, N)$ for the Special Scheme (3.49), (3.47) in the Case of Problem (3.10)

$\varepsilon \backslash N$	4	16	64	256	1024
1	3.773e − 1	3.307e − 1	1.321e − 1	3.709e − 2	9.546e − 3
2^{-2}	4.760e − 1	1.474e − 1	3.817e − 2	9.615e − 3	2.408e − 3
2^{-4}	1.007e − 1	3.821e − 2	9.629e − 3	2.409e − 3	6.024e − 4
2^{-6}	2.722e − 2	7.376e − 3	1.620e − 3	4.175e − 4	1.305e − 4
2^{-8}	7.563e − 3	2.253e − 3	5.582e − 4	1.361e − 4	3.308e − 5
2^{-10}	1.938e − 3	5.870e − 4	1.479e − 4	3.680e − 5	9.149e − 6
2^{-12}	4.873e − 4	1.482e − 4	3.746e − 5	9.360e − 6	2.337e − 7
$\bar{Q}(N)$	4.760e − 1	3.307e − 1	1.321e − 1	3.709e − 2	9.546e − 3

TABLE XXVIII
Table of Errors of the Normalized Flux $Q(\varepsilon, N)$ for the Special Scheme (3.50), (3.47) in the Case of Problem (3.14)

$\varepsilon \backslash N$	4	16	64	256	1024
1	1.675e − 1	3.658e − 2	8.844e − 3	2.193e − 3	5.470e − 4
2^{-2}	8.878e − 2	1.728e − 2	3.903e − 3	9.482e − 4	2.353e − 4
2^{-4}	2.524e − 2	5.094e − 2	6.211e − 3	1.095e − 3	2.444e − 4
2^{-6}	9.922e − 3	7.228e − 2	1.431e − 2	2.137e − 3	3.520e − 4
2^{-8}	1.153e − 2	7.228e − 2	1.431e − 2	2.137e − 3	3.520e − 4
2^{-10}	1.163e − 2	7.228e − 2	1.431e − 2	2.137e − 3	3.520e − 4
2^{-12}	1.163e − 2	7.228e − 2	1.431e − 2	2.137e − 3	3.520e − 4
$\bar{Q}(N)$	1.675e − 1	7.288e − 2	1.431e − 2	2.193e − 3	5.470e − 4

obvious from the tables that the discrete diffusion fluxes converge to the actual fluxes $P(x, t; U(\cdot))$ and $P(x, t; W(\cdot))$ when N increases and $N_0 = N$.

In Table XXIX the errors $Q_{1024,1024}(\varepsilon, N)$ and $\bar{Q}_{1024,1024}(N)$ are given for problem (3.15). We see from this table that, as N increases and $N_0 = N$, the error $Q_{1024,1024}(\varepsilon, N)$ tends to zero, and the discrete diffusion flux $P^{h+}(x, t; v_{N^*,N_0^*}(\cdot))$ converges to the flux $P(x, t; v_{N^*,N_0^*}(\cdot))$ ε-uniformly. The analysis of Tables XXVII–XXIX demonstrates the ε-uniform convergence of the grid fluxes for each component of the representation (3.13) and, consequently, for the total grid flux for the boundary value problem (3.7), (3.9).

Finally, we have also analyzed the convergence of the discrete fluxes for the Dirichlet boundary value problem (2.16). Here we used the special finite difference scheme. The grid fluxes were considered separately for each component in the representation (2.24). In Tables XXX and XXXI, we give the errors $Q(\varepsilon, N)$, $\bar{Q}(N)$ computed from the solutions of the difference schemes (2.79), (2.77) and (2.80), (2.77) corresponding to problems (2.20) and (2.25). Table XXXII shows the

TABLE XXIX
Table of Errors of the Normalized Flux $Q_{1024,1024}(\varepsilon, N)$ for the Special Scheme (3.51), (3.47) in the Case of Problem (3.15)

$\varepsilon \backslash N$	4	16	64	256
1	1.258e − 01	3.549e − 02	9.153e − 03	2.306e − 03
2^{-2}	5.353e − 04	2.816e − 04	8.831e − 05	2.342e − 05
2^{-4}	3.980e − 32	3.979e − 32	3.464e − 32	1.585e − 32
2^{-6}	0.000e − 00	0.000e − 00	0.000e − 00	0.000e − 00
$\bar{Q}(N)$	1.258e − 01	3.549e − 02	9.153e − 03	2.306e − 03

TABLE XXX

Table of Errors of the Normalized Flux $Q(\varepsilon, N)$ for the Special Scheme (2.79), (2.77) in the Case of Problem (2.20)

$\varepsilon \backslash N$	4	16	64	256	1024
1	7.240e − 1	3.218e − 1	9.282e − 2	2.413e − 2	6.095e − 3
2^{-2}	2.729e − 1	7.402e − 2	1.909e − 2	4.807e − 3	1.204e − 3
2^{-4}	1.024e − 1	4.590e − 2	1.587e − 2	4.295e − 3	1.095e − 3
2^{-6}	8.890e − 2	3.619e − 2	1.385e − 2	4.077e − 3	1.076e − 3
2^{-8}	8.983e − 2	3.612e − 2	1.384e − 2	4.075e − 3	1.076e − 3
2^{-10}	9.009e − 2	3.611e − 2	1.384e − 2	4.075e − 3	1.076e − 3
2^{-12}	9.015e − 2	3.611e − 2	1.384e − 2	4.075e − 3	1.076e − 3
$\bar{Q}(N)$	7.240e − 1	3.218e − 1	9.282e − 2	2.413e − 2	6.095e − 3

TABLE XXXI

Table of Errors of the Normalized Flux $Q(\varepsilon, N)$ for the Special Scheme (2.80), (2.77) in the Case of Problem (2.25)

$\varepsilon \backslash N$	4	16	64	256	1024
1	1.620e − 1	6.123e − 2	2.362e − 2	9.946e − 3	4.496e − 3
2^{-2}	4.516e − 1	1.328e − 1	4.345e − 2	1.535e − 2	5.904e − 3
2^{-4}	7.867e − 1	4.332e − 1	1.228e − 1	3.321e − 2	1.086e − 2
2^{-6}	7.727e − 1	5.505e − 1	2.428e − 1	8.507e − 2	2.701e − 2
2^{-8}	7.690e − 1	5.505e − 1	2.428e − 1	8.507e − 2	2.701e − 2
2^{-10}	7.680e − 1	5.505e − 1	2.428e − 1	8.507e − 2	2.701e − 2
2^{-12}	7.678e − 1	5.505e − 1	2.428e − 1	8.507e − 2	2.701e − 2
$\bar{Q}(N)$	7.867e − 1	5.505e − 1	2.428e − 1	8.507e − 2	2.701e − 2

errors $Q_{1024,1024}(\varepsilon, N)$, $\bar{Q}_{1024,1024}(N)$ obtained from the solution of the difference scheme (2.81), (2.77) corresponding to problem (2.26).

The results demonstrate that, as N increases with $N_0 = N$, the discrete fluxes related to each component of the representation (2.24) converge to the actual fluxes ε-uniformly, and that the total grid flux for the

TABLE XXXII

Table of Errors of the Normalized Flux $Q_{1024,1024}(\varepsilon, N)$ for the Special Scheme (2.81), (2.77) in the Case of Problem (2.26)

$\varepsilon \backslash N$	4	16	64	256
1	3.934e − 02	1.176e − 02	3.268e − 03	1.033e − 03
2^{-2}	4.777e − 04	3.054e − 04	1.066e − 04	3.701e − 05
2^{-4}	5.963e − 32	1.591e − 31	3.514e − 31	2.499e − 31
2^{-6}	0.000e − 00	0.000e − 00	0.000e − 00	0.000e − 00
$\bar{Q}(N)$	3.934e − 02	1.176e − 02	3.268e − 03	1.033e − 03

difference scheme (2.78), (2.77) also converges to the actual flux for the boundary value problem (2.16) ε-uniformly.

Thus, we see that the newly constructed finite difference schemes are indeed effective and that they allow us to approximate the solution and the normalized diffusion fluxes ε-uniformly for both Dirichlet and Neumann boundary value problems with singular perturbations.

IV. DIFFUSION EQUATIONS WITH CONCENTRATED SOURCES

In Sections II and III, we considered boundary value problems modeling the diffusion process for some substance in a homogeneous material. We have studied problems in which the unknown function takes a prescribed value on the boundary (in Section II), or the diffusion flux of the unknown function is given on the boundary (in Section III). The right-hand sides of the equations were supposed to be smooth. Boundary layers appear in these problems for small values of the parameter.

In this section, we consider the singularly perturbed diffusion equation when linear combinations of the solution and its diffusion flux are given on the domain boundary. Such boundary conditions make it possible to realize any of the boundary conditions considered in Sections II and III. Moreover, concentrated sources act inside the domain. These sources lead to the appearance of interior layers. Thus, in addition to the computational problems accompanying the solution of the boundary value problems in Sections II and III, there arise new problems due to the presence of these interior layers.

A. Mathematical Formulation of the Problems

We begin our consideration with a simple example that brings us to the singularly perturbed boundary value problems considered in this section. Suppose that it is required to find the function $C(y, \tau)$, which is the distribution of temperature in a homogeneous material, or in a layer of solid material with a thickness $2L$. As in Section II.A, we assume that the distributed heat sources of density $F(y, \tau)$ act inside the material. Besides these sources, in the middle part of the material a concentrated source of the strength $Q(\tau)$ is situated at $y = 0$. In a simplified variant, the temperature distribution in a layer of material is described by the heat equation

$$D\frac{\partial^2}{\partial y^2}C(y, \tau) - \frac{\partial}{\partial \tau}C(y, \tau) = -F(y, \tau)$$
$$-L < y < L \qquad 0 < \tau \leq \vartheta \qquad y \neq 0 \tag{4.1}$$

(see Section II.A). Here D is the coefficient of heat conductivity, and ϑ is a typical duration of the process. The following conditions are fulfilled at $y = 0$:

$$C(-0, \tau) = C(+0, \tau) \tag{4.2}$$

$$\lambda \frac{\partial}{\partial y} C(+0, \tau) - \lambda \frac{\partial}{\partial y} C(-0, \tau) = -Q(\tau) \qquad 0 < \tau \leq \vartheta \tag{4.3}$$

Here $C(a-0, \tau)$ [or $C(a+0, \tau)$] denotes the limit of the function $C(y, \tau)$ for $y \to a$ when $y < a$ (or, respectively, when $y > a$)

$$C(a-0, \tau) = \lim_{\substack{\xi \to a \\ \xi < a}} C(\xi, \tau) \qquad C(a+0, \tau) = \lim_{\substack{\xi \to a \\ \xi > a}} C(\xi, \tau)$$

where a is an arbitrary number in the interval $(-L, L)$. In the same way the derivatives $(\partial/\partial y)C(a-0, \tau)$, $(\partial/\partial y)C(a+0, \tau)$ are defined; in formulas (4.2) and (4.3) $a = 0$. Condition (4.2) means that the temperature on the left of the concentrated source $C(-0, \tau)$ is equal to that on the right of the source $C(+0, \tau)$, that is, the temperature in the material is continuous when it crosses the source point $y = 0$. Condition (4.3) can be written in the form

$$P^H(+0, \tau) - P^H(-0, \tau) = -Q(\tau) \qquad 0 < \tau \leq \vartheta$$

where $P^H(y, \tau) \equiv \lambda(\partial/\partial y)C(y, \tau)$ is the heat flux. Thus, condition (4.3) means that when crossing the point $y = 0$, where the concentrated source is located, the heat flux varies by an amount equal to the source strength. Conditions (4.2), (4.3) are called conjugation conditions at the point where the concentrated source acts. These conditions are usually written in the form

$$[C(y, \tau)] = 0 \qquad \lambda\left[\frac{\partial}{\partial y} C(y, \tau)\right] = -Q(\tau) \qquad y = 0 \qquad 0 < \tau \leq \vartheta \tag{4.4}$$

where the symbol $[v(a, \tau)]$ denotes the jump of the function $v(y, \tau)$ at the point $y = a$

$$[v(a, \tau)] = v(a+0, \tau) - v(a-0, \tau)$$

On the left and right boundaries of the layer under consideration the following condition is prescribed

$$\alpha(y)C(y,\tau)+\nu(y)\lambda\frac{\partial}{\partial n}C(y,\tau)=\Psi(y,\tau) \qquad y=-L, L \qquad 0<\tau\leq\vartheta \tag{4.5}$$

Here $(\partial/\partial n)$ is the derivative along the outward normal to the surface of the set G, the coefficients $\alpha(y)$, $\nu(y)$, $y=-L, L$ satisfy the relations

$$0\leq\alpha(y),\ \nu(y)<\infty \qquad \alpha(y)+\nu(y)>0 \qquad y=-L, L$$

The right-hand side in Eq. (4.5) characterizes the intensity of the surface heat sources, and the left-hand side shows where this heat is spent. The first term in the left-hand side of (4.5) characterizes the part of the heat flux that passes from the surface to the surrounding medium by means of heat transfer, while the second term describes the part of the heat flux that passes from the surface into the material by heat conductivity. At the initial instant the temperature distribution in the material is known

$$C(y,0)=C^0(y) \qquad -L\leq y\leq L \tag{4.6}$$

By solving problem (4.1), (4.4), (4.5), (4.6) we find the function $C(y,\tau)$, that is, the distribution of the temperature in the material at the time τ.

For the problem of substance diffusion in the layer in the presence of concentrated substance sources, Eqs. (4.1), (4.6) hold, while Eqs. (4.4) and (4.5) are replaced by the equations

$$[C(y,\tau)]=0 \qquad D\left[\frac{\partial}{\partial y}C(y,\tau)\right]=-Q(\tau) \qquad y=0 \qquad 0<\tau\leq\vartheta \tag{4.7}$$

$$\alpha(y)C(y,\tau)+\nu(y)D\frac{\partial}{\partial n}C(y,\tau)=\Psi(y,\tau) \qquad y=-L, L \qquad 0<\tau\leq\vartheta \tag{4.8}$$

Equations (4.1), (4.7), (4.8), (4.6) allow us to find the distribution of substance concentration in the material for $0\leq\tau\leq\vartheta$.

In the boundary value problems (4.1), (4.4), (4.5), (4.6) and (4.1), (4.7), (4.8), (4.6), we change to new variables defined by relations (2.6)

$$x=L^{-1}y \qquad t=\vartheta^{-1}\tau$$

If we take into account the notation

$$\varepsilon^2 = D\vartheta L^{-2}$$

we arrive at the problem

$$\begin{gathered}
\varepsilon^2 \frac{\partial^2}{\partial x^2} u(x,t) - \frac{\partial}{\partial t} u(x,t) = f(x,t) \qquad -1<x<1 \qquad 0<t\le 1 \qquad x \ne 0 \\
[u(x,t)] = 0 \qquad \varepsilon\left[\frac{\partial}{\partial x} u(x,t)\right] = q(t) \qquad x = 0 \\
\beta(x)u(x,t) + (1-\beta(x))\varepsilon \frac{\partial}{\partial n} u(x,t) = \psi(x,t) \qquad x = -1, 1 \qquad 0<t\le 1 \\
u(x,0) = \varphi(x) \qquad -1 \le x \le 1
\end{gathered} \tag{4.9}$$

Here $\beta(x)$, $f(x,t)$, $q(t)$, $\psi(x,t)$, $\varphi(x)$ are given functions, $0 \le \beta(x) \le 1$. These functions and the function $u(x,t)$, which is the solution of problem (4.9), are defined by the relations

$$u(x,t) = C_*^{-1} C(y(x), \tau(t)) \qquad f(x,t) = -\vartheta C_*^{-1} F(y(x), \tau(t))$$
$$\varphi(x) = C_*^{-1} C^0(y(x))$$

and

$$\beta(x) = \vartheta^{1/2}\alpha(y(x))[\vartheta^{1/2}\alpha(y(x)) + D^{1/2}\nu(y(x))]^{-1}$$
$$q(t) = -\vartheta^{1/2} C_*^{-1} D^{-1/2} Q(\tau(t))$$
$$\psi(x,t) = \vartheta^{1/2} C_*^{-1} [\vartheta^{1/2}\alpha(y(x)) + D^{1/2}\nu(y(x))]^{-1} \Psi(y(x), \tau(t))$$

for substance diffusion, and by the relations

$$\beta(x) = \vartheta^{1/2} D^{1/2}\alpha(y(x))[\vartheta^{1/2} D^{1/2}\alpha(y(x)) + \lambda\nu(y(x))]^{-1}$$
$$q(t) = -\lambda^{-1}\vartheta^{1/2} C_*^{-1} D^{1/2} Q(\tau(t))$$
$$\psi(x,t) = \vartheta^{1/2} C_*^{-1} D^{1/2} [\vartheta^{1/2} D^{1/2}\alpha(y(x)) + \lambda\nu(y(x))]^{-1} \Psi(y(x), \tau(t))$$

for heat transfer.

Problem (4.9) is a Dirichlet or Neumann problem under the conditions $\beta(-1) = \beta(1) = 1$ or $\beta(-1) = \beta(1) = 0$, respectively. For other values of $\beta(x)$, problem (4.9) is a problem with mixed boundary conditions.

In the case when $q(t) \equiv 0$ for $0 < t \le 1$, problem (4.9) is equivalent to the problem

$$\varepsilon^2 \frac{\partial^2}{\partial x^2} u(x, t) - \frac{\partial}{\partial t} u(x, t) = f(x, t) \qquad -1 < x < 1, \qquad 0 < t \leq 1$$

$$\beta(x) u(x, t) + (1 - \beta(x)) \varepsilon \frac{\partial}{\partial n} u(x, t) = \psi(x, t) \qquad x = -1, 1 \qquad 0 < t \leq 1$$

$$u(x, 0) = \varphi(x) \qquad -1 \leq x \leq 1 \tag{4.10}$$

When $\beta(-1) = \beta(1) = 1$, we arrive at the problem with prescribed values of the solution on the material surface, which was considered in Section II. In the case of $\beta(-1) = \beta(1) = 0$, problem (4.10) is the problem with given diffusion flux on the boundary, which was studied in Section III.

On the set

$$G = D \times (0, T] \tag{4.11a}$$

where D is the segment $[d_0, d_1]$ on the axis x, we consider the following singularly perturbed equation of parabolic type with a concentrated source at the point $x = d^*$

$$L_{(4.12)} u(x, t) \equiv \left\{ \varepsilon^2 a(x, t) \frac{\partial^2}{\partial x^2} - p(x, t) \frac{\partial}{\partial t} - c(x, t) \right\} u(x, t) = f(x, t)$$
$$(x, t) \in G^* \tag{4.12a}$$

Here,

$$G^* = D^* \times (0, T] \qquad D^* = \{x: d_0 < x < d_1, x \neq d^*\}$$
$$d_0 < d^* < d_1 \tag{4.11b}$$

On the set $S^* = \{x = d^*\} \times (0, T]$ the following conjugation conditions (or the interface conditions) are given:

$$[u(x, t)] = 0 \qquad l^*_{(4.12)} u(x, t) \equiv \varepsilon \left[\frac{\partial}{\partial x} u(x, t) \right] = q(t) \qquad (x, t) \in S^* \tag{4.12b}$$

On the lateral boundary $S^L = \{x: x = d_0, d_1\} \times (0, T]$ the condition

$$l_{(4.12)} u(x, t) \equiv \left\{ \beta(x) + (1 - \beta)(x)) \varepsilon \frac{\partial}{\partial n} \right\} u(x, t) = \psi(x, t)$$
$$(x, t) \in S^L \tag{4.12c}$$

is satisfied. At the initial instant, that is, on the set $S^0 = \bar{D} \times \{t = 0\}$, the function $u(x, t)$ takes the given values

$$u(x, t) = \varphi(x) \qquad (x, t) \in S^0 \tag{4.12d}$$

All the functions are assumed to be sufficiently smooth, moreover, $a(x, t) \geq a_0 > 0$, $c(x, t) \geq 0$, $p(x, t) \geq p_0 > 0$, $(x, t) \in \bar{G}$.

We want to find the solution for the boundary value problem (4.12), (4.11), that is, the function $u(x, t)$, $(x, t) \in \bar{G}$ satisfying Eq. (4.12a) on G^*, conjugation conditions (4.12b) on S^* and boundary conditions (4.12c) and (4.12d) on the sets S^L and S^0, respectively.

The quantity

$$P(x, t) = \varepsilon \frac{\partial}{\partial x} u(x, t) \tag{4.13}$$

is called a normalized diffusion flux. This function has a discontinuity on the set S^*. It is convenient to represent the set $\bar{G}$ as a sum of the two sets $\bar{G}^1$ and $\bar{G}^2$

$$\bar{G} = \bar{G}^1 \cup \bar{G}^2 \qquad G^1 = D^1 \times (0, T] \qquad G^2 = D^2 \times (0, T] \tag{4.11c}$$

where $D^1 = \{x: d_0 < x < d_*\}$, $D^2 = \{x: d^* < x < d_1\}$. Note that, generally speaking, the function $P(x, t)$ is not continuous on each set $\bar{G}^j$, $j = 1, 2$. It has discontinuities on the set $S_0^{L*} = \gamma^{L*} \times \{t = 0\}$, where γ^{L*} is the set of the ends of the segment $\bar{D}$ and the point $x = d^*$ where the concentrated source acts. We consider the function $P(x, t)$ on the sets $\bar{G}_0^j$, $j = 1, 2$, where

$$\bar{G}_0^j = \bar{G}^j \backslash S_0^{L*} \qquad j = 1, 2$$

When the problem data satisfy the special compatibility conditions on the set S_0^{L*}, the function $P(x, t)$ is continuous on the sets $\bar{G}^j$, $j = 1, 2$.

When the parameter tends to zero, in a neighborhood of the sets S^L and S^* there appear boundary and interior layers, respectively. Note that the function $P(x, t)$, $(x, t) \in \bar{G}^j$, $j = 1, 2$ remains bounded uniformly with respect to the parameter.

We also consider a two-dimensional problem on the rectangle D

$$D = \{x: d_{0s} < x_s < d_{1s}, s = 1, 2\} \tag{4.14a}$$

The set D is considered to be composed of two rectangles

$$\bar{D} = \bar{D}^1 \cup \bar{D}^2 \qquad D^1 = \{x: d_{01} < x_1 < d_1^*, x \in D\}$$
$$D^2 = \{x: d_1^* < x_1 < d_{11}, x \in D\} \qquad d_{01} < d_1^* < d_{11} \tag{4.14b}$$

mutually crossing on the interval $\Gamma^* = \{x: x_1 = d_1^*, x \in D\}$. We introduce the sets

$$G = D \times (0, T] \qquad G^j = D^j \times (0, T] \qquad j = 1, 2$$
$$\bar{G} = \bar{G}^1 \cup \bar{G}^2 \qquad S^* = \Gamma^* \times (0, T] \tag{4.14c}$$

On the set $\bar{G}$ we consider the boundary value problem for the singularly perturbed equation of parabolic type with a concentrated source on S^*

$$L_{(4.15)}u(x,t) \equiv \left\{ \varepsilon^2 \sum_{s=1,2} a_s(x,t) \frac{\partial^2}{\partial x_s^2} - p(x,t)\frac{\partial}{\partial t} - c(x,t) \right\} u(x,t) = f(x,t)$$
$$(x,t) \in G^* \tag{4.15a}$$

Here $G^* = D^* \times (0, T]$, $D^* = \{x: d_{0s} < x_s < d_{1s},\ s = 1, 2,\ x_1 \neq d_1^*\}$ is a rectangular domain with the excluded interval Γ^* on which the concentrated source acts. The following conjugation conditions are fulfilled on the set S^*

$$[u(x,t)] = 0 \qquad l^*_{(4.15)}u(x,t) \equiv \varepsilon \left[\frac{\partial}{\partial x_1} u(x,t) \right] = q(t) \qquad (x,t) \in S^* \tag{4.15b}$$

The jump $[v(x,t)]$, $(x,t) \in S^*$ is defined by the relation

$$[v(x,t)] = v(x_1 + 0, x_2, t) - v(x_1 - 0, x_2, t) \qquad (x,t) \in S^*$$

On the lateral boundary $S^L = \Gamma \times (0, T]$, $\Gamma = \bar{D} \setminus D$ and on the set $S^0 = \bar{D} \times \{t = 0\}$, that is, the lower base of the set $\bar{G}$, the function $u(x,t)$ satisfies the conditions

$$l_{(4.15)}u(x,t) \equiv \beta(x)u(x,t) + (1 - \beta(x))\varepsilon \frac{\partial}{\partial n} u(x,t) = \psi(x,t)$$
$$(x,t) \in S^L \tag{4.15c}$$

$$u(x,t) = \varphi(x) \qquad (x,t) \in S^0 \tag{4.15d}$$

The coefficients and the right-hand side of Eq. (4.15a) are assumed to be sufficiently smooth on the set $\bar{G}$, besides, $a_s(x,t) \geq a_0 > 0$, $c(x,t) \geq 0$, $p(x,t) \geq p_0 > 0$, $(x,t) \in \bar{G}$; the function $q(t)$ is sufficiently smooth for $0 < t \leq T$. We describe the properties of the functions $\beta(x)$ and $\psi(x,t)$, $\varphi(x)$.

We define the sets S^{Lj}, S^{0j} as follows:

$$S^{Lj} = S^L \cap S^j \qquad S^{0j} = S^0 \cap S^j \qquad j = 1, 2$$

where $S^j = \bar{G}^j \backslash G^j$ is the boundary of the set G^j. The set of corner points of both of the rectangles D^1 and D^2 is denoted by Γ^c. Suppose $S^c = \Gamma^c \times (0, T]$. We assume that the function $\beta(x)$ is continuous on every side of the rectangle D, however, it is not continuous on the boundary Γ. This function can have discontinuities in the corner points of D. We assume that $\beta(x)$, $x \in \Gamma$ satisfies the following condition: either $\beta(x) = 1$ or $\beta(x) < 1$ on every side of the rectangle D. The sides of the rectangle D on which $\beta(x) = 1$ are denoted by Γ^D. Dirichlet conditions are given on these sides. We denote the other sides by Γ^N; the sets Γ^D and Γ^N are assumed to be closed. Neumann or mixed boundary conditions are given on the set Γ^N. The functions $\varphi(x)$ and $\psi(x,t)$ are sufficiently smooth, respectively, on the sets S^{0j} and on each side forming the set S^{Lj}, $j = 1, 2$. The function $\psi(x,t)$ is continuous on the set $S^D = \Gamma^D \times (0, T]$ and coincides with the function $\varphi(x)$ only on the intersection $\bar{S}^D \cap S^0$. However, the function $\psi(x,t)$ is not assumed to be continuous on the set S^{Nc}. Here $S^{Nc} = \Gamma^{Nc} \times (0, T]$, Γ^{Nc} is a set of corner points belonging to the set Γ^N, $\Gamma^{Nc} = \Gamma^c \cap \Gamma^N$.

It is required to find the solution of problem (4.15), (4.14). We define the normalized diffusion fluxes $P_s(x,t)$ by the relations

$$P_s(x,t) \equiv \varepsilon \frac{\partial}{\partial x_s} u(x,t) \qquad s = 1, 2 \tag{4.16}$$

Note that the function $P_1(x,t)$ is not continuous on S^*, where the concentrated source acts. Moreover, generally speaking, the functions $P_s(x,t)$ have discontinuities on the set $S_0^{cL*} = S^c \cup S_0^{L*}$, where $S_0^{L*} = \gamma^{L*} \times \{t = 0\}$, γ^{L*} is the sum of the sets Γ and Γ^*, $\gamma^{L*} = \Gamma \cup \Gamma^*$. We consider the functions $P_s(x,t)$, $s = 1, 2$ on the sets $\bar{G}_0^j = \bar{G}^j \backslash S_0^{cL*}$, $j = 1, 2$. The functions $P_s(x,t)$, $s = 1, 2$ are continuous on $\bar{G}^j$, $j = 1, 2$ under special conditions imposed on the problem data on the set S_0^{cL*}.

When the parameter tends to zero, boundary and interior layers appear in a neighborhood of the sets S^L and S^*, respectively.

B. Classical and Special Finite Difference Schemes

First we construct classical finite difference schemes for the boundary value problems (4.12), (4.11) and (4.15), (4.14). We use rectangular grids; the distribution of nodes in the space grids is assumed to be arbitrary.

We construct a finite difference scheme for problem (4.12), (4.11). On the set $\bar{G}_{(4.11)}$ we introduce the grid

$$\bar{G}_h = \bar{D}_h \times \bar{\omega}_0 = \bar{\omega}_1 \times \bar{\omega}_0 \tag{4.17}$$

where $\bar{\omega}_1$ and $\bar{\omega}_0$ are grids on the sets $[d_0, d_1]$ and $[0, T]$, respectively, $\bar{\omega}_1$ is a grid with an arbitrary distribution of nodes, the point $x = d^*$ belongs to the grid $\bar{\omega}_1$, the grid $\bar{\omega}_0$ is uniform; $N + 1$ and $N_0 + 1$ are the number of nodes in the grids $\bar{\omega}_1$ and $\bar{\omega}_0$, respectively. On the grid $\bar{G}_h$ we consider the finite difference scheme

$$\begin{aligned} \Lambda_{(4.18)} z(x, t) = f(x, t) \qquad & (x, t) \in G_h^* \\ \lambda^*_{(4.18)} z(x, t) = q(t) \qquad & (x, t) \in S_h^* \\ \lambda_{(4.18)} z(x, t) = \psi(x, t) \qquad & (x, t) \in S_h^L \\ z(x, t) = \varphi(x) \qquad & (x, t) \in S_h^0 \end{aligned} \tag{4.18}$$

Here

$$G_h^0 = G^0 \cap \bar{G}_h \qquad G^0 \subseteq \bar{G} \tag{4.19}$$

where G^0 is one of the sets G^*, S^*, S^L, S^0

$$\Lambda_{(4.18)} z(x, t) \equiv \{\varepsilon^2 a(x, t)\delta_{\bar{x}\hat{x}} - p(x, t)\delta_{\bar{t}} - c(x, t)\} z(x, t) \qquad (x, t) \in G_h^*$$

$$\lambda^*_{(4.18)} z(x, t) \equiv \varepsilon\{\delta_x - \delta_{\bar{x}}\} z(x, t) \qquad (x, t) \in S_h^*$$

$$\lambda_{(4.18)} z(x, t) \equiv \beta(x) z(x, t) + (1 - \beta(x))\varepsilon\delta_n z(x, t)$$

$$\delta_n z(x, t) \equiv \begin{cases} -\delta_x z(x, t) & x = d_0 \\ \delta_{\bar{x}} z(x, t) & x = d_1 \end{cases}$$

The difference scheme (4.18), (4.17) allows us to find the function $z(x, t)$, $(x, t) \in \bar{G}_h$. Let us construct an approximation of the function $P_{(4.13)}(x, t)$. On the sets $\bar{G}_0^j$ we introduce the grid

$$\bar{G}_{0h}^{j-} \qquad j = 1, 2 \tag{4.20a}$$

Here $\bar{G}^{j-}_{0h}$ is a set of nodes (x, t) from $\bar{G}^{j}_{0h} = \bar{G}^{j}_{0} \cap \bar{G}_h$ for which the operator δ_x is defined, that is, a set of nodes $(x^i, t) \in \bar{G}^{j}_{0h}$ such that $(x^i, t), (x^{i+1}, t) \in \bar{G}^{j}_{0h}$. In a similar way, we define the grids

$$\bar{G}^{j-}_{h} \qquad j = 1, 2 \tag{4.20b}$$

on the sets $\bar{G}^j$. To approximate the function $P(x, t)$, $(x, t) \in \bar{G}^{j}_{0}$, we use the grid function

$$P^{h+}(x, t) \equiv \varepsilon \delta_x z(x, t) \qquad (x, t) \in \bar{G}^{j-}_{0h} \qquad j = 1, 2 \tag{4.20c}$$

If the function $P(x, t)$ is continuous, then, in order to approximate it, we apply the function $P^{h+}(x, t)$ on the grid $\bar{G}^{j-}_{h}$, $j = 1, 2$.

Now we construct a finite difference scheme for problem (4.15), (4.14). The following grid is constructed on the set $\bar{G}_{(4.14)}$

$$\bar{G}_h = \bar{D}_h \times \bar{\omega}_0 = \bar{\omega}_1 \times \bar{\omega}_2 \times \bar{\omega}_0 \tag{4.21}$$

where $\bar{\omega}_s$ is a grid on the set $[d_{0s}, d_{1s}]$ on the axis x_s, $s = 1, 2$ (generally speaking, a nonuniform grid), the point $x_1 = d^*_1$ belongs to the grid $\bar{\omega}_1$; $\bar{\omega}_0 = \bar{\omega}_{(4.17)}$ is a uniform grid; $N_s + 1$ are the number of nodes in the grid $\bar{\omega}_s$. On the grid $\bar{G}_h$, for the approximation of the boundary value problem (4.15), (4.14) we apply the finite difference scheme

$$\begin{aligned} \Lambda_{(4.22)} z(x, t) &= f(x, t) \qquad (x, t) \in G^*_h \\ \lambda^*_{(4.22)} z(x, t) &= q(t) \qquad (x, t) \in S^*_h \\ \lambda_{(4.22)} z(x, t) &= \psi(x, t) \qquad (x, t) \in S^L_h \backslash S^c_h \\ z(x, t) &= \varphi(x) \qquad (x, t) \in S^0_h \end{aligned} \tag{4.22a}$$

Here the sets G^*_h, S^*_h, S^L_h, S^0_h are defined by relation (4.19) and

$$\Lambda_{(4.22)} z(x, t) \equiv \Big\{ \varepsilon^2 \sum_{s=1,2} a_s(x, t) \delta_{\overline{xs}\widehat{xs}} - p(x, t) \delta_{\bar{t}} - c(x, t) \Big\} z(x, t) \quad (x, t) \in G^*_h$$

$$\lambda^*_{(4.22)} z(x, t) \equiv \varepsilon \{ \delta_{x1} - \delta_{\overline{x1}} \} z(x, t) \qquad (x, t) \in S^*_h$$

$$\lambda_{(4.22)} z(x, t) \equiv \beta(x) z(x, t) + (1 - \beta(x)) \varepsilon \delta_n z(x, t)$$

$$\delta_n z(x, t) \equiv \begin{cases} -\delta_{xs} z(x, t) & x_s = d_{0s} \\ \delta_{\overline{xs}} z(x, t) & x_s = d_{1s} \end{cases}$$

The finite difference equations (4.22) enable us to find the function

$z(x, t)$ for $(x, t) \in \bar{G}_h \backslash S_h^c$. Now we state relations which allow us to find the function $z(x, t)$ for $(x, t) \in S_h^c$. We assume that

$$z(x, t) = \psi(x, t) \qquad (x, t) \in S_h^c \cap S^D \tag{4.22b}$$

Suppose that $(x^0, t) \in S_h^c$ and $(x^0, t) \not\in S^D$, and that this node (x^0, t) is one of the four nodes (x_1^i, x_2^k, t), (x_1^{i+1}, x_2^k, t), (x_1^i, x_2^{k+1}, t), $(x_1^{i+1}, x_2^{k+1}, t) \in \bar{G}_h^j$, $j = 1, 2$. The function $z(x^0, t)$ is determined from the relation

$$z(x_1^i, x_2^k, t) + z(x_1^{i+1}, x_2^{k+1}, t) - z(x_1^{i+1}, x_2^k, t) - z(x_1^i, x_2^{k+1}, t) = 0 \tag{4.22c}$$

Relation (4.22c) means that the values of the grid function $z(x, t)$ at the above-mentioned nodes are located on one plane. Formulas (4.22b) and (4.22c) define the function $z(x, t)$ on the set $S_h^c \cap \bar{G}_h^j$. The function $z(x^0, t)$ defined by relation (4.22c) on $\bar{G}_h^j$ is denoted by $z^j(x^0, t)$. If, for the node (x, t), the point x is a corner of the rectangle D, we suppose that

$$z(x, t) = z^j(x, t) \qquad (x, t) \in S_h^c \backslash S^D \tag{4.22d}$$

otherwise,

$$z(x, t) = 2^{-1}[z^1(x, t) + z^2(x, t)] \qquad (x, t) \in S_h^c \backslash S^D \tag{4.22e}$$

Thus, the required finite difference scheme [scheme (4.22), (4.21)] has been constructed. The approximate solution $z(x, t)$ is defined on the set $\bar{G}_h$.

Now we construct approximations of the diffusion fluxes $P_s(x, t)$. On the sets $\bar{G}_0^j$ we construct the grids

$$\bar{G}_{0hs}^{j-} \qquad j, s = 1, 2 \tag{4.23a}$$

$\bar{G}_{0hs}^{j-}$ is a set of nodes (x, t) from $\bar{G}_{0h}^j = \bar{G}_0^j \cap \bar{G}_h$ for which the operator δ_{xs} is defined. For example, the set $\bar{G}_{0h1}^{2-}$ is an assembly of nodes $(x_1^i, x_2, t) \in \bar{G}_{0h}^2$, such that (x_1^i, x_2, t), $(x_1^{i+1}, x_2, t) \in \bar{G}_{0h}^2$. In a similar way, we define the grids

$$\bar{G}_{hs}^{j-} \qquad j, s = 1, 2 \tag{4.23b}$$

on the sets $\bar{G}^j$. To approximate the diffusion fluxes $P_s(x, t)$, $(x, t) \in \bar{G}_0^j$, we apply the grid functions

$$P_s^{h+}(x, t) \equiv \varepsilon \delta_{xs} z(x, t) \qquad (x, t) \in \bar{G}_{0hs}^{j-} \qquad j, s = 1, 2 \tag{4.23c}$$

If the functions $P_s(x, t)$ are continuous on $\bar{G}^j$, in order to approximate them, we use the grid functions $P_s^{h+}(x, t)$ on $\bar{G}_{hs}^{j-}$, $j, s = 1, 2$.

The boundary value problems from Sections II and III are particular cases of the problems considered in this section. Therefore, computational problems arising for the classical finite difference schemes described in Sections II and III remain acute for the boundary value problems in this section.

Note, in particular, one feature in the behavior of the approximate solutions of boundary value problems with a concentrated source. It follows from the results of Section II that, in the case of the Dirichlet problem, the solution of the classical finite difference scheme is bounded ε-uniformly, and even though the grid solution does not converge ε-uniformly, it approximates qualitatively the exact solution ε-uniformly. But now, in the case of a Dirichlet boundary value problem with a concentrated source, the behavior of the approximate solution differs sharply from what was said above. For example, in the case of a Dirichlet boundary value problem with a concentrated source acting in the middle of the segment $\bar{D} = [-1, 1]$, when the equation coefficients are constant, the right-hand side and the boundary function are equal to zero, the solution is equivalent to the solution of the problem on $[0, 1]$ with a Neumann condition at $x = 0$. It follows that the solution of the classical finite difference scheme for the Dirichlet problem with a concentrated source is not bounded ε-uniformly, and that it does not approximate the exact solution uniformly in ε, even qualitatively.

We give some estimates of the solution and its derivatives for problem (4.12), (4.11). Suppose that, for a fixed value of the parameter, the solution of the boundary value problem is sufficiently smooth on each set $\bar{G}^j$ and has the continuous derivatives

$$\frac{\partial^{k+k_0}}{\partial x^k \partial t^{k_0}} u(x, t) \qquad (x, t) \in \bar{G}^j \qquad 0 \le k + 2k_0 \le K \qquad j = 1, 2 \quad (4.24)$$

where $K = 4$. The solution of the boundary value problem on each subdomain $\bar{G}^j$ can be represented as a sum of two functions

$$u(x, t) = U(x, t) + V(x, t) \qquad (x, t) \in \bar{G}^j \qquad j = 1, 2 \qquad (4.25a)$$

For the functions $U(x, t)$ and $V(x, t)$, that is, for the regular and singular parts of the solution, and for their derivatives, the following estimates are fulfilled

$$\left| \frac{\partial^{k+k_0}}{\partial x^k \partial t^{k_0}} U(x, t) \right| \leq M$$

$$\left| \frac{\partial^{k+k_0}}{\partial x^k \partial t^{k_0}} V(x, t) \right| \leq M\varepsilon^{-k} \exp(-m\varepsilon^{-1} r(x, \Gamma^j)) \tag{4.25b}$$

$$(x, t) \in \bar{G}^j \qquad 0 \leq k + 2k_0 \leq 4 \qquad j = 1, 2$$

where $r(x, \Gamma^j)$ is the distance from the point x to the set $\Gamma^j = \bar{D}^j \backslash D^j$; m is a sufficiently small arbitrary number.

We see that these estimates on the subsets $\bar{G}^j$ are similar to the estimates of the solutions and their derivatives considered in Sections II.C and III.C. Therefore, to construct ε-uniformly convergent schemes, we can apply the results of Sections II.C and III.C.

Now we construct a special grid $\bar{G}_h$, which condenses in the boundary and interior layers. Suppose that

$$\bar{G}_h = \bar{D}_h \times \bar{\omega}_0 = \bar{\omega}_1^* \times \bar{\omega}_0 \equiv \bar{G}_h(x_0, d_1, d^*) \tag{4.26a}$$

The grid $\bar{D}_h$ can be defined by the relations

$$\bar{D}^j \cap \bar{D}_h = \bar{D}_h^j \qquad j = 1, 2$$

$$\begin{aligned} \bar{D}_h^1 &= \bar{D}_h^1(\varepsilon, N, d_0, d^*, m) \\ &= \bar{\omega}_{*(2.58)}(\sigma_{(2.58)}(\varepsilon, 2^{-1}N, d_0, d^*, m), N_* = 8^{-1}N, 2^{-1}N, d_0, d^*) \\ \bar{D}_h^2 &= \bar{D}_h^2(\varepsilon, N, d^*, d_1, m) \\ &= \bar{\omega}_{*(2.58)}(\sigma_{(2.58)}(\varepsilon, 2^{-1}N, d^*, d_1, m), N_* = 8^{-1}N, 2^{-1}N, d^*, d_1) \end{aligned} \tag{4.26b}$$

where $m = m_{(4.26)}$ is an arbitrary number, $\bar{\omega}_0 = \bar{\omega}_{0(4.17)}$

Let us discuss the finite difference scheme (4.18), (4.26). The solution of scheme (4.18), (4.26) converges to the solution of the boundary value problem (4.12), (4.11) ε-uniformly, that is,

$$\max_{\bar{G}_h} |u(x, t) - z(x, t)| \leq \lambda(N, N_0) \qquad \bar{G}_h = \bar{G}_{h(4.26)}$$

where $\lambda(N, N_0) \to 0$ when $N, N_0 \to \infty$. Now we estimate the error in the case when estimates (4.25) are fulfilled for the solution of the boundary value problem. According to estimates (4.25), the difference operator $\Lambda_{(4.18)}$ approximates the differential operator $L_{(4.12)}$ on the solution of the boundary value problem ε-uniformly on every subdomain G^j

$$\max_{G_h^j} |L_{(4.12)}u(x,t) - \Lambda_{(4.18)}u(x,t)| \le M[N^{-1}\ln N + N_0^{-1}] \qquad j=1,2 \tag{4.27a}$$

The grid operators $\lambda^*_{(4.18)}$ and $\lambda_{(4.18)}$ also approximate the differential operators $l^*_{(4.12)}$ and $l_{(4.12)}$ on the solution of the boundary value problem ε-uniformly.

$$\begin{aligned} &\max_{S_h^*} |l^*_{(4.12)}u(x,t) - \lambda^*_{(4.18)}u(x,t)| \le MN^{-1}\ln N \\ &\max_{S_h^L} |l_{(4.12)}u(x,t) - \lambda_{(4.18)}u(x,t)| \le MN^{-1}\ln N \end{aligned} \tag{4.27b}$$

The estimates (4.27) imply the ε-uniform convergence of the solution of the finite difference scheme (4.18), (4.26)

$$\max_{\bar{G}_h} |u(x,t) - z(x,t)| \le M[N^{-1}\ln N + N_0^{-1}] \qquad \bar{G}_h = \bar{G}_{h(4.26)}$$

If, for a fixed value of the parameter ε, the solution of the boundary value problem has the continuous derivatives (4.24), where $K=4$, the computed normalized diffusion flux $P^{h+}(x,t)$ converges to the actual flux $P(x,t)$ ε-uniformly on $\bar{G}_h^{j-}$

$$\max_{\bar{G}_h^{j-}} |P(x,t) - P^{h+}(x,t)| \le M[N^{-1}\ln N + N_0^{-1}]$$

$$\bar{G}_h^{j-} = \bar{G}_{h(4.26)}^{j-} \qquad j=1,2$$

In the case of problem (4.15), (4.14) we write a special grid on which the solution of the finite difference scheme (4.22) converges ε-uniformly. Suppose that

$$\bar{G}_h = \bar{D}_h \times \bar{\omega}_0 = \bar{\omega}_1^* \times \bar{\omega}_2^* \times \bar{\omega}_0 \tag{4.28a}$$

where $\bar{\omega}_0 = \bar{\omega}_{0(4.17)}$. The grids $\bar{\omega}_1^*$ and $\bar{\omega}_2^*$ are defined by the relations

$$[d_{01}, d_1^*] \cap \bar{\omega}_1^* = \bar{\omega}_1^{*1} \qquad [d_1^*, d_{11}] \cap \bar{\omega}_1^* = \bar{\omega}_1^{*2}$$

$$\bar{\omega}_1^{*1} = \bar{\omega}_{*(2.58)}(\sigma_{(2.58)}(\varepsilon, 2^{-1}N_1, d_{01}, d_1^*, m), N_* = 8^{-1}N_1, 2^{-1}N_1, d_{01}, d_1^*)$$

$$\bar{\omega}_1^{*2} = \bar{\omega}_{*(2.58)}(\sigma_{(2.58)}(\varepsilon, 2^{-1}N_1, d_1^*, d_{11}, m), N_* = 8^{-1}N_1, 2^{-1}N_1, d_1^*, d_{11})$$

$$\bar{\omega}_2^* = \bar{\omega}_{*(2.58)}(\sigma_{(2.58)}(\varepsilon, N_2, d_{02}, d_{12}, m), N_* = 4^{-1}N_2, N_2, d_{02}, d_{12})$$

where $m = m_{(4.28)}$ is an arbitrary number.

The solution of the finite difference scheme (4.22), (4.28) converges to the solution of the boundary value problem ε-uniformly

$$\max_{\bar{G}_h} |u(x, t) - z(x, t)| \leq \lambda(N, N_0) \qquad \bar{G}_h = \bar{G}_{h(4.28)}$$

where $\lambda(N, N_0) \to 0$ when $N, N_0 \to \infty$, $N = \min[N_1, N_2]$. If the solution of the boundary value problem (4.15), (4.14) is sufficiently smooth on each subset $\bar{G}^j$, the solution of the finite difference scheme and the computed normalized diffusion fluxes converge ε-uniformly to the solution of the boundary value problem and the actual fluxes, respectively. For the solution of the finite difference scheme (4.22), (4.28) the estimate

$$\max_{\bar{G}_h} |u(x, t) - z(x, t)| \leq M[N^{-1}\ln N + N_0^{-1}] \qquad \bar{G}_h = \bar{G}_{h(4.28)}$$

is valid, and, for the computed normalized diffusion fluxes $P_s^{h+}(x, t)$, the following estimate is fulfilled

$$\max_{\bar{G}_{hs}^{j-}} |P_s(x, t) - P_s^{h+}(x, t)| \leq M[N^{-1}\ln N + N_0^{-1}]$$

$$\bar{G}_h^j = \bar{G}_{h(4.28)}^j \qquad j, s = 1, 2$$

C. Numerical Experiments with the Classical and Special Finite Difference Schemes

Now we formulate a problem and thereafter study classical and new special finite difference schemes to solve it. Let the segment $\bar{D}$ be divided into two parts $\bar{D}^1$ and $\bar{D}^2$

$$\bar{D} = \bar{D}^1 \cup \bar{D}^2 \tag{4.29a}$$

where $D = (-1, 1)$, $D^1 = (-1, 0)$, $D^2 = (0, 1)$. The concentrated source is assumed to act at the point $x = 0$. On the domain $\bar{G}$

$$\bar{G} = \bar{G}^1 \cup \bar{G}^2 \qquad G^j = D^j \times (0, T] \qquad j = 1, 2 \tag{4.29b}$$

where $T = 1$, we consider the following boundary value problem with a concentrated source

$$
\begin{aligned}
L_{(4.30)}u(x,t) &= f(x,t) \qquad (x,t)\in G^j \\
l^*_{(4.30)}u(x,t) &= q(t) \qquad [u(x,t)]=0 \qquad (x,t)\in S^* \\
l_{(4.30)}u(x,t) &= \psi(x,t) \qquad (x,t)\in S^L \\
u(x,t) &= \varphi(x) \qquad (x,t)\in S^0
\end{aligned}
\tag{4.30}
$$

Here

$$S^* = \{x=0\}\times(0,T] \qquad S^L=\{x=-1,1\}\times(0,T] \qquad S^0=\bar{D}\times\{t=0\}$$

$$L_{(4.30)}u(x,t) \equiv \left\{\varepsilon^2\frac{\partial^2}{\partial x^2}-\frac{\partial}{\partial t}\right\}u(x,t) \qquad (x,t)\in G^j$$

$$f(x,t)=\begin{cases}-1 & (x,t)\in\bar{G}^1\\ -t & (x,t)\in\bar{G}^2\end{cases}$$

$$l^*_{(4.30)}u(0,t)\equiv\varepsilon\left(\frac{\partial}{\partial x}u(+0,t)-\frac{\partial}{\partial x}u(-0,t)\right)=-2\frac{1}{\sqrt{\pi}}t^{1/2}-\frac{4}{3\sqrt{\pi}}t^{3/2}$$

$$\equiv q(t) \qquad 0<t\le T$$

$$l_{(4.30)}u(x,t)\equiv\left\{\begin{matrix}u(x,t) & x=-1\\ \varepsilon\dfrac{\partial}{\partial n}u(x,t) & x=1\end{matrix}\right\}=\left\{\begin{matrix}0 & x=-1\\ \dfrac{2}{3\sqrt{\pi}}t^{3/2} & x=1\end{matrix}\right\}$$

$$\equiv\psi(x,t) \qquad (x,t)\in S^L$$

$$\varphi(x)=0 \qquad x\in\bar{D}$$

The function $f(x,t)$ is continuous on every set $\bar{G}^j$ and it has a discontinuity on the set S^*. The Dirichlet boundary condition is given on the left end of the segment D, while the Neumann boundary condition is realized on the right end of it. The graph of the problem solution is shown in Fig. 19 for $\varepsilon=1.0$, 0.1, and 0.01.

The solution of the boundary value problem is continuous on $\bar{G}$ and sufficiently smooth on every subdomain $\bar{G}^j$. The solution can be represented in the form

$$u(x,t)=U(x,t)+W(x,t)+v(x,t) \qquad (x,t)\in\bar{G}^j \qquad j=1,2 \tag{4.31}$$

Here $U(x,t)$, $W(x,t)$ are the principal regular and singular terms of the solution expansion, and $v(x,t)$ is the remainder term of the expansion.

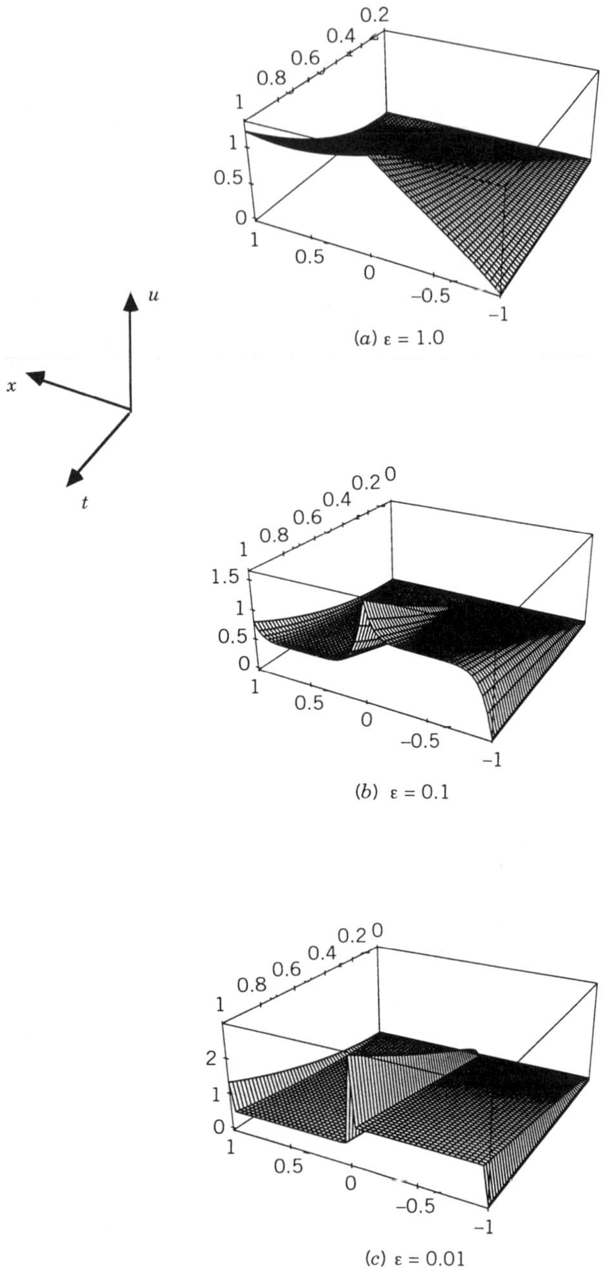

Figure 19. Solution of problem (4.30) for $\varepsilon = 1.0$, 0.1, and 0.01.

The components of the representation (4.31) are defined by the relations

$$U(x,t)=\begin{cases} t & (x,t)\in\bar{G}^1 \\ 2^{-1}t^2 & (x,t)\in\bar{G}^2 \end{cases}$$

$$W(x,t)=\begin{cases} -W_{(2.23)}(x+1,t)+2^{-1}W_{(3.12)}(-x,t) & (x,t)\in\bar{G}^1 \\ W_{(2.23)}(x,t)+4^{-1}W_{(3.12)}(-x+1,t) & (x,t)\in\bar{G}^2 \end{cases}$$

$$v(x,t)=v_0(x,t)+v_1(x,t) \qquad (x,t)\in\bar{G}^j \qquad j=1,2$$

$$v_1(x,t)=\begin{cases} 4^{-1}W_{(3.12)}(-x+1,t) & (x,t)\in\bar{G}^1 \\ -W_{(2.23)}(x+1,t) & (x,t)\in\bar{G}^2 \end{cases}$$

The function $v_0(x,t)$, $(x,t)\in\bar{G}$ is the solution of the boundary value problem

$$L_{(4.32)}v_0(x,t)\equiv\left\{\varepsilon^2\frac{\partial^2}{\partial x^2}-\frac{\partial}{\partial t}\right\}v_0(x,t)=0 \qquad (x,t)\in G$$

$$\begin{aligned} l_{(4.32)}v_0(x,t)&\equiv\left\{\begin{matrix} v_0(x,t) & x=-1 \\ \varepsilon\dfrac{\partial}{\partial x}v_0(x,t) & x=1 \end{matrix}\right\} \\ &=\left\{\begin{matrix} -2^{-1}W_{(3.12)}(1,t)-4^{-1}W_{(3.12)}(2,t) & x=-1 \\ -\varepsilon\dfrac{\partial}{\partial x}W_{(2.23)}(1,t)+\varepsilon\dfrac{\partial}{\partial x}W_{(2.23)}(2,t) & x=1 \end{matrix}\right\} \\ &\equiv\psi_{(4.32)}(x,t) \qquad (x,t)\in S^L \end{aligned} \tag{4.32}$$

$$v_0(x,t)=0 \qquad (x,t)\in S^0$$

We use finite difference schemes to solve problem (4.30). On the set $\bar{G}$, we construct the rectangular grid

$$\bar{G}_h=\bar{G}_{h(4.17)}=\bar{\omega}_1\times\bar{\omega}_0 \tag{4.33}$$

where $\bar{\omega}_1$ is, generally speaking, a nonuniform grid on the segment $\bar{D}_{(4.29)}$, the point $x=0$ belongs to $\bar{\omega}_1$. Problem (4.30) is approximated by the finite difference scheme

$$\begin{aligned} \Lambda_{(4.34)} z(x,t) &= f(x,t) \qquad (x,t) \in G_h^j \\ \lambda^*_{(4.34)} z(x,t) &= q(t) \qquad (x,t) \in S_h^* \\ \lambda_{(4.34)} z(x,t) &= \psi(x,t) \qquad (x,t) \in S_h^L \\ z(x,t) &= \varphi(x) \qquad (x,t) \in S_h^0 \end{aligned} \tag{4.34}$$

Here

$$\Lambda_{(4.34)} z(x,t) \equiv \{\varepsilon^2 \delta_{\bar{x}\hat{x}} - \delta_{\bar{t}}\} z(x,t) \qquad (x,t) \in G_h^j$$

$$\lambda^*_{(4.34)} z(x,t) \equiv \lambda^*_{(4.18)} z(x,t) \qquad (x,t) \in S_h^*$$

$$\lambda_{(4.34)} z(x,t) \equiv \begin{Bmatrix} z(x,t) & x=-1 \\ \varepsilon \delta_{\bar{x}} z(x,t) & x=1 \end{Bmatrix} \qquad (x,t) \in S_h^L$$

The grid equations are considered either on the uniform grids

$$\bar{G}_h = \bar{G}_{h(4.35)} \tag{4.35}$$

or on the special grids

$$\bar{G}_h = \bar{G}_{h(4.36)} \equiv \bar{G}_{h(4.26)}(d_0 = -1, d_1 = 1, d^* = 0) \tag{4.36}$$

To analyze the errors in the approximate solution obtained with the finite difference schemes (4.34), (4.35) and (4.34), (4.36), one can use the method described in Sections II and III. In this method, we use the analytical representation of the functions $U(x, t)$, $W(x, t)$, $v_1(x, t)$ and the numerical solution of problem (4.32) on a fine grid (possibly, a uniform grid) to compute the function $v_0(x, t)$.

We can do this as follows. We know that the grid solution of problem (4.34), (4.36) converges to the solution of the boundary value problem ε-uniformly when $N, N_0 \to \infty$. The following estimate is valid:

$$\max_{\bar{G}_h} |u(x,t) - z(x,t)| \leq \lambda_1(N, N_0) \tag{4.37}$$

where $\lambda_1(N, N_0)$, generally speaking, is a function unknown to us, and also $\lambda_1(N, N_0) \to 0$ uniformly with respect to the parameter ε when N, $N_0 \to \infty$. We take the function $\bar{u}(x,t) = \bar{u}_{N^*, N_0^*}(x,t)$, $(x,t) \in \bar{G}$ as the exact solution of problem (4.30). This function is constructed by linear interpolation with respect to x and t from the values of the grid function

$z_{N^*,N_0^*}(x,t) \in \bar{G}_{h(4.36)}(N^*, N_0^*)$, that is, the solution of problem (4.34) on the fine grid $\bar{G}_{h(4.36)}(N=N^*,\ N_0=N_0^*)$. According to the estimate (4.37), the function $\bar{u}_{N^*,N_0^*}(x,t)$ can be made close to the function $u(x,t)$ for sufficiently large N^*, N_0^*.

We are interested in the quantity

$$\max_{\bar{G}_h} |u(x,t) - z(x,t)|$$

that is, in the error in the approximate solution $z(x,t)$ obtained by the scheme (4.34), (4.35) or (4.34), (4.35) or (4.34), (4.36) for various values of ε and N, N_0. This error is close to the value

$$\max_{\bar{G}_h} |\bar{u}(x,t) - z(x,t)|$$

for sufficiently large N^*, N_0^*. Let us introduce the values $E_{N^*,N_0^*}(\varepsilon, N, N_0)$ and $\bar{E}_{N^*,N_0^*}(N, N_0)$

$$\begin{aligned} E_{N^*,N_0^*}(\varepsilon, N, N_0) &= E(\varepsilon, N, N_0; N^*, N_0^*) \equiv \max_{\bar{G}_h} |\bar{u}(x,t) - z(x,t)| \\ \bar{E}_{N^*,N_0^*}(N, N_0) &= \bar{E}(N, N_0; N^*, N_0^*) \equiv \max_{\varepsilon} E_{N^*,N_0^*}(\varepsilon, N, N_0) \end{aligned} \tag{4.38a}$$

We set

$$E_{N^*,N_0^*}(\varepsilon, N) = E_{N^*,N_0^*}(\varepsilon, N, N) \qquad \bar{E}_{N^*,N_0^*}(N) = \bar{E}_{N^*,N_0^*}(N, N) \tag{4.38b}$$

In Tables XXXIII and XXXIV we give the values of $E_{N^*,N_0^*}(\varepsilon, N)$ and $\bar{E}_{N^*,N_0^*}(N)$ computed for various values of the parameter ε and the number of nodes N, $\varepsilon = 2^{-12} \div 1.0$, $N = 8 \div 512$, $N^* = N_0^* = 1024$ in the case of the finite difference schemes (4.34), (4.35) and (4.34), (4.36), respectively.

One can see from Table XXXIII that the approximate solution on the uniform grid converges for a fixed value of the parameter, although it does not converge ε-uniformly. The error in the approximate solution increases without bound when $\varepsilon N \to 0$. We recall that the maximum modulus of the solution proper varies between 1.30 and 1.55

$$1.30 \le \max_{\bar{G}} |u(x,t;\varepsilon)| \le 1.55$$

It follows from Table XXXIV that the approximate solution on the special grid converges both for a fixed value of the parameter and ε-uniformly.

TABLE XXXIII
Table of Errors $E_{1024,1024}(\varepsilon, N)$ for the Classical Scheme (4.34), (4.35) in the Case of Problem (4.30)

$\varepsilon \backslash N$	4	16	64	512
1	2.688e − 1	6.182e − 2	1.379e − 2	1.965e − 3
2^{-2}	6.377e − 1	1.357e − 1	2.944e − 2	4.164e − 3
2^{-4}	3.169e + 0	5.907e − 1	1.139e − 1	1.554e − 2
2^{-6}	1.432e + 1	3.104e + 0	5.475e − 1	7.880e − 2
2^{-8}	5.942e + 1	1.429e + 1	3.095e + 0	5.447e − 1
2^{-10}	2.400e + 2	5.940e + 1	1.428e + 1	3.093e + 0
2^{-12}	9.621e + 2	2.399e + 2	5.939e + 1	1.428e + 1
$\bar{E}(N)$	9.621e + 2	2.399e + 2	5.939e + 1	1.428e + 1

We are also interested in the quantity

$$\max_{j,\bar{G}_h^{j-}} |P(x,t) - P^{h+}(x,t)|$$

that is, in the error in the computed normalized diffusion flux obtained by the finite difference schemes (4.34), (4.35) or (4.34), (4.36). Since the solution of the boundary value problem (4.30) is sufficiently smooth on every subdomain $\bar{G}^j$, the function $P^{h+}(x,t)$ computed with the use of the solution of scheme (4.34), (4.36) converges to the function $P(x,t)$ ε-uniformly when $N, N_0 \to \infty$

$$\max_{j,\bar{G}_h^{j-}} |P(x,t) - P^{h+}(x,t)| \leq \lambda_2(N, N_0)$$

where $\lambda_2(N, N_0) \to 0$ uniformly with respect to the parameter ε when $N, N_0 \to \infty$.

TABLE XXXIV
Table of Errors $E_{1024,1024}(\varepsilon, N)$ for the Special Scheme (4.34), (4.36) in the Case of Problem (4.30)

$\varepsilon \backslash N$	8	32	128	512
1	2.688e − 1	6.182e − 2	1.379e − 2	1.965e − 3
2^{-2}	6.377e − 1	1.357e − 1	2.944e − 2	4.164e − 3
2^{-4}	1.984e + 0	5.907e − 1	1.139e − 1	1.554e − 2
2^{-6}	1.862e + 0	8.436e − 1	2.342e − 1	3.869e − 2
2^{-8}	1.846e + 0	8.436e − 1	2.342e − 1	3.869e − 2
2^{-10}	1.842e + 0	8.436e − 1	2.342e − 1	3.869e − 2
2^{-12}	1.841e + 0	8.436e − 1	2.342e − 1	3.869e − 2
$\bar{E}(N)$	1.984e + 0	8.435e − 1	2.342e − 1	3.869e − 2

We assume that the function $\bar{P}(x,t)=\bar{P}(x,t;N^*,N_0^*)$, $(x,t)\in\bar{G}^j$, $j=1,2$ is the exact normalized diffusion flux $P(x,t)$. The function $\bar{P}(x,t)$ can be defined by the relation

$$\bar{P}(\xi,\tau)=\varepsilon\delta_x z_{N^*,N_0^*}(x,t) \qquad \text{for} \qquad x^i\le\xi<x^{i+1} \qquad x^{i+1}\ne 0,1$$
$$x^i\le\xi\le x^{i+1} \qquad x^{i+1}=0,1 \qquad t^k\le\tau<t^{k+1}$$
$$(x,t)=(x^i,t^k)\in\bar{G}_h^{j-} \qquad (\xi,\tau)\in\bar{G}^j$$
$$\bar{G}_h^j=\bar{G}^j_{h(4.36)}(N=N^*,N_0=N_0^*) \qquad j=1,2$$

We introduce the quantities $Q_{N^*,N_0^*}(\varepsilon,N,N_0)$ and $\bar{Q}_{N^*,N_0^*}(N,N_0)$

$$Q_{N^*,N_0^*}(\varepsilon,N,N_0)=Q(\varepsilon,N,N_0;N^*,N_0^*)$$
$$\equiv \max_{j,\bar{G}_h^{j-}}|\bar{P}(x,t)-P^{h+}(x,t)| \tag{4.39a}$$
$$\bar{Q}_{N^*,N_0^*}(N,N_0)=\bar{Q}(N,N_0;N^*,N_0^*)\equiv\max_\varepsilon Q(\varepsilon,N,N_0)$$

We set

$$Q_{N^*,N_0^*}(\varepsilon,N)=Q_{N^*,N_0^*}(\varepsilon,N,N) \qquad \bar{Q}_{N^*,N_0^*}(N)=\bar{Q}_{N^*,N_0^*}(N,N) \tag{4.39b}$$

In Tables XXXV and XXXVI we give the values of $Q_{N^*,N_0^*}(\varepsilon,N)$ and $\bar{Q}_{N^*,N_0^*}(N)$ computed for various values of ε and N in the case of finite difference schemes (4.34), (4.35) and (4.34), (4.36), respectively.

It follows from Table XXXV that, in the case of scheme (4.34), (4.35), the computed normalized diffusion flux $P^{h+}(x,t)$ converges to the actual flux $P(x,t)$ for a fixed value of the parameter ε, but it does not converge

TABLE XXXV

Table of Errors of the Normalized Flux $Q_{1024,1024}(\varepsilon,N)$ for the Classical Scheme (4.34), (4.35) in the Case of Problem (4.30)

$\varepsilon\backslash N$	8	32	128	512
1	2.633e − 1	6.893e − 2	1.630e − 2	3.622e − 3
2^{-2}	6.264e − 1	2.193e − 1	5.446e − 2	7.969e − 3
2^{-4}	8.877e − 1	6.058e − 1	1.962e − 1	3.088e − 2
2^{-6}	1.018e + 0	8.817e − 1	5.694e − 1	1.013e − 1
2^{-8}	1.065e + 0	1.018e + 0	8.803e − 1	5.683e − 1
2^{-10}	1.076e + 0	1.065e + 0	1.018e + 0	8.799e − 1
2^{-12}	1.079e + 0	1.076e + 0	1.065e + 0	1.018e + 0
$\bar{Q}(N)$	1.079e + 0	1.076e + 0	1.065e + 0	1.018e + 0

TABLE XXXVI
Table of Errors of the Normalized Flux $Q_{1024,1024}(\varepsilon, N)$ for the Special Scheme (4.34), (4.36) in the Case of Problem (4.30)

$\varepsilon\backslash N$	8	32	128	512
1	2.633e − 1	6.893e − 2	1.630e − 2	3.622e − 3
2^{-2}	6.264e − 1	2.193e − 1	5.446e − 2	7.969e − 3
2^{-4}	8.703e − 1	6.058e − 1	1.962e − 1	3.088e − 2
2^{-6}	8.589e − 1	6.938e − 1	3.201e − 1	4.849e − 2
2^{-8}	8.541e − 1	6.938e − 1	3.201e − 1	4.849e − 2
2^{-10}	8.527e − 1	6.938e − 1	3.201e − 1	4.849e − 2
2^{-12}	8.524e − 1	6.938e − 1	3.201e − 1	4.849e − 2
$\bar{Q}(N)$	8.703e − 1	6.938e − 1	3.201e − 1	4.849e − 2

ε-uniformly. The error $Q_{N^*,N_0^*}(\varepsilon, N)$ increases up to a constant exceeding 1 when $\varepsilon N \to 0$.

As is seen from Table XXXVI, the function $P^{h+}(x, t)$ in the case of scheme (4.34), (4.36) converges to the function $P(x, t)$ both for a fixed value of the parameter and ε-uniformly when $N \to \infty$, $N_0 = N$.

The results given in Tables XXXIII–XXXVI agree with the theory. Finite difference schemes on space grids with an arbitrary distribution of nodes, in particular, on uniform space grids, can cause errors in the approximate solutions that are many times greater than the maximum absolute value of the actual solutions. Besides, for small values of ε, the errors in the computed fluxes tend to some constant exceeding unity when $N, N_0 \to \infty$. We can see that the application of finite difference schemes on special condensing grids allows us to find approximate solutions and computed normalized diffusion fluxes that are ε-uniformly convergent when $N, N_0 \to \infty$.

Thus, for singularly perturbed diffusion equations with mixed boundary conditions and concentrated sources, we have constructed new finite difference schemes that enable us to find solutions and normalized diffusion fluxes to an ε-uniform accuracy.

V. APPLICATION TO HEAT TRANSFER IN SOME TECHNOLOGIES

Many processes of heat and mass transfer, for example, fast-running processes, lead to the investigation of singularly perturbed boundary value problems with a perturbation parameter ε. For example, those problems arise in the analysis of heat and mass transfer for mechanical working of materials, in particular, metals. The use of classical methods for the numerical solution of such problems (see, e.g., [1, 11, 12]) leads us to errors that exceed the exact solution by many orders of magnitude. In this connection, the problem consists in developing special numerical methods for which the error in the solution does not depend on the parameter value ε, that is, ε-uniformly convergent methods. Such methods are developed, for example, in [4, 9, 13–15].

In this section, we consider the processes of heat and mass transfer that take place under the condition of plastic shear of a material in Section V.A), in hot die-forming (in V.B) and hot rolling (in V.C). All these processes are characterized by the small duration of the process and, therefore, lead to boundary value problems for singularly perturbed equations, that is, equations with a small parameter multiplying the highest order derivatives. To solve the problem in the case of plastic shear, we use a special finite difference scheme developed in Section IV. Thus, in some sense, the results given in Section V.A are illustrative. Here we give a description of physical processes that accompany heat transfer under plastic shear. We also give the mathematical formulation of the heat transfer problem. The results of the numerical analysis of the process are given and discussed.

In Section V.B we consider the singularly perturbed problem that arises in the modeling of heat transfer in die-forming. At the initial instant two bodies with different temperature are brought into contact. Thus, we come to a problem with a discontinuous initial function. In Section V.C the problem of heat transfer in hot rolling is considered. In this singularly perturbed problem, the nature of the boundary condition changes on the interface of the contiguous bodies. For the numerical analysis of such heat processes we construct special, ε-uniformly convergent, finite difference schemes. These schemes can be analyzed fairly well by means of the methods given in the previous sections.

By using these special difference schemes we investigate heat transfer numerically for some parameters that are typical for material processing.

The construction and investigation of special difference schemes for a particular singularly perturbed boundary value problem with a discontinuous initial condition were examined in [16–19].

A. Plastic Shear in a Material

The analysis of heat exchange processes, in the case of the plastic shear of a material, leads us to singularly perturbed boundary value problems with a concentrated source. Problems such as these were considered in Section IV, where it was shown that classical difference schemes give rise to errors, which exceed the exact solution by many orders of magnitude if the perturbation parameter is sufficiently small. Besides, a special finite difference scheme, which allows us to approximate both the solution and the diffusion fluxes to ε-uniform accuracy, was constructed and investigated. We use this difference scheme in this section to analyze the process of heat transfer under plastic shear. The technological parameters given here are typical of metal working: forming, machining, and so on. In Section V.A.1, we describe heat exchange under plastic shear. In Section V.A.2, we give the mathematical formulation of the problem. In Section V.A.3, we make a numerical investigation and compare the results that have been obtained by the classical difference scheme and the special one developed in the previous sections. If we use grids with the number of partitioning intervals $N_0 = 100$ and $N = 100$, respectively, to the time and space variables, the error in the computed maximal temperature is 71% in the case of the classical scheme and 2.02% in the case of the special scheme.

1. Description of Heat Exchange under Plastic Shear

In products made by metals, forming a defect may occur, which is detected on macrosections, in the form of thin stripes in which the metal structure is different from that of the surrounding parts of the material. It is known that the properties of the metal in these stripes are much worse. Therefore products with such stripes on the macrosections of samples (either a lot or a group of lots) are often rejected.

The mechanism of the formation of such stripes can be the following. Under plastic deformation the material can reach some definite limit of hardening (here this problem is not discussed in detail). Then this material behaves as an ideally plastic material. From the theory of ideal plasticity, it is known that such a state is the necessary condition for the loss of the stability of homogeneous plastic flow, that is, layers of large plastic shear appear. Such layers are modeled as the surfaces of the shear. That is to say, the parts of the material separated by these surfaces slip relative to each other with a resistance to the plastic flow that is equal to the plastic shear stress for this material. The disposition of the shear surface for die-forming is shown in Fig. 20.

As a result of this shear, heat is liberated on the shear surface. Here the heating can be so great that phase transformations in the metal can occur, which become apparent on the macrosections in the form of stripes. Sometimes these stripes are called adiabatic shear stripes. Of course, the shear on these surfaces is not adiabatic. It is important to know the temperature on the surfaces in order to exert an influence on the deformation conditions and prevent this defect.

In the narrow region of the relatively large shear, heat is liberated due to the slip of the shifting parts. A diagram of the shear line in the body and the principal temperature distribution near the shear are given in Fig.

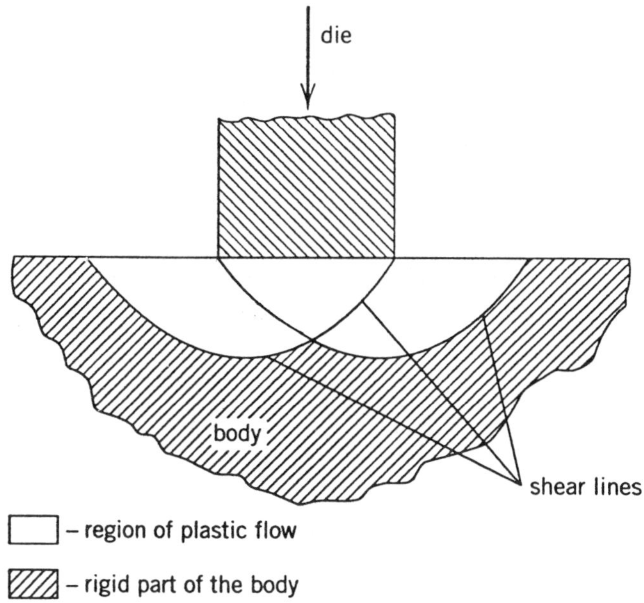

Figure 20. "Die-body" interaction scheme.

21. For simplicity, we assume that the body being deformed consists of two parts shifting relative to each other as solid bodies. Suppose that the heat transfer takes place only along the normal to the shear surface. The heat produced due to the slip is concentrated on the shear surface. Assume that the body under deformation is sufficiently thick and consists of two parts. We assume also that these parts have the same thickness L (m). The duration of metal working is equal to ϑ (s), the velocity of the relatively moving layers v (m/s) is uniform in time. Suppose that the thermal coefficients of the material are constant. At the initial instant, the temperature of the body under deformation is assumed to be constant along the cross section

$$C(y,0)=C^0 \qquad -L<y<L$$

Here $C(y,\tau)$ is the temperature of the body at the point y at the moment τ. The shear surface corresponds to the coordinate plane $y=0$. Since the body is sufficiently thick and the duration of working (e.g., die-forming) is sufficiently short, the temperature of the material on the body surface practically does not vary during the process. So we may assume that the

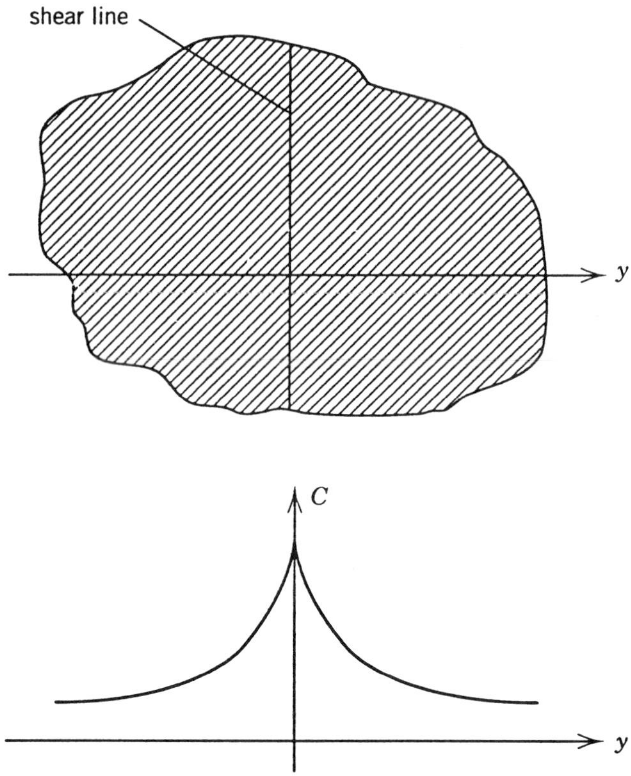

Figure 21. Shear line scheme and near-shear temperature distribution.

temperature on the body surface

$$C(-L, \tau) = C_{-L}(\tau) \qquad C(L, \tau) = C_L(\tau)$$

is constant for all values of time and is equal to the temperature of this body at the initial instant of working

$$C_{-L}(\tau) = C_L(\tau) = C^0 \qquad 0 < \tau \leq \vartheta$$

We want to find the maximal temperature in the body

$$R_1 = \max_{y,\tau} C(y, \tau) \qquad -L \leq y \leq L \qquad 0 \leq \tau \leq \vartheta \tag{5.1a}$$

the thickness of the heated regions in the neighborhood of the intensive shear and the maximal gradient of the temperatures

$$R_2 = \max_{y,\tau} \left| \frac{\partial}{\partial y} C(y, \tau) \right| \qquad -L \leq y \leq L \qquad 0 \leq \tau \leq \vartheta \tag{5.1b}$$

Note that phase transformations in metal occur in a certain temperature interval. The intensity of such phase transitions depends on the rate of material cooling in the phase transition temperature interval. In the problem described above, immediately after the end of the shear and before the moment 2ϑ, we want to find the largest rate of cooling in passing through the given temperature of the phase transition C^P, that is the value

$$R_3 = \max_{\tau} \left\{ -\frac{\partial}{\partial \tau} C(y^P(\tau), \tau) \right\} \qquad \vartheta \leq \tau \leq 2\vartheta \tag{5.1c}$$

where $y^P(\tau)$ is the surface in the body on which the temperature is equal to that of a phase transition

$$C(y^P(\tau), \tau) = C^P \qquad \vartheta \leq \tau \leq 2\vartheta$$

If we know the temperatures and heat fluxes in the narrow region of intensive shear, this knowledge may help in analyzing and solving important theoretical problems. For example, we are able to study the problem concerning the loss of stability of homogeneous deformation and its transition to the local narrow region. It is theoretically proved that ideally plastic material (i.e., material that ceases to harden with the growth of deformation) can exhibit localized, discontinuous or unstable flow. Before the ideally plastic state, two competitive processes take place in the material. They are further hardened by deformation and the drop of resistance to deformation on account of heat liberation. Therefore, it is important to know what process data will not result in unstable flows.

2. *Problem Formulation*

The distribution of temperatures in the material is described by the equations

$$D^H \frac{\partial^2}{\partial y^2} C(y, \tau) - \frac{\partial}{\partial \tau} C(y, \tau) = 0 \qquad -L < y < L \qquad y \neq 0 \qquad 0 < \tau \leq \vartheta$$

$$C(-0, \tau) = C(+0, \tau) \qquad \lambda \frac{\partial}{\partial y} C(+0, \tau) - \lambda \frac{\partial}{\partial y} C(-0, \tau) = -Q(\tau)$$

$$0 < \tau \leq \vartheta \tag{5.2}$$

$$C(y, \tau) = C^0 \qquad y = -L, L \qquad 0 \le \tau \le \vartheta$$

$$C(y, 0) = C^0 \qquad -L \le y \le L$$

Here D^H, λ are the coefficients of temperature conductivity and heat conduction of the material, $Q(\tau)$ is the density of heat sources (on the boundary surface of the shearing layers) caused by slipping. Problem (5.2) is a Dirichlet problem with a concentrated source.

Equations (5.2) are transformed to the dimensionless form

$$
\begin{gathered}
L_{(5.3)} u(x, t) = 0 \qquad (x, t) \in G^j \qquad j = 1, 2 \\
\ell^*_{(5.3)} u(x, t) = -q(t) \qquad [u(x, t)] = 0 \qquad (x, t) \in S^* \\
u(x, t) = 0 \qquad (x, t) \in S
\end{gathered}
\tag{5.3}
$$

Here the dimensionless variables x, t; the functions $u(x, t)$, $q(t)$; and the sets G^j, S^*, S are defined by the relations

$$
\begin{gathered}
G^j = D_j \times (0, T] \qquad G = D \times (0, T] \qquad D_1 = (-1, 0) \\
D_2 = (0, 1) \qquad D = (-1, 1) \\
S^* = \{x = 0\} \times (0, T] \qquad S = \bar{G} \setminus G \\
x = L^{-1} y \qquad t = 2^{-1} \vartheta^{-1} T \tau \\
u(x, t) = C_*^{-1} \{ C(Lx, 2\vartheta T^{-1} t) - C^0 \} \qquad (x, t) \in \bar{G}
\end{gathered}
\tag{5.4}
$$

$$
q(t) = \begin{cases} \sqrt{2} \lambda^{-1} (D^H)^{1/2} \vartheta^{1/2} T^{-1/2} C_*^{-1} Q(2\vartheta T^{-1} t) & 0 < t \le 2^{-1} T \\ 0 & 2^{-1} T < t \le T \end{cases}
$$

$$L_{(5.3)} u(x, t) \equiv \left\{ \varepsilon^2 \frac{\partial^2}{\partial x^2} - \frac{\partial}{\partial t} \right\} u(x, t) \qquad (x, t) \in G^j$$

$$\ell^*_{(5.3)} u(x, t) \equiv \varepsilon \left(\frac{\partial}{\partial x} u(x + 0, t) - \frac{\partial}{\partial x} u(x - 0, t) \right) \qquad (x, t) \in S^*$$

$$\varepsilon^2 = 2 D^H L^{-2} \vartheta T^{-1}$$

The value C_* is the scaling factor for temperature; it is chosen below. For convenience, the value T is assumed to be equal to unity. If ε turns out to be greater than 1, then we choose the value T so that the following inequality holds:

$$\varepsilon \le 1$$

Thus, problem (5.3) is a boundary value problem for a singularly perturbed equation with a concentrated source, which was considered in Section IV.

The derivative (or the gradient) $(\partial/\partial y)C(y,\tau)$ and the heat flux

$$P^H(y,\tau) = \lambda \frac{\partial}{\partial y} C(y,\tau)$$

are defined by the formulas

$$\begin{aligned} \frac{\partial}{\partial y} C(y,\tau) &= C_* L^{-1} \frac{\partial}{\partial x} u(L^{-1}y, 2^{-1}\vartheta^{-1}T\tau) \\ P^H(y,\tau) &= M_P P(L^{-1}y, 2^{-1}\vartheta^{-1}T\tau) \end{aligned} \tag{5.5}$$

where $M_P = 2^{-1/2} C_* \lambda (D^H)^{-1/2} \vartheta^{-1/2} T^{1/2}$

$$P(x,t) = \varepsilon \frac{\partial}{\partial x} u(x,t)$$

The function $P(x,t)$ is called the normalized diffusion flux. It is expressed in terms of the first-order derivative of the function $u(x,t)$, which is the solution of the boundary value problem in dimensionless form.

In the case of problem (5.3), we want to find the solution of the boundary value problem, as well as the normalized diffusion flux and the first-order derivative with respect to t.

3. *Numerical Investigation of the Heat Exchange Problem*

Note that, in the case of problem (5.3), an interior layer appears as the parameter ε tends to zero. Boundary layers do not appear. In the problem under consideration the parameter ε is characterized by some value ε_0. It is required to decide: whether ε_0 is a small value, and, therefore, the special finite difference scheme should be applied, or the value ε_0 is not too small, and, therefore, we can use a finite difference scheme on a uniform grid. The answer depends on N, that is, the number of nodes of the space grid. According to the results in Section IV (see Section IV.C), if $N\varepsilon_0 \le 1$, we should expect large errors to occur in the case of uniform grids. If $N\varepsilon_0 \gg 1$, then good accuracy of the approximate solution on uniform grids can be expected.

For the solution of problem (5.3), we use finite difference schemes either on the grids

$$\bar{G}_h = \bar{\omega}_1 \times \bar{\omega}_0 \tag{5.6}$$

where $\bar{\omega}_1$ and $\bar{\omega}_0$ are uniform grids on the segments $\bar{D}=[-1,1]$ and $[0,T]$, respectively, with step sizes $h=2N^{-1}$ and $h_0=TN_0^{-1}$, $N+1$ and N_0+1 are the numbers of nodes in the grids $\bar{\omega}_1$ and $\bar{\omega}_0$, or on the special grids

$$\bar{G}_h=\bar{\omega}_1\times\bar{\omega}_0 \tag{5.7a}$$

where $\bar{\omega}_0$ is a uniform grid, $\bar{\omega}_0=\bar{\omega}_{0(5.6)}$. The grid $\bar{\omega}_1=\bar{\omega}_{1(5.7)}$ condensing in the neighborhood of the point $x=0$, where the concentrated source acts, is constructed in the following way. The segment $\bar{D}=[-1,1]$ is divided into three parts $[-1,-\sigma]$, $[-\sigma,\sigma]$, $[\sigma,1]$. The step size of the grid is constant on each part and equal to $h_{(1)}$ on the segment $[-\sigma,\sigma]$ and $h_{(2)}$ on the segments $[-1,-\sigma]$, $[\sigma,1]$. We set $h_{(1)}=4\sigma N^{-1}$, $h_{(2)}=4(1-\sigma)N^{-1}$, where $N+1$ is the number of nodes in the grid $\bar{\omega}_1$. The value σ is defined by the relation

$$\sigma=\sigma(\varepsilon,N)=\min[2^{-1},m^{-1}\varepsilon\ln N] \tag{5.7b}$$

where m is an arbitrary number. For definiteness, m is assumed to be equal to 1. This relation completes the construction of the grid $\bar{\omega}_{1(5.7)}$. Corresponding to problem (5.3) we introduce the finite difference scheme

$$\begin{aligned}\Lambda_{(5.8)}z(x,t)&=0 \qquad (x,t)\in G_h^j\\ \lambda^*_{(5.8)}z(x,t)&=-q(t) \qquad (x,t)\in S_h^*\\ z(x,t)&=0 \qquad (x,t)\in S_h\end{aligned} \tag{5.8}$$

Here

$$G_h^j=G^j\cap\bar{G}_h \qquad S_h^*=S^*\cap\bar{G}_h \qquad S_h=S\cap\bar{G}_h$$

$$\Lambda_{(5.8)}z(x,t)\equiv\{\varepsilon^2\delta_{\bar{x}\hat{x}}-\delta_{\bar{t}}\}z(x,t) \qquad (x,t)\in G_h^j$$

$$\lambda^*_{(5.8)}z(x,t)\equiv\varepsilon\{\delta_x z(x,t)-\delta_{\bar{x}}z(x,t)\} \qquad (x,t)\in S_h^*$$

$\delta_{\bar{x}\hat{x}}z(x,t)$ and $\delta_x z(x,t)$, $\delta_{\bar{x}}z(x,t)$, $\delta_{\bar{t}}z(x,t)$ are second- and first-order finite differences on nonuniform grids

$$\delta_x z(x,t)=(x^{i+1}-x^i)^{-1}(z(x^{i+1},t)-z(x^i,t))$$

$$\delta_{\bar{x}}z(x,t)=(x^i-x^{i-1})^{-1}(z(x^i,t)-z(x^{i-1},t))$$

$$\delta_{\bar{x}\hat{x}}z(x,t)=2(x^{i+1}-x^{i-1})^{-1}(\delta_x z(x,t)-\delta_{\bar{x}}z(x,t))$$

$$x = x^i \qquad x^{i-1}, x^i, x^{i+1} \in \bar{\omega}_1$$

$$\delta_{\bar{t}} z(x, t) = (t^j - t^{j-1})^{-1}(z(x, t^j) - z(x, t^{j-1}))$$

$$t = t^j \qquad t^{j-1}, t^j \in \bar{\omega}_0$$

The solution of the difference scheme (5.8), (5.7) converges ε-uniformly to the solution of problem (5.3).

We solve problems (5.8), (5.6) and (5.8), (5.7) for $N_0 = N = 100$. The subdivision of the segments $\bar{D}$ and $[0, T]$ into 100 parts seems to be sufficient according to common sense.

We are interested in an industrial process with the following characteristics. We assume that the ingot is steel, $L = 1$ m, $\vartheta = 1$ s, $D^H = 2 \times 10^{-5}\ \text{m}^2 \cdot \text{s}^{-1}$, $\lambda = 63$ Wt/(m · K), $C^0 = 600$ °C, $T = 1$. The tangential stress under slipping τ_s is taken as $5 \times 10^7\ \text{N} \cdot \text{m}^{-2}$. Then, for $v = 10^{-1}\ \text{m} \cdot \text{s}^{-1}$, on an element with the area $10^{-4}\ \text{m}^2$ in the slipping plane, the following heat is liberated per unit time: $\tau_s v = 5 \times 10^2$ J, that is, $Q(\tau) = 5 \times 10^6$ Wt/m^2. For this problem, the parameter ε defined by the relation

$$\varepsilon^2 = 2D^H L^{-2} \vartheta T$$

then takes the value

$$\varepsilon_0 = 6.3 \times 10^{-3}$$

We choose $C_* = \sqrt{2}\lambda^{-1}(D^H)^{1/2}\vartheta^{1/2}T^{-1/2}\max_\tau Q(\tau)$ $[C_* = 5.02 \times 10^2\ °\text{C}]$. Then, $\max q(t) = 1$. In this case, the maximum of the function $u(x, t)$ will be a quantity of order unity.

The solution of problem (5.8) on the special grid (5.7) for $N = 100$ and $N_0 = N$ is given in Fig. 22. From the figures, one can see that in the neighborhood of the point $x = 0$ the solution of problem (5.8), (5.7) varies sharply displaying the character of an interior layer.

For comparison, the approximate solution was computed for the case of the difference scheme (5.8) on the uniform grid (5.6). The solution of problem (5.8), (5.6) for $N = 100$ and $N_0 = N$ is given in Fig. 23. If we compare the figures, we can see that the results computed with finite difference schemes (5.8), (5.6) and (5.8), (5.7) differ sharply. For example, the maximum values differ by a factor greater than 3

$$\max_{\bar{G}_h} z_{(5.8, 5.7)}(x, t) = 0.477$$

$$\max_{\bar{G}_h} z_{(5.8, 5.6)}(x, t) = 1.663$$

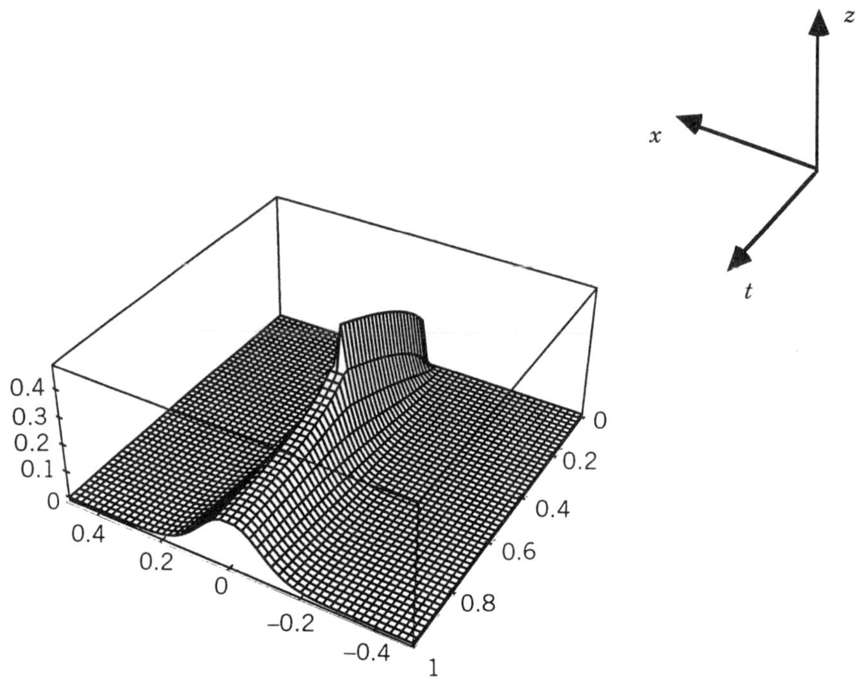

Figure 22. Solution of problem (5.8) on special grid (5.7) in the subinterval $-0.5 \leq x \leq 0.5$ for $\varepsilon = \varepsilon_0 = 6.3 \times 10^{-3}$.

where $z_{(5.8,5.7)}(x, t)$ and $z_{(5.8,5.6)}(x, t)$ are the solutions of problems (5.8), (5.7) and (5.8), (5.6), respectively. By taking into account the formula [see (5.4)]

$$C(y, \tau) = C_* u(x(y), t(\tau)) + C^0 = C_* u(L^{-1}y, 2^{-1}\vartheta^{-1}T\tau) + C^0$$

for the value $R_{1(5.1)}$, we find that

$$R_1 = 839\,°\mathrm{C}$$

in the case of the finite difference scheme (5.8), (5.7) and

$$R_1 = 1435\,°\mathrm{C}$$

in the case of the finite difference scheme (5.8), (5.6).

We would like to evaluate the error in the value of R_1. From the theory considered in Section IV it is known that the solution of the finite difference scheme (5.8), (5.7) converges ε-uniformly for $N, N_0 \to \infty$. Let

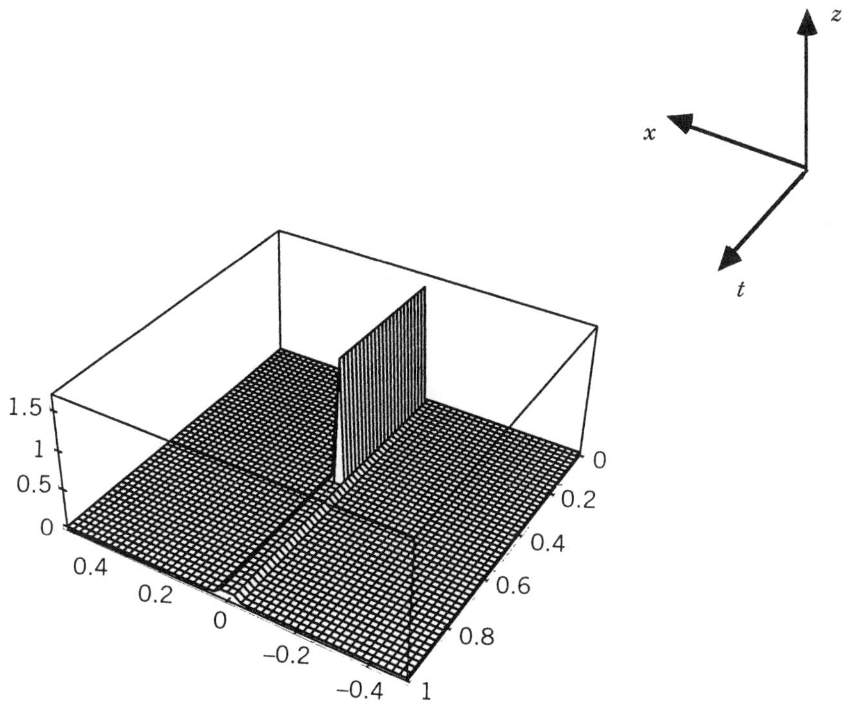

Figure 23. Solution of problem (5.8) on the uniform grid (5.6) in the subinterval $-0.5 \le x \le 0.5$ for $\varepsilon = \varepsilon_0 = 6.3 \times 10^{-3}$.

$z_{N,N_0}(x, t)$ be the solution of problem (5.8), (5.6) with the number of nodes in the grids $\bar{\omega}_1$ and $\bar{\omega}_0$ equal to $N+1$ and N_0+1, respectively. The error

$$E_{N,N_0} \equiv \max_{\bar{G}_h} |u(x, t) - z_{N,N_0}(x, t)|$$

is close to the quantity

$$E^*_{N,N_0} \equiv \max_{\bar{G}_h} |\bar{z}_{N^*,N^*_0}(x, t) - z_{N,N_0}(x, t)| \qquad \bar{G}_h = \bar{G}_h(N, N_0)$$

if the values of N^* and N^*_0 considerably exceed the values of N and N_0, respectively. Here $\bar{z}_{N^*,N^*_0}(x, t)$, $(x, t) \in \bar{G}$ is a continuous function constructed by linear interpolation in the variables x and t from the values of the grid function $z_{N^*,N^*_0}(x, t)$. Computations made by scheme (5.8),

(5.7) for $N = N_0 = 100$ and $N^* = N_0^* = 200$ give

$$E^*_{N,N} \equiv \max_{\bar{G}_h} |\bar{z}_{N^*,N^*}(x,t) - z_{N,N}(x,t)| = 0.033$$

$$\bar{G}_h = \bar{G}_h(N = N_0 = 100)$$

Thus, we take $E_{N,N} = 0.033$. Consequently, the error in $R_1 = 839\,°\mathrm{C}$ is of the same order as the value $C_* E^*_{N,N} = 17\,°\mathrm{C}$.

A graph of the function $C(y, \tau)$ constructed from the solution of problem (5.8), (5.7), is given in Fig. 24. We see that the principal growth in the temperature for $\tau \leq \vartheta$ takes place in the region $|y| \leq \varepsilon_0 L = 0.63$ cm.

The maximal gradient $R_{2(5.1)}$ is found from the equation

$$R_2 = C_* L^{-1} \max_{x,t} \left| \frac{\partial}{\partial x} u(x,t) \right| \qquad |x| \leq 1 \qquad 0 \leq t \leq T$$

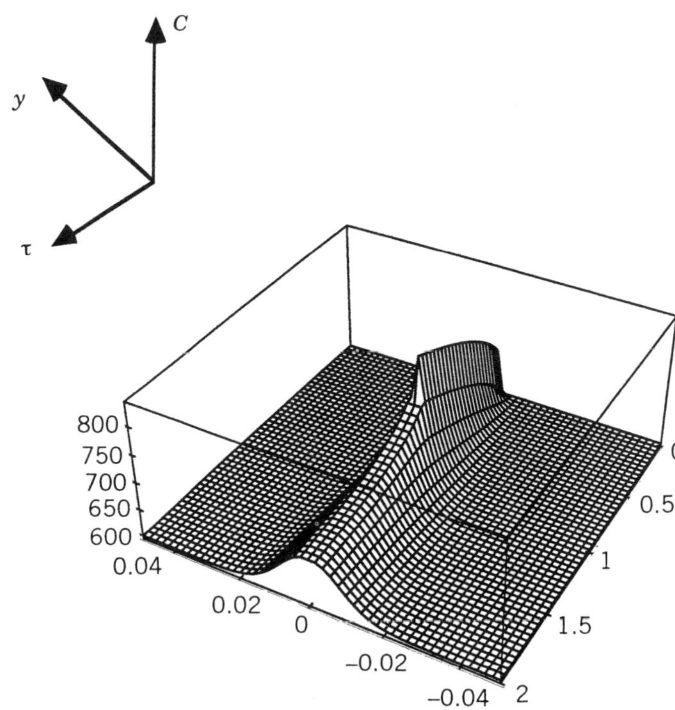

Figure 24. Graph of the temperature $C(y, \tau)\,°\mathrm{C}$ computed from the solution of problem (5.8), (5.7) for $-0.05\,\mathrm{m} \leq y \leq 0.05\,\mathrm{m}$, $0 \leq \tau \leq 2\,\mathrm{s}$.

and in this problem it is attained for $y = 0$ $(x = 0)$; R_2 is given by the relation

$$R_2 = 2^{-1} C_* L^{-1} \varepsilon_0^{-1} q$$

and it equals 3.98×10^4 (K m^{-1}). The highest rate of cooling for $\vartheta \leq \tau \leq 2\vartheta$ (or $2^{-1}T \leq t \leq T$) and for $C^P = 700\,°\text{C}$ is defined by the formula

$$R_{3(5.1)} = 2^{-1} C_* \vartheta^{-1} T \max_{2^{-1} \leq t \leq 1} \left\{ -\frac{\partial}{\partial t} u(x^P(t), t) \right\}$$

$$u(x^P(t), t) = C_*^{-1}\{700 - C^0\} = 0.199$$

and it equals 55.7 °C/s for $y = 0.0$ m, $\tau = 1.5$ s, with $C^0 = 600\,°\text{C}$.

B. Hot Die-Forming

In this section, we consider heat transfer in the case when, at the initial instant, two bodies with different temperatures are brought into contact. Such problems arise, for example, in hot die-forming. For the short duration of the process, heat transfer can be described by a singularly perturbed equation, that is, an equation with a small parameter multiplying the highest order derivatives. For small values of the perturbation parameter the error in the approximate solution obtained by the classical scheme can be commensurate with the solution itself. Moreover, the error in the heat fluxes, in particular, on the interface of the bodies in contact, can exceed the actual heat flux by many orders of magnitude. Therefore, for the numerical analysis of the die-forming process, we need to develop special difference schemes, the accuracy of which does not depend on the value of the parameter.

In Section V.B.1, we describe heat transfer in hot die-forming and give a mathematical formulation of the problem. In Section V.B.2, we construct a special finite difference scheme. This scheme allows us to approximate ε-uniformly both the problem solution and the amount of heat that passes through the "die-body" interface. In Section V.B.3 we numerically analyze the problem of heat exchange between the die and the hot body with technological parameters typical for die-forming.

1. Description of Heat Exchange for Hot Die-Forming: Problem Formulation

As a rule, die-forming consists of stages when the die acts on the body and pauses between successive actions. During one of these stages, the die and the body, which have different temperatures, are brought into contact for a short time. Schematically, this stage is shown in Fig. 20.

During the pause the die is subjected to cooling before forming the next article. This step is repeated many times. Here two principal questions arise. One of them concerns the temperatures in the contact area between the die and the body, and the other concerns the heat fluxes that pass through the interface. High temperatures near the working surface of the die cause lower abrasion and crushing strength. In the long run, lower strength causes loss of accuracy in the die-forming. On the other hand, excessive decrease in the body temperature leads to the appearance of defects on the surface of the product.

The heat fluxes, which the die and body exchange, determine the duration of die cooling and the cooling agent consumption.

Suppose that we want to find the distribution of temperature in the die and the hot body, which come into contact with each other during the process of die-forming. For simplicity, we assume that the die and the body are in contact along a plane, and heat transfer in this system proceeds in the direction normal to this plane. The scheme of mutual location of the die and the body in contact, and the qualitative distribution of temperatures in the interface region, are presented in Fig. 25.

The thermal characteristics of the die and the body are assumed to be constant. The body and the die have thickness L_1 and L_2, respectively, and occupy the domains $-L_1 \le y \le 0$ and $0 \le y \le L_2$. The distribution of temperatures in the body and the die can be defined by the heat equations

$$
\begin{aligned}
D_1^H \frac{\partial^2}{\partial y^2} C(y,\tau) - \frac{\partial}{\partial \tau} C(y,\tau) &= 0 \qquad -L_1 < y < 0 \\
D_2^H \frac{\partial^2}{\partial y^2} C(y,\tau) - \frac{\partial}{\partial \tau} C(y,\tau) &= 0 \qquad 0 < y < L_2 \qquad 0 < \tau \le \vartheta
\end{aligned}
\tag{5.9a}
$$

Here D_1^H and D_2^H are temperature conductivity coefficients for the body and the die, respectively, ϑ is the duration of the process. At the initial instant the temperatures of the body and the die have the given values

$$
C(y,0) = \begin{cases} C_M^0(y) & -L_1 \le y < 0 \\ C_S^0(y) & 0 < y \le L_2 \end{cases}
\tag{5.9b}
$$

moreover, on the contact surface $y = 0$ the temperature of this "body-die" system is discontinuous:

$$
C_M^0 \ne C_S^0
$$

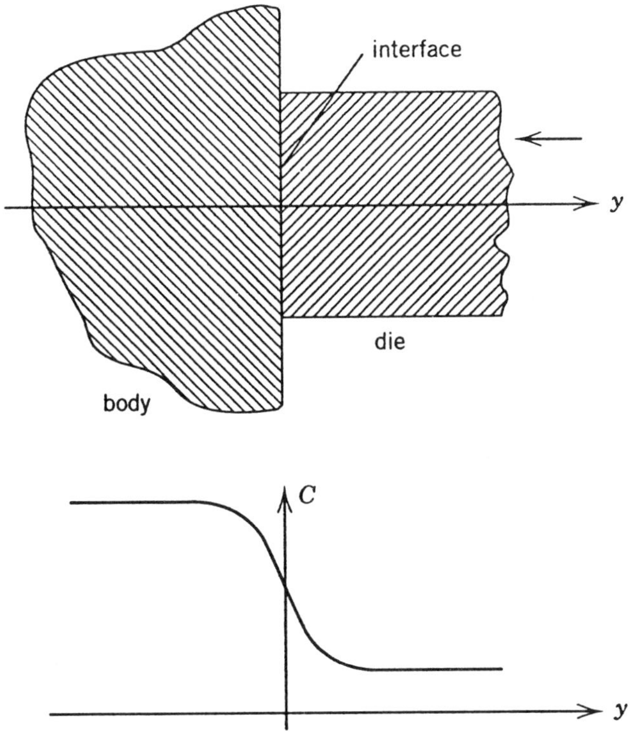

Figure 25. Scheme of the "die-body" disposition in contact, and the temperature distribution near the interface.

where $C_M^0 = C_M^0(-0)$, $C_S^0 = C_S^0(+0)$. The temperature jump at the point of discontinuity is determined by the relation

$$C(+0,0) - C(-0,0) = C_S^0 - C_M^0$$

On the surfaces of the body and the die, adjacent to the surrounding medium, that is, for $y = -L_1, L_2$, we observe convective heat exchange

$$\lambda_1 \frac{\partial}{\partial y} C(-L_1, \tau) = \alpha_1 C(-L_1, \tau)$$
$$-\lambda_2 \frac{\partial}{\partial y} C(L_2, \tau) = \alpha_2 C(L_2, \tau) \qquad 0 < \tau \leq \vartheta \tag{5.9c}$$

Here α_1 and α_2 are the coefficients of convective heat emission in the surrounding medium, λ_1 and λ_2 are the coefficients of heat conductivity of the body and the die, respectively; the temperature of the surrounding medium is assumed to be zero. For simplicity, we assume that there is no thermal resistance on the interface. Then

$$C(-0, \tau) = C(+0, \tau)$$
$$\lambda_1 \frac{\partial}{\partial y} C(-0, \tau) = \lambda_2 \frac{\partial}{\partial y} C(+0, \tau) \qquad 0 < \tau \leq \vartheta \tag{5.9d}$$

We want to find the temperature distribution in the "body-die" system during forming, and the total amount of heat transferred from the metal body into the die.

The quantity Q^H of heat transferred from the body to the die during the length of time ϑ, is determined by the relation

$$Q^H = -\lambda_2 \int_0^{\vartheta} \frac{\partial}{\partial y} C(+0, \eta)\, d\eta \tag{5.10}$$

Q^H is the quantity of heat (J m^{-2}) passing through the unit (1 m^2) surface of the die at $y = 0$ during the length of time ϑ.

Equations (5.9) are transformed to the dimensionless form

$$L^1_{(5.11)} u(x, t) \equiv \left\{ \varepsilon_1^2 \frac{\partial^2}{\partial x^2} - \frac{\partial}{\partial t} \right\} u(x, t) = 0 \qquad (x, t) \in G^1$$
$$L^2_{(5.11)} u(x, t) \equiv \left\{ \varepsilon_2^2 \frac{\partial^2}{\partial x^2} - \frac{\partial}{\partial t} \right\} u(x, t) = 0 \qquad (x, t) \in G^2$$
$$u(+0, t) - u(-0, t) = 0$$
$$\beta^0 \varepsilon_2 \frac{\partial}{\partial x} u(+0, t) - \varepsilon_1 \frac{\partial}{\partial x} u(-0, t) = 0 \qquad 0 < t < T \tag{5.11}$$
$$\varepsilon_1 \frac{\partial}{\partial x} u(-1, t) - \gamma_1 u(-1, t) = 0$$
$$\varepsilon_2 \frac{\partial}{\partial x} u(1, t) + \gamma_2 u(1, t) = 0 \qquad 0 < t \leq T$$
$$u(0, t) = \varphi(x) \qquad -1 \leq x \leq 1 \qquad x \neq 0$$

Here

$$x = L_1^{-1} y \qquad y < 0 \qquad x = L_2^{-1} y \qquad y > 0 \qquad t = \vartheta^{-1} T \tau$$
$$G^1 = D_1 \times (0, T] \qquad G^2 = D_2 \times (0, T] \qquad D_1 = (-1, 0) \qquad D_2 = (0, 1)$$
$$\varepsilon_j^2 = D_j^H \vartheta T^{-1} L_j^{-2} \qquad \gamma_j = \alpha_j \lambda_j^{-1} \vartheta^{1/2} T^{-1/2} (D_j^H)^{1/2} \qquad j = 1, 2$$

$$\beta^0 = \lambda_1^{-1}\lambda_2(D_1^H)^{1/2}(D_2^H)^{-1/2} \tag{5.12}$$

$$u(x,t) = \begin{cases} C_*^{-1}C(L_1x, \vartheta T^{-1}t) & -1 \le x < 0 \\ C_*^{-1}C(L_2x, \vartheta T^{-1}t) & 0 < x \le 1 \quad 0 \le t \le T \quad (x,t) \ne (0,0) \end{cases}$$

$$\varphi(x) = \begin{cases} C_*^{-1}C_M^0(L_1x) & -1 \le x < 0 \\ C_*^{-1}C_S^0(L_2x) & 0 < x \le 1 \end{cases}$$

and the scaling factor C_* is found by the relation

$$C_* = \max_{D_j, j} \varphi(x)$$

It is convenient to assume that the value T is equal to unity. If ε_1 and/or ε_2 exceed unity, then we choose the value T so that the following relations are fulfilled:

$$\varepsilon_1, \varepsilon_2 \le 1$$

For the chosen value of C_* the maximum of the function $u(x,t)$ does not exceed unity.

When the values L_1, L_2 are sufficiently large and/or the values ϑ and D_1^H, D_2^H are sufficiently small, the parameters $\varepsilon_1, \varepsilon_2$ can take any small values. It can happen that the values of ε_1 and ε_2 are not coupled and, generally speaking, take different values. Depending on the problem formulation, we assume that the values of $\varepsilon_1, \varepsilon_2$ are independent of each other and take arbitrary values in the half-interval $(0, 1]$.

In what follows it is convenient to rewrite problem (5.11) in the following form:

$$\begin{gathered} L_{(5.13)}u(x,t) \equiv \left\{\varepsilon^2 \frac{\partial^2}{\partial x^2} - \frac{\partial}{\partial t}\right\} u(x,t) = 0 \qquad (x,t) \in G \setminus S^* \\ [u(x,t)] = 0 \qquad \left[\beta\varepsilon \frac{\partial}{\partial x} u(x,t)\right] = 0 \qquad (x,t) \in S^* \\ \varepsilon \frac{\partial}{\partial n} u(x,t) + \gamma u(x,t) = 0 \qquad (x,t) \in S^1 \\ u(x,t) = \varphi(x) \qquad (x,t) \in S^0 \setminus S_0^* \end{gathered} \tag{5.13}$$

Here

$$G = D \times (0, T] \qquad D = (-1, 1) \qquad D = D_1 \cup D^0 \cup D_2 \qquad D_1 = (-1, 0)$$

$$D_2 = (0, 1) \qquad D^0 = \{x = 0\} \qquad S^* = \{x = 0\} \times (0, T] \qquad S_0^* = (0, 0)$$

$$S^1 = \{x = -1, 1\} \times (0, T] \qquad S^0 = \bar{D} \times \{t = 0\}$$

$$\varepsilon = \varepsilon(x) = \varepsilon_j \qquad \beta = \beta(x) = \beta_j \qquad \gamma = \gamma(x) = \gamma_j \qquad x \in \bar{D}_j \qquad j = 1, 2$$

$$\beta_1 = 1 \qquad \beta_2 = \beta^0$$

$\partial/\partial n$ is the derivative along the normal outward to the lateral boundary S^1; $[v(x, t)]$, $(x, t) \in S^*$ is the jump of the function $v(x, t)$ when crossing the set S^*

$$[v(x, t)] = v(x + 0, t) - v(x - 0, t) \qquad (x, t) \in S^*$$

generally speaking, the initial function $\varphi(x)$ is discontinuous at $x = 0$; the parameters ε_1 and ε_2, that is, the coefficients at the highest derivatives in (5.13), take any values in the half-interval $(0, 1]$.

Problem (5.13) is a singularly perturbed boundary value problem for a parabolic equation with discontinuous coefficients at the highest order derivative and with a discontinuous initial condition.

The jump of the function $\varphi(x)$ at the point $x = 0$ is defined by the relation

$$[\varphi(0)] = \varphi(+0) - \varphi(-0) = C_*^{-1}(C_S^0 - C_M^0)$$

The discontinuous initial function $\varphi(x)$ is redefined at the point of discontinuity by

$$\varphi(0) = 2^{-1}\{\varphi(+0) + \varphi(-0)\}$$

The function $u(x, t)$ with a discontinuity of the first kind on the set S_0^* is defined on this set by the relation

$$u(x, t) = \varphi(x) \qquad (x, t) \in S_0^*$$

The derivative $(\partial/\partial y)C(y, \tau)$ and the heat flux $P^H(y, \tau)$ have discontinuities on the set S^*:

$$\frac{\partial}{\partial y} C(+0, \tau) \neq \frac{\partial}{\partial y} C(-0, \tau) \qquad P^H(+0, \tau) \neq P^H(-0, \tau)$$

where

$$P^H(y, \tau) = \lambda \frac{\partial}{\partial y} C(y, \tau)$$

$$\lambda = \lambda(y) = \lambda_j \qquad \lambda_j = \lambda_1 \quad \text{for} \quad y<0 \qquad \lambda_j = \lambda_2 \quad \text{for} \quad y>0$$

The derivative $(\partial/\partial y)C(y, \tau)$, the heat flux $P^H(y, \tau)$ and the value of Q^H are defined by the formulas

$$\frac{\partial}{\partial y} C(y, \tau) = C_* L^{-1} \frac{\partial}{\partial x} u(L^{-1}y, \vartheta^{-1}T\tau)$$

$$P^H(y, \tau) = C_* \lambda (D^H)^{-1/2} \vartheta^{-1/2} T^{1/2} P(L^{-1}y, \vartheta^{-1}T\tau)$$

$$Q^H = -M_Q \int_0^T P(+0, \eta)\, d\eta$$

Here

$$P(x, t) = \varepsilon \frac{\partial}{\partial x} u(x, t)$$

is the normalized diffusion flux; the function $P(x, t)$ and the derivative $(\partial/\partial x)u(x, t)$ are continuous on $\bar{G}^j$ for $(x, t) \not\in S_0^*$; $\bar{G}^j = \bar{D}_j \times (0, T]$;

$$L = L(x) = L_j \qquad \lambda = \lambda(x) = \lambda_j \qquad D^H = D^H(x) = D_j^H$$

$$x \in \bar{D}_j \qquad j = 1, 2$$

and the constant M_Q is determined by

$$M_Q = C_* \lambda_2 (D_2^H)^{-1/2} \vartheta^{1/2} T^{-1/2}$$

We call the quantity

$$Q(x_0) \equiv -\int_0^T P(x_0, \eta)\, d\eta \tag{5.14}$$

the total amount of a normalized substance passing in the positive direction of the axis x through the cross section $x = x_0$ during the process.

Thus, in the case of problem (5.13) we want to find the solution of the boundary value problem and the value $Q = Q(0)$, which is the total amount of a normalized substance passing through the section $x = 0$ during the time T.

When the parameter ε [$\varepsilon = \varepsilon(x), x \in \bar{D}_j$, $j = 1, 2$] tends to zero, interior and boundary layers appear in the neighborhood of the sets S^* and S^1, respectively. Even for $\varepsilon = 1$, the solution of the boundary value problem (5.13) is not smooth in the neighborhood of the set S_0^*; the derivatives of the solution with respect to x and t increase without bound when the point (x, t) tends to the point $(0, 0)$.

2. *Finite Difference Scheme for Problem (5.13)*

The use of classical finite difference schemes for solving problem (5.13) gives rise to some difficulties. These difficulties are caused by the loss of smoothness in the neighborhood of the discontinuity of the initial function, and by the unbounded increase of the derivatives with respect to x in the neighborhood of the sets S^*, S^1, as the parameter tends to zero.

We describe a finite difference scheme, the solution of which approximates the solution of the boundary value problem ε-uniformly on the whole grid set in $\bar{G}$.

To solve the problem, we use the method of additive representation of singularities (see, e.g., [20]). The solution of problem (5.13) is represented as a sum of continuous and discontinuous parts:

$$u(x, t) = U(x, t) + W(x, t) \qquad (x, t) \in \bar{G} \tag{5.15a}$$

Here the function $W(x, t)$, which has a discontinuity at the point $(0, 0)$, is defined by the relation

$$W(x, t) = \left\{ \begin{array}{ll} [\varphi(0)](1 + \beta_1 \beta_2^{-1})^{-1} \operatorname{erf}\left(\dfrac{x}{2\varepsilon_1 t^{1/2}}\right) & x \leq 0 \\ [\varphi(0)](1 + \beta_1^{-1} \beta_2)^{-1} \operatorname{erf}\left(\dfrac{x}{2\varepsilon_2 t^{1/2}}\right) & x > 0 \end{array} \right\} \qquad (x, t) \in \bar{G} \tag{5.15b}$$

the function $\operatorname{erf}(\xi)$ is the error function

$$\operatorname{erf}(\xi) = \frac{2}{\sqrt{\pi}} \int_0^{\xi} \exp(-\alpha^2)\, d\alpha$$

The function $U(x, t)$, being continuous on $\bar{G}$, is the solution of the boundary value problem

$$\begin{gathered} L_{(5.13)}U(x,t)=0 \qquad (x,t)\in G\backslash S^* \\ [U(x,t)]=0 \qquad \left[\beta\varepsilon\frac{\partial}{\partial x}U(x,t)\right]=0 \qquad (x,t)\in S^* \\ \varepsilon\frac{\partial}{\partial n}U(x,t)+\gamma U(x,t)=\psi(x,t) \qquad (x,t)\in S^1 \\ U(x,t)=\varphi_1(x) \qquad (x,t)\in S^0 \end{gathered} \tag{5.16}$$

Here

$$\psi(x,t)=-\,\varepsilon\frac{\partial}{\partial n}W(x,t)-\gamma W(x,t)$$

$$\varphi_1(x)=\varphi(x)-W(x,0)$$

The function $\varphi_1(x)$ is continuous.

We study finite difference schemes for the solution of problem (5.16), (5.15). First, we consider a classical scheme approximating problem (5.16). On the set $\bar{G}$ we introduce the rectangular grid

$$\bar{G}_h=\bar{\omega}_1\times\bar{\omega}_0 \tag{5.17}$$

where $\bar{\omega}_1$ is a grid with an arbitrary distribution of nodes, the point $x=0$ belongs to the grid $\bar{\omega}_1$; $\bar{\omega}_0$ is a uniform grid; $N+1$ and N_0+1 are the numbers of nodes in the grids $\bar{\omega}_1$ and $\bar{\omega}_0$, respectively. The value h (the maximal step-size of the grid $\bar{\omega}_1$) satisfies the relation: $h\le MN^{-1}$, the constant M is independent of ε, N, N_0.

To solve problem (5.16), we use the difference scheme

$$\begin{gathered} \Lambda_{(5.18)}z(x,t)\equiv\{\varepsilon^2\delta_{\bar{x}\hat{x}}-\delta_{\bar{t}}\}z(x,t)=0 \qquad (x,t)\in G_h^j \qquad j=1,2 \\ \beta_1\varepsilon_1\delta_{\bar{x}}z(x,t)=\beta_2\varepsilon_2\delta_x z(x,t) \qquad (x,t)\in S_h^* \\ -\varepsilon_1\delta_x z(x,t)+\gamma_1 z(x,t)=\psi(x,t) \qquad (x,t)\in S_h^1 \qquad x=-1 \\ \varepsilon_2\delta_{\bar{x}}z(x,t)+\gamma_2 z(x,t)=\psi(x,t) \qquad (x,t)\in S_h^1 \qquad x=1 \\ z(x,t)=\varphi_1(x) \qquad (x,t)\in S_h^0 \end{gathered} \tag{5.18}$$

Here

$$G_h^j=G^j\cap\bar{G}_h \qquad S_h^*=S^*\cap\bar{G}_h \qquad S_h^1=S^1\cap\bar{G}_h \qquad S_h^0=S^0\cap\bar{G}_h$$

$$\varepsilon=\varepsilon(x)=\varepsilon_{(5.12)}(x)$$

The approximate solution of problem (5.13) is given by

$$Z(x, t) = z(x, t) + W(x, t) \qquad (x, t) \in \bar{G}_h \tag{5.19}$$

where $z(x, t)$ is the solution of problem (5.18), (5.17).

Note that the approximate solution $Z_{(5.19)}(x, t)$, which is obtained with the classical finite difference scheme on the grid with an arbitrary distribution of nodes, does not converge ε-uniformly. This is due to the unsatisfactory approximation of the function $U(x, t)$ by the function $z(x, t)$ in the neighborhood of the boundary laycr.

For problem (5.16) we construct a special grid, on which the solution of the finite difference scheme (5.18) approximates the function $U(x, t)$ ε-uniformly.

On the set $\bar{G}$, we introduce a grid condensing in the neighborhood of the boundary layers

$$\bar{G}_h = \bar{\omega}_1^* \times \bar{\omega}_0 \tag{5.20}$$

where $\bar{\omega}_1^*$ is a piecewise uniform grid, $\bar{\omega}_0 = \bar{\omega}_{0(5.17)}$. Let us describe the grid $\bar{\omega}_1^*$. The interval $[-1, 1]$ is divided into four parts: $[-1, -1+\sigma_1]$, $[-1+\sigma_1, 0]$, $[0, 1-\sigma_2]$, $[1-\sigma_2, 1]$. The step size of the grid on each part is constant and equals $h_1 = 4\sigma_1 N^{-1}$ and $h_2 = 4\sigma_2 N^{-1}$ on the subintervals $[-1, -1+\sigma_1]$ and $[1-\sigma_2, 1]$, respectively, whereas it equals $h_3 = 4(1-\sigma_1)N^{-1}$ and $h_4 = 4(1-\sigma_2)N^{-1}$ on the subintervals $[-1+\sigma_1, 0]$ and $[0, 1-\sigma_2]$, respectively. Suppose that

$$\sigma_1 = \sigma_1(\varepsilon_1, N) = \min[2^{-1}, m^{-1}\varepsilon_1 \ln N]$$

$$\sigma_2 = \sigma_2(\varepsilon_2, N) = \min[2^{-1}, m^{-1}\varepsilon_2 \ln N]$$

where m is an arbitrary number. This relation completes the construction of the grid $\bar{G}_h$.

The solution of problem (5.18), (5.20) approximates the solution of problem (5.16) ε-uniformly. We determine the approximate solution of problem (5.13) by the relation

$$Z(x, t) = z(x, t) + W(x, t) \qquad (x, t) \in \bar{G}_h \tag{5.21}$$

where $z(x, t)$ is the solution of problem (5.18), (5.20). This function $Z_{(5.21)}(x, t)$ converges to the solution of problem (5.13) ε-uniformly for N, $N_0 \to \infty$.

Let us describe the algorithm for computing the value of $Q^H_{(5.10)}$. This value is found from the function $P(+0, t)$, which is the normalized diffusion flux to the right from the point $x = 0$. The quantity $Q_{(5.14)}(0)$ is approximated by the quantity Q^h

$$Q^h = Q_1^h + Q_2 \tag{5.22}$$

where

$$Q_1^h = -\varepsilon_2 h_0 \sum_{k=1}^{N_0} \delta_x z(0, t_k)$$

$$Q_2 = -\varepsilon_2 \int_0^T \frac{\partial}{\partial x} W(+0, t)\, dt = -2\pi^{-1/2}[\varphi(0)](1 + \beta_1^{-1}\beta_2)^{-1} T^{1/2}$$

$t_k = kh_0$, $h_0 = TN_0^{-1}$, h_0 is the step size of the grid $\bar{\omega}_0$, $z(x, t)$ is the solution of problem (5.18), (5.20).

The quantity Q^{Hh}, approximating $Q^H_{(5.10)}$, is determined by the formula

$$Q^{Hh} = M_Q Q^h$$

The quantities $Q^h_{(5.22)}$ and Q^{Hh} approximate $Q_{(5.14)}(0)$ and $Q^H_{(5.10)}$ ε-uniformly.

3. *Numerical Investigation of the Heat Transfer Problem*

We give the parameters of the technological process, for which we solve the heat transfer problem for hot die-forming. Suppose that the body and the die are steel, $D_1^H = D_2^H = 2 \times 10^{-5}\ \mathrm{m}^2 \cdot \mathrm{s}^{-1}$, $\lambda_1 = \lambda_2 = 63\ \mathrm{Wt/(m \cdot K)}$; the thickness of the body and the die, and their initial temperatures have the following values, respectively: $L_1 = 0.01$ m, $L_2 = 0.50$ m, $C_M^0 = 600\ °\mathrm{C}$, $C_S^0 = 50\ °\mathrm{C}$. The process duration ϑ is 0.1 s. The coefficient of heat transfer from the surface to the surrounding medium α is assumed to be equal to $10\ \mathrm{Wt/(m^2 \cdot K)}$. This value of α corresponds mainly to convective heat exchange for the die.

For these parameters we obtain

$$\begin{gathered} \varepsilon_1 = 1.41 \times 10^{-1} \qquad \varepsilon_2 = 2.83 \times 10^{-3} \qquad \gamma_1 = \gamma_2 = 2.24 \times 10^{-4} \\ \beta_2 = 1 \qquad C_* = 600\ °\mathrm{C} \qquad T = 1 \end{gathered} \tag{5.23}$$

Note that no boundary layers appear for these special parameters of the process. Therefore, we may use the special scheme on the uniform grid

$$\bar{G}_h = \bar{\omega}_1 \times \bar{\omega}_0 \tag{5.24}$$

that is a grid with the uniform distribution of nodes in x and t.

To solve problem (5.13), (5.23), we use the finite difference scheme (5.18), (5.24). We use grids with $N = N_0 = 100$.

The solution of problem (5.13), (5.23), constructed with the special scheme on the uniform grid, that is, scheme (5.18), (5.24), (5.19), is represented in Fig. 26. As is seen from the figure, the solution of the problem changes sharply thus exhibiting the character of an interior layer. As a result, the value of Q^{Hh} is equal to 1.28×10^6 Wt/(m$^2 \cdot$ s).

To estimate the accuracy of the quantity Q^{Hh}, computations were carried out for $N = N_0 = 200$. The error in Q^{Hh} was found to be 2.1%.

A graph of the function $C(y, \tau)$ for $-L_1 \le y \le 0.1L_2$, constructed from the solution of the grid problem (5.18), (5.24), (5.19), is presented in Fig. 27.

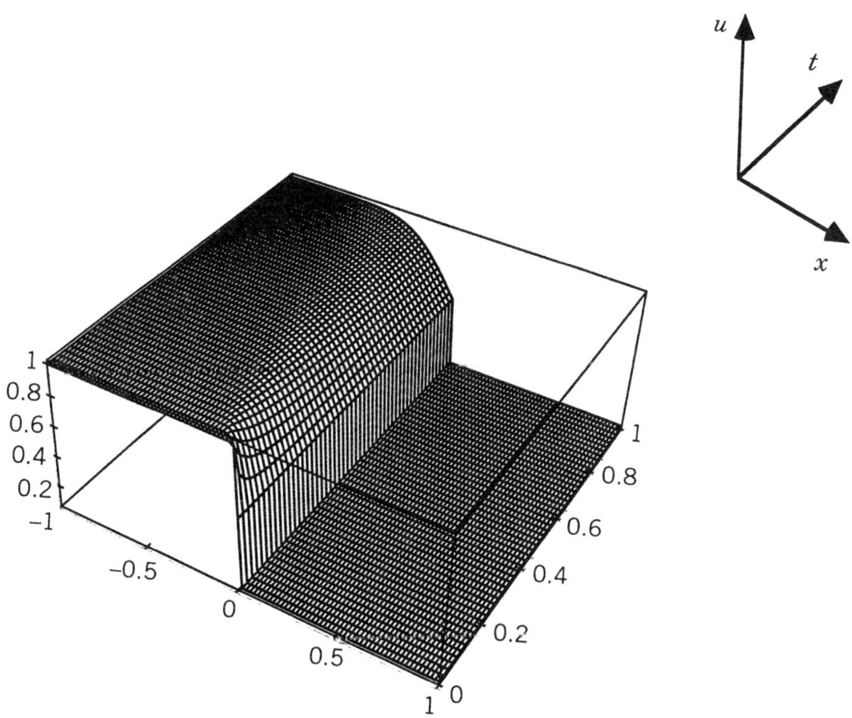

Figure 26. Solution of problem (5.13), (5.23) constructed with the special scheme (5.18), (5.19) on the uniform grid (5.24).

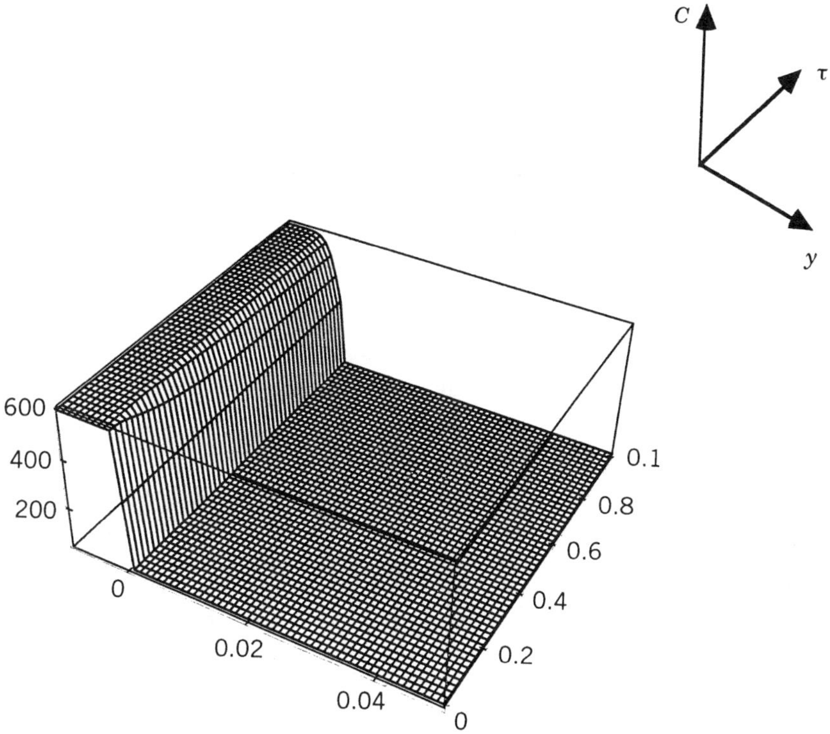

Figure 27. Graph of the temperature $C(y,\tau)$ °C, for $-L_1 \le y \le 0.1L_2$ m, $0 \le \tau \le 0.1$ s, computed from the solution of problem (5.18), (5.19), (5.24).

For comparison, the approximate solution of problem (5.13), (5.23) was also computed with the classical finite difference scheme

$$\begin{aligned}
\Lambda_{(5.25)}z(x,t) \equiv \{\varepsilon^2\delta_{x\bar{x}} - \delta_{\bar{t}}\}z(x,t) = 0 \qquad & (x,t) \in G_h^j \qquad j = 1, 2 \\
\beta_1\varepsilon_1\delta_{\bar{x}}z(x,t) = \beta_2\varepsilon_2\delta_x z(x,t) \qquad & (x,t) \in S_h^* \\
-\varepsilon_1\delta_x z(x,t) + \gamma_1 z(x,t) = 0 \qquad & (x,t) \in S_h^1 \qquad x = -1 \\
\varepsilon_2\delta_{\bar{x}}z(x,t) + \gamma_2 z(x,t) = 0 \qquad & (x,t) \in S_h^1 \qquad x = 1 \\
z(x,t) = \varphi(x) \qquad & (x,t) \in S_h^0
\end{aligned} \tag{5.25}$$

Here $\bar{G}_h = \bar{G}_{h(5.24)}$. The quantities Q^h and Q^{Hh} were calculated by the formulas

$$Q^h = -\varepsilon_2 h_0 \sum_{k=1}^{N_0} \delta_x z(0, t_k) \qquad Q^{Hh} = M_Q Q^h \tag{5.26}$$

where $z(x, t)$ is the solution of problem (5.25), (5.24), $t_k = kh_0$, $t_k \in \bar{\omega}_0$. The computed solution approximates the solution of problem (5.13), (5.23) qualitatively well. However, the value of $Q^{Hh}_{(5.26)}$ obtained with the use of the classical scheme (5.25), (5.24) is equal to 5.10×10^4 $\mathrm{Wt/(m^2 \cdot s)}$.

Note that the change of the coefficient α by an order of magnitude does not significantly influence the results obtained.

C. Hot Rolling

Let us consider heat transfer when two bodies with different initial temperature are periodically brought into contact. In this case, on the interface of these bodies the nature of the boundary condition varies. At the instants when the two bodies come into contact, the temperature on the interface is discontinuous. At subsequent times the temperature and heat fluxes on the interface are continuous. In the case when the bodies do not touch each other, the condition of heat exchange with the surrounding medium is given on the surfaces of the bodies. The solutions of such boundary value problems have restricted smoothness. These solutions have discontinuities at the instants when the bodies come into contact, and their first-order space derivatives have discontinuities on the interface of the bodies at all times when the two bodies are in contact. In processes of small duration, heat transfer is described by a singularly perturbed equation. For small values of the perturbation parameter, in the neighborhood of the interface, a transient (interior) layer appears when the bodies are in contact, and a boundary layer appears when they are not. Therefore, the use of classical finite difference schemes for solving such problems leads to large errors.

The above-described problems arise, for example, in the modeling of heat exchange processes for hot rolling. One of these heat-transfer processes is considered in this section. In Section V.C.1 we give a description of the heat exchange process and the mathematical formulation of the problem. In Section V.C.2 we construct a special finite difference scheme. Numerical investigation of heat transfer for rolling is made in Section V.C.3.

1. *Description of Heat Transfer for Hot Rolling: Problem Formulation*

We want to find the distribution of temperature during rolling in the roller and in a layer of the hot material near the contact surfaces, and to find the heat fluxes on the contact boundaries.

The thermal process in rolling is similar to that in die-forming (see Section V.B). The temperatures of the roller and the die coming into contact with the material, differ from the temperature of the material. Now we discuss the principal differences. The thickness of the material in die-forming can be much smaller than the thickness of the die. Therefore, we obtain a singularly perturbed problem with discontinuous coefficients which differ essentially in different subdomains. The "material-die" system exchanges heat during the whole process. Thus, on the contact surface, the heat exchange condition remains valid during the process. In rolling, the thickness of the material and the radius of the roller are comparable. So we obtain a singularly perturbed problem with discontinuous coefficients, which are of the same order of magnitude in the different subdomains. The material and the roller exchange heat only on the contact surface. Outside of this surface they exchange heat with the surrounding medium. Thus, alternation in the kind of heat exchange conditions takes place on the material and roller boundaries.

Now we describe heat transfer in rolling. We consider a roller and a material layer (or, in short, a body) in the orthogonal (normal) cross section. Suppose that they come into contact on a region of width ℓ, the roller radius is R and the body thickness is L_1. The scheme of mutual location of the roller and the body, and the qualitative distribution of temperatures in the region of contact, are presented in Fig. 28.

At the initial instant (at the beginning of rolling) the roller and the body have a given distribution of temperatures. The "roller-body" contact is assumed to be ideal, that is, the temperatures of the roller and the body are equal on the contact surface, the heat fluxes orthogonal to the roller and body surfaces are equal too. The body is infinitely long and moves with constant velocity v.

We state a few simplifications under which we study this problem. We are interested in the region of sharpest variation of temperatures, in particular, we would like to know the distribution of temperatures in the neighborhood of the contact surfaces during the contact and after it. We are also interested in the quantity $Q^H(\tau)$, that is, the amount of heat that passes into the roller through its surface at the moment τ. The value ϑ, that is, the duration of the rolling process, is assumed not to be very large so that the deep interior parts of the roller have no time to become very hot. The value R is comparable to the value L_1. For simplicity, we consider the roller to be a tube with outside radius R and inside one R_0.

We make the problem even simpler. Let us unroll the tube in a plane. We consider the roller as a plane layer, periodic in length, with period $L = 2\pi R$ and width $L_2 = R - R_0$. Thus, the body and the roller occupy the regions $[-L_1, 0]$ and $[0, L_2]$, respectively, along the axis y_2. The sizes of the body and the roller along y_1 are unbounded.

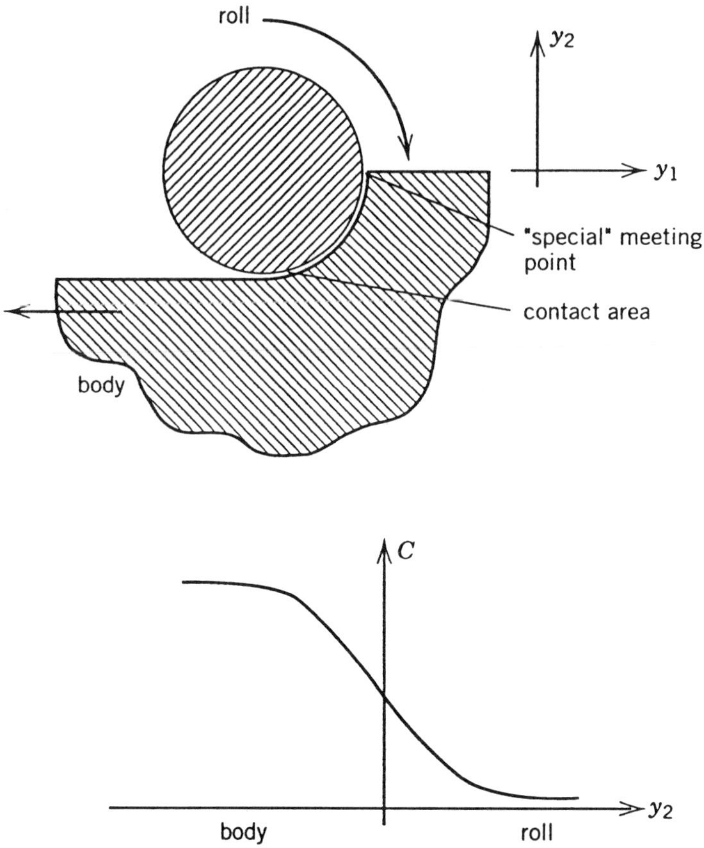

Figure 28. "Roll-body" disposition scheme and the temperature distribution in the contact region.

The roller and body surfaces have two sides each. Every point of the roller, which is situated on its outward rolling side, is defined by the distance between this point and some fixed point on the roller surface. Every point of the body surface rolled by the roller is also defined by the distance between the point and some fixed point on the body surface. At the beginning of the process the surfaces of the roller and the body come into contact in the region $[0, \ell]$ on the axis y_1 for $y_2 = 0$. Then this region of width ℓ moves with velocity v in the positive direction of the axis y_1. The thermal characteristics of the roller and the body are assumed to be constant. We are also interested in the distribution of temperatures with depth, that is, along the axis y_2, for the roller in a fixed cross section $y_1^R = y_1^0$, where $0 < y_1^0 \leq 2\pi R$, and for the body in the sections $y_1^n =$

$y_1^0 + 2n\pi R$, $n = 0, 1, 2, \ldots$ that come into contact with the section $y_1^R = y_1^0$ as the roller rotates. Here n is the number of revolutions of the roller.

The distribution of temperature in the body and the roller satisfies the heat equation

$$\begin{gathered} D_1^H\left(\frac{\partial^2}{\partial y_1^2} C(y,\tau) + \frac{\partial^2}{\partial y_2^2} C(y,\tau)\right) - \frac{\partial}{\partial \tau} C(y,\tau) = 0 \\ -\infty < y_1 < \infty \qquad -L_1 < y_2 < 0 \\ D_2^H\left(\frac{\partial^2}{\partial y_1^2} C(y,\tau) + \frac{\partial^2}{\partial y_2^2} C(y,\tau)\right) - \frac{\partial}{\partial \tau} C(y,\tau) = 0 \\ 0 < y_1 < L \qquad 0 < y_2 < L_2 \qquad 0 < \tau \le \vartheta \end{gathered} \tag{5.27a}$$

Here D_1^H, D_2^H are temperature conductivity coefficients for the body and the roller, respectively, ϑ is the duration of the process. Since the roller is a periodic layer along y_1,

$$\begin{gathered} C(y_1, y_2, \tau) = C(y_1 + L, y_2, \tau) \\ \frac{\partial}{\partial y_1} C(y_1, y_2, \tau) = \frac{\partial}{\partial y_1} C(y_1 + L, y_2, \tau) \\ 0 < y_2 < L_2 \qquad 0 < \tau \le \vartheta \end{gathered} \tag{5.27b}$$

We consider the temperature of the roller and the body at the beginning of the process to be constant and equal to C_R^0 and C_B^0, respectively,

$$\begin{gathered} C(y, 0) = C_B^0 \qquad -\infty < y_1 < \infty \qquad -L_1 \le y_2 \le 0 \\ C(y, 0) = C_R^0 \qquad 0 \le y_1 \le L \qquad 0 \le y_2 \le L_2 \end{gathered} \tag{5.27c}$$

During rolling the temperatures on the surfaces of the roller for $y_2 = L_2$ and of the body for $y_2 = -L_1$ are assumed to be constant and equal to their initial temperatures

$$\begin{gathered} C(y_1, -L_1, \tau) = C_B^0 \qquad -\infty < y_1 < \infty \\ C(y_1, L_2, \tau) = C_R^0 \qquad 0 \le y_1 \le L \end{gathered} \tag{5.27d}$$

The parts of the roller and body surfaces that are in contact during rolling, that is, at $y_2 = 0$, satisfy the conditions

$$
\begin{gathered}
C(y_1 - v\tau, -0, \tau) = C(y_1 - nL - v\tau, +0, \tau) \\
\lambda_1 \frac{\partial}{\partial y_2} C(y_1 - v\tau, -0, \tau) = \lambda_2 \frac{\partial}{\partial y_2} C(y_1 - nL - v\tau, +0, \tau) \qquad (5.27\text{e}) \\
0 < y_1 - v\tau < \ell \qquad 0 < \tau \le \vartheta
\end{gathered}
$$

Here, $n = n(y_1)$ are nonnegative integers satisfying the relation

$$
0 \le y_1 - nL < \ell \qquad 0 \le y_1 < \infty \tag{5.28}
$$

$n = n(y_1)$ is the number of complete revolutions of the roller when the body is removed at the distance y_1 from the beginning of the process; λ_1, λ_2 are heat conductivity coefficients. The parts of the surfaces for $y_2 = 0$, which are not in contact, transfer heat to the surrounding medium by convective heat exchange

$$
\begin{gathered}
\lambda_1 \frac{\partial}{\partial y_2} C(y_1 - v\tau, -0, \tau) + \alpha_1 C(y_1 - v\tau, -0, \tau) = 0 \\
\lambda_2 \frac{\partial}{\partial y_2} C(y_1 - nL - v\tau, +0, \tau) - \alpha_2 C(y_1 - nL - v\tau, +0, \tau) = 0 \qquad (5.27\text{f}) \\
-\infty < y_1 - v\tau < 0 \qquad \ell < y_1 - v\tau < \infty \qquad 0 < \tau \le \vartheta
\end{gathered}
$$

Here α_1, α_2 are the heat transfer coefficients, the temperature of medium is assumed to be zero.

Let us make another simplification of problem (5.27). We assume that heat exchange takes place only along the axis y_2, and that it is absent along the axis y_1. In this case, only Eqs. (5.27a), (5.27b) change. They take the form

$$
\begin{gathered}
D_1^H \frac{\partial^2}{\partial y_2^2} C(y, \tau) - \frac{\partial}{\partial \tau} C(y, \tau) = 0 \qquad -\infty < y_1 < \infty \qquad -L_1 < y_2 < 0 \\
D_2^H \frac{\partial^2}{\partial y_2^2} C(y, \tau) - \frac{\partial}{\partial \tau} C(y, \tau) = 0 \qquad 0 \le y_1 \le L \qquad 0 < y_2 < L_2 \qquad (5.29) \\
C(y_1, y_2, \tau) = C(y_1 + L, y_2, \tau) \qquad 0 < y_2 < L_2 \qquad 0 < \tau \le \vartheta
\end{gathered}
$$

Equations (5.29), (5.27c)–(5.27f) allow us to find the temperatures in the body and the roller, namely, the function

$$C(y,\tau) \qquad -\infty < y_1 < \infty \qquad -L_1 \le y_2 \le 0 \qquad 0 \le \tau \le \vartheta$$

$$\text{for } y_1 \notin [0, \ell],\ y_2 = 0,\ \tau = 0 \qquad y_1 - v\tau \ne 0, \ell,\ y_2 = 0,\ 0 < \tau \le \vartheta$$

and the function

$$C(y,\tau) \qquad 0 \le y_1 \le L \qquad 0 \le y_2 \le L_2 \qquad 0 \le \tau \le \vartheta$$

$$\text{for } y_1 \notin [0, \ell],\ y_2 = 0,\ \tau = 0 \qquad y_1 - v\tau - nL \ne 0, \ell$$

$$y_2 = 0 \qquad 0 < \tau \le \vartheta \qquad n = n_{(5.28)}$$

The function $C(y, \tau)$ includes the variable y_1 as a parameter. The quantity $Q^H(\tau)$, that is, the amount of heat passing through the roller surface at $y_2 = 0$ during the time τ, is determined by the relation

$$Q^H(\tau) \equiv -\lambda_2 \int_0^\tau \int_0^L \frac{\partial}{\partial y_2} C(y_1, +0, \eta)\, dy_1\, d\eta \tag{5.30}$$

The quantity of heat $Q^H(\tau)$ ($\mathrm{J\,m^{-1}}$) passes through the surface between the roller cross sections spaced 1 m apart; these cross sections are orthogonal to the roller axis.

It is convenient to rewrite problem (5.29), (5.27c)–(5.27f) in dimensionless form.

We change to the variables x, t

$$x_1 = L^{-1} y_1 \qquad x_2 = L_1^{-1} y_2,\ y_2 < 0 \qquad x_2 = L_2^{-1} y_2,\ y_2 > 0 \qquad t = b^{-1} L^{-1} v\tau$$

Here b is the dimensionless velocity of the body chosen below. Suppose

$$u(x,t) = C_*^{-1} C(y(x), \tau(t))$$

where

$$C_* = \max_{y, \tau = 0} C(y, \tau)$$

The domains occupied by the body and the roller in these new variables are $\bar{H}^-$ and $\bar{H}^+$, respectively, where

$$H^- = \{(x,t)\colon -\infty < x_1 < \infty,\ -1 < x_2 < 0,\ 0 < t \le T\}$$

$$H^+ = \{(x,t)\colon 0 \le x_1 \le 1,\ 0 < x_2 < 1,\ 0 < t \le T\} \qquad T = b^{-1} L^{-1} v\vartheta$$

The surfaces of the body and the roller at $x_2 = 0$ are denoted by F^- and

F^+, respectively,

$$F^- = \{(x,t)\colon -\infty < x_1 < \infty,\ x_2 = 0,\ 0 < t \le T\}$$

$$F^+ = \{(x,t)\colon 0 \le x_1 \le 1,\ x_2 = 0,\ 0 < t \le T\}$$

We will also denote points (x,t) from the sets $\bar{H}^-$ and $\bar{H}^+$ by (x^-,t), (x^+,t), respectively. The body and the roller have mutual contacts over the sets $\bar{F}_c^- \subset F^-$ and $\bar{F}_c^+ \subset F^+$. Here

$$F_c^- = \{(x,t)\colon 0 < x_1 < \infty,\ x_2 = 0,\ t_0^-(x_1) < t \le t_1^-(x_1),\ 0 < t \le T\}$$

$$t_1^-(x_1) \equiv b^{-1}x_1 \qquad t_0^-(x_1) \equiv \max[0, b^{-1}(x_1 - d)] \qquad 0 \le x_1 < \infty$$

at the moment $t_0^-(x_1)$ the body surface comes into contact with the roller at the point x_1, and at the moment $t_1^-(x_1)$ this contact is interrupted; $d = L^{-1}\ell$. Let us describe the set F_c^+. On the semiaxis $t \ge 0$, we introduce the grid $\bar{\omega}_t$ with the step size $h_t = b^{-1}$, that is, the nondimensional time of one turn of the roller; $\bar{\omega}_t \ne \bar{\omega}_0$. Let $t_n = nh_t$, $n = 0, 1, 2, \ldots$. Suppose

$$t_1^+(x_1) \equiv t_n + b^{-1}x_1 \qquad 0 \le x_1 \le 1 \qquad n = 0, 1, 2, \ldots$$

$$t_0^+(x_1) \equiv \begin{cases} 0 & 0 \le x_1 \le d \qquad n = 0 \\ t_1^+(x_1) - b^{-1}d & d < x_1 \le 1 \qquad n = 0 \\ t_1^+(x_1) - b^{-1}d & 0 \le x_1 \le 1 \qquad n = 1, 2, \ldots \end{cases}$$

Then

$$F_c^+ = \{(x,t)\colon 0 < x_1 \le 1,\ x_2 = 0,\ t_0^+(x_1) < t \le t_1^+(x_1),\ 0 < t \le T\}$$

We put the points $(x_1^-, 0, t) \in F_c^-$ into correspondence with the points $(x_1^+, 0, t) \in F_c^+$ [the latter are in contact with the points $(x_1^-, 0, t)$] where

$$x_1^+ = x_1^+(x_1^-) = x_1^- - n(x_1^-)$$

Here $n(a)$ is the greatest integer not exceeding a value a; $n(x_1^-)$ is the number of full revolutions of the roller when the body moves a distance x_1^-.

We define the sets $F_1^-, F_1^+, \bar{H}$:

$$F_1^- = F^- \backslash \bar{F}_c^- \qquad F_1^+ = F^+ \backslash \bar{F}_c^+ \qquad \bar{H} = \bar{H}^- \cup \{\bar{H}^+ \backslash F_c^+\}$$

$\bar{H}$ is the domain of definition for the function $u(x,t)$, F_1^- and F_1^+ are the sets over which the body and the roller are in contact with the surrounding medium at $x_2 = 0$. The set $\bar{H}$ has a "cut" where F_1^- and F_1^+ are different sides of this cut. We want to find the function $u(x, t)$ for $(x, t) \in \bar{H}$; the function $u(x, t)$ is continuous on the sets F_c^- and F_c^+, that is, on the "body-roller" interface

$$u(x_1^-, -0, t) = u(x_1^+(x_1^-), +0, t) \qquad (x_1^-, 0, t) \in F_c^- \qquad (x_1^+, 0, t) \in F_c^+$$

Generally speaking, on different sides of the cut this function takes different values

$$u(x_1^-, -0, t) \neq u(x_1^+(x_1^-), +0, t) \qquad (x_1^-, 0, t) \in F_1^- \qquad (x_1^+, 0, t) \in F_1^+$$

We do not complete the definition of the function $u(x, t)$ on different sides of the cut.

Now we define the boundary function $\varphi(x, t)$, $(x, t) \in S$, where $S = S^0 \cup S^1$, S^0 and S^1 are the base and the external (lateral) surface of the set H

$$S^0 = \{(x, t)\colon -\infty < x_1 < \infty,\ -1 \le x_2 \le 0 \text{ and } 0 \le x_1 \le 1,\ 0 \le x_2 \le 1 \text{ at } t = 0\}$$

$$S^1 = \{(x, t)\colon -\infty < x_1 < \infty,\ x_2 = -1 \text{ and } 0 \le x_1 \le 1,\ x_2 = 1 \text{ at } 0 < t \le T\}$$

Suppose

$$\varphi(x, t) = \begin{cases} C_*^{-1} C_B^0 & (x, t) \in \{S^0 \cup S^1\} \cap \bar{H}^- \\ C_*^{-1} C_R^0 & (x, t) \in \{S^0 \cup S^1\} \cap \bar{H}^+ \end{cases}$$

the function $\varphi(x, t)$, $(x, t) \in S^0 \cup S^1$ has a discontinuity at $0 < x_1 < d$, $x_2 = 0$, $t = 0$, $(x, t) \in S^0$; the cut of the set $\bar{H}$ passes along the set $\{(x, t)\colon d < x_1 \le 1,\ x_2 = 0,\ t = 0\}$, $(x, t) \in S^0$.

Problem (5.29), (5.27c)–(5.27f) in the new variables takes the form

$$L^1_{(5.31)} u(x, t) \equiv \left\{ \varepsilon_1^2 \frac{\partial^2}{\partial x_2^2} - \frac{\partial}{\partial t} \right\} u(x, t) = 0 \qquad (x, t) \in H^-$$

$$L^2_{(5.31)} u(x, t) \equiv \left\{ \varepsilon_2^2 \frac{\partial^2}{\partial x_2^2} - \frac{\partial}{\partial t} \right\} u(x, t) = 0 \qquad (x, t) \in H^+$$

$$u(x_1^+, x_2, t) = u(1 + x_1^+, x_2, t) \qquad (x, t) \in S^2$$

$$u(x_1^-, -0, t) = u(x_1^+(x_1^-), +0, t) \tag{5.31a}$$

$$\varepsilon_1 \frac{\partial}{\partial x_2} u(x_1^-, -0, t) = \beta \varepsilon_2 \frac{\partial}{\partial x_2} u(x_1^+(x_1^-), +0, t) \quad (x^-, t) \in F_c^-,\ (x^+, t) \in F_c^+$$

$$\varepsilon_1 \frac{\partial}{\partial x_2} u(x_1^-, -0, t) + \gamma_1 u(x_1^-, -0, t) = 0 \qquad (x^-, t) \in F_1^-$$

$$\varepsilon_2 \frac{\partial}{\partial x_2} u(x_1^+, +0, t) - \gamma_2 u(x_1^+, +0, t) = 0 \qquad (x^+, t) \in F_1^+$$

$$u(x, t) = \varphi(x, t) \qquad (x, t) \in S^0 \cup S^1$$

Here

$$S^2 = \{(x, t): x_1 = 0,\ 0 < x_2 < 1,\ 0 < t \le T\}$$

$$\begin{gathered} \varepsilon = \varepsilon(x) \qquad \varepsilon = \varepsilon_1 \qquad x_2 < 0 \qquad \varepsilon = \varepsilon_2 \qquad x_2 > 0 \\ \varepsilon_i = b^{1/2} L^{1/2} v^{-1/2} L_i^{-1} (D_i^H)^{1/2} \\ \gamma = \gamma(x) \qquad \gamma = \gamma_1 \qquad x_2 < 0 \qquad \gamma = \gamma_2 \qquad x_2 > 0 \\ \gamma_i = b^{1/2} L^{1/2} v^{-1/2} \alpha_i \lambda_i^{-1} (D_i^H)^{1/2} \qquad i = 1, 2 \\ \beta = \lambda_1^{-1} \lambda_2 (D_1^H)^{1/2} (D_2^H)^{-1/2} \end{gathered} \tag{5.31b}$$

Note that the parameters ε_1 and ε_2 are defined by the value bLv^{-1}, where Lv^{-1} is the duration of one revolution of the roller. The value b is usually chosen to be equal to unity. If it happens that either ε_1 or ε_2 exceeds unity then we choose the value b so that the following inequalities are fulfilled

$$\varepsilon_1, \varepsilon_2 \le 1$$

The quantity $Q^H_{(5.30)}(\tau)$ is determined by the relation

$$Q^H(\tau) = -M_Q \int_0^t \int_0^1 P_2(x_1, 0, \eta)\, dx_1\, d\eta \qquad t = t(\tau) = b^{-1} L^{-1} v\tau$$

where $M_Q = C_* \lambda_2 (D_2^H)^{-1/2} b^{1/2} L_2 L^{1/2} v^{-1/2}$, $P_2(x_1, 0, \eta) = P_2(x_1, +0, \eta)$ is the normalized diffusion flux through the roller surface. The normalized diffusion fluxes in the roller and the body are found by the formulas

$$\begin{aligned} P_2(x, t) &\equiv \varepsilon_2 \frac{\partial}{\partial x_2} u(x, t) \qquad (x, t) \in \bar{H}^+ \\ P_1(x, t) &\equiv \varepsilon_1 \frac{\partial}{\partial x_2} u(x, t) \qquad (x, t) \in \bar{H}^- \end{aligned} \tag{5.32}$$

We now discuss a particular problem. We want to find the distribution of roller temperatures with thickness (i.e., along the radius) in the cross section ξ^0 and of body temperatures with thickness in the cross sections

$$\xi_n = \xi_n(\xi^0, n) = \xi^0 + n \qquad n = 0, 1, 2, \dots \qquad \xi^0 \in (0, 1]$$

The cross-sections ξ^0 and ξ_n are associated with the roller and the body. The distribution of temperatures in the cross-section ξ^0 of the roller is denoted by $u^+(x, t; \xi^0)$. Here $x \in [0, 1]$, x is the distance to the surface of the roller. The distribution of temperatures in the cross-section ξ_n of the body is denoted by $u_n^-(x, t; \xi^0)$. Here $x \in [-1, 0]$, $-x$ is the distance to the body surface in contact with the roller. The following relations hold:

$$u^+(x, t; \xi^0) = u(\xi^0, x, t)$$

$$u_n^-(x, t; \xi^0) = u(\xi_n, x, t) \qquad \xi_n = \xi_n(\xi^0, n)$$

where $u(\xi^0, x, t)$, $u(\xi_n, x, t)$ are the components of the solution for problem (5.31). For $n = 0, 1, 2, \dots$ the roller cross-section ξ^0 and one of the body cross-sections ξ_n come into contact at the instants

$$t_0^n \equiv \max[0, b^{-1}(\xi_n - d)]$$

At

$$t_1^n \equiv b^{-1}\xi_n$$

this contact is interrupted; d is the width of the contact region. We are also interested in the quantity

$$q^H(t; \xi^0) \equiv -M_q \int_0^t P_2(0, \eta; \xi^0)\, d\eta \tag{5.33}$$

where $q^H(t; \xi^0)$ is the quantity of heat passing through 1 m^2 of the surface in the section ξ^0, $M_q = M_Q L^{-1} = C_* \lambda_2 (D_2^H)^{-1/2} b^{1/2} L_2 L^{-1/2} v^{-1/2}$, $P_2(0, t; \xi^0) = P_2(+0, t; \xi^0)$ is the normalized diffusion flux through the roller surface. The normalized diffusion fluxes in the roll and the body are determined by the relations

$$P_2(x, t; \xi^0) = P_2(x, t; \xi^0; u^+(\cdot; \xi^0)) \equiv \varepsilon_2 \frac{\partial}{\partial x} u^+(x, t; \xi^0) \qquad (x, t) \in \bar{G}^+$$

$$P_1(x, t; \xi^0) = P_1(x, t; \xi^0; u_n^-(\cdot; \xi^0)) \equiv \varepsilon_1 \frac{\partial}{\partial x} u_n^-(x, t; \xi^0)$$

$$(x, t) \in \bar{G}^- \qquad n = 0, 1, 2, \dots \tag{5.34a}$$

$$P_{2(5.34)}(x, t; \xi^0; u^+(\cdot; \xi^0)) = P_{2(5.32)}(\xi^0, x, t)$$

$$P_{1(5.34)}(x, t; \xi^0; u_n^-(\cdot; \xi^0)) = P_{1(5.32)}(\xi_n(\xi^0), x, t)$$

The quantities $q^H(t; \xi^0)$ and $Q^H(\tau)$ are related as follows:

$$Q^H(\tau) = Q^H(bLv^{-1}t) = L \int_0^1 q^H(t; \xi^0)\, d\xi^0$$

We call the value

$$q(x_0, t; \xi^0) \equiv -\int_0^t P_2(x_0, \eta; \xi^0; u^+(\cdot; \xi^0))\, d\eta \qquad x_0 \in \bar{D}_2 \quad (5.34b)$$

the total amount of normalized substance passing in the positive direction of the axis x through the cross-section $x = x_0$ at time t.

We rewrite problem (5.31) in a more convenient form for numerical solution. We introduce a number of notations. Let

$$G^- = D_1 \times (0, T] \qquad D_1 = (-1, 0) \qquad S^- = \bar{G}^- \backslash G^-$$

$$G^+ = D_2 \times (0, T] \qquad D_2 = (0, 1) \qquad S^+ = \bar{G}^+ \backslash G^+$$

$$S_0^- = \bar{D}_1 \times \{t = 0\} \qquad S_0^+ = \bar{D}_2 \times \{t = 0\}$$

$$S^{L-} = \{x = -1\} \times (0, T] \qquad S^{L+} = \{x = 1\} \times (0, T]$$

$$S^I = \{x = 0\} \times (0, T] \qquad S^{I-} = S^- \backslash \{S_0^- \cup S^{L-}\}$$

$$S^{I+} = S^+ \backslash \{S_0^+ \cup S^{L+}\}$$

S^I and S^{I+} are the right and left faces of the sets G^- and G^+. The function

$$u(x, t; \xi^0) = \{u^+(x, t; \xi^0),\ (x, t) \in \bar{G}^+,\ u_n^-(x, t; \xi^0),\ (x, t) \in \bar{G}^-,$$
$$n = 0, 1, 2, \ldots\} \qquad (x, t) \in \bar{H}$$

is to be found, where

$$\bar{H} = \bar{H}(\xi^0) \equiv \bar{G}^+ \cup \Big\{\bigcup_n \bar{G}_n^-\Big\} \qquad \bar{G}_n^- = \{(x, t; n) \colon (x, t) \in \bar{G}^-\}$$
$$n = 0, 1, 2, \ldots$$

The set $\bar{H}$ consists of the union of the set $\bar{G}^+$ and the interrelated sets $\bar{G}_n^-$, $n = 0, 1, 2, \ldots$. The sets $\bar{G}^+$ and $\bar{G}_n^-$ have a common part at $x = 0$,

$t \in (t_0^n, t_1^n]$. The function $u(x, t; \xi^0)$ defined in $\bar{H}$ is the union of the function $u^+(x, t; \xi^0)$ and the functions $u_n^-(x, t; \xi^0)$, $n = 0, 1, 2, \ldots$, which are continuous across the set $x = 0$, $t \in (t_0^n, t_1^n]$.

On the set S, which is the "external" boundary of the set $\bar{H}$, where

$$S = \{S_0^+ \cup S^{L+}\} \cup \Big\{ \bigcup_n \{S_0^- \cup S^{L-}\}_n,\ n = 0, 1, 2, \ldots \Big\}$$

$$\{S_0^- \cup S^{L-}\}_n = \{(x, t; n): (x, t) \in S_0^- \cup S^{L-}\}$$

we have the given function $\Phi(x, t) = \Phi(x, t; \xi^0)$, $(x, t) \in S$:

$$\Phi(x, t) = \begin{cases} \Phi^0(x, t) = \varphi(\xi^0, x, t), & (x, t) \in S_0^+ \cup S^{L+} \\ \Phi_n(x, t) = \varphi(\xi_n, x, t), & (x, t) \in \{S_0^- \cup S^{L-}\}_n \qquad n = 0, 1, 2, \ldots \end{cases}$$

Generally speaking,

$$\Phi^0(0, 0) \neq \Phi_n(0, 0) \qquad n = 0, 1, 2, \ldots$$

The function $u(x, t; \xi^0)$ is found from the solution of the problem

$$L_{(5.35)} u^+(x, t; \xi^0) \equiv \left\{ \varepsilon_2^2 \frac{\partial^2}{\partial x^2} - \frac{\partial}{\partial t} \right\} u^+(x, t; \xi^0) = 0 \qquad (x, t) \in G^+$$

$$L_{(5.35)} u_n^-(x, t; \xi^0) \equiv \left\{ \varepsilon_1^2 \frac{\partial^2}{\partial x^2} - \frac{\partial}{\partial t} \right\} u_n^-(x, t; \xi^0) = 0 \qquad (x, t) \in G^-$$

$$\varepsilon_2 \frac{\partial}{\partial x} u^+(x, t; \xi^0) - \gamma_2 u^+(x, t; \xi^0) = 0 \qquad (x, t) \in S^{I+}$$

$$t \notin \Big\{ \bigcup_n (t_0^n, t_1^n],\ n \geq 0 \Big\} \tag{5.35a}$$

$$\varepsilon_1 \frac{\partial}{\partial x} u_x^-(x, t; \xi^0) + \gamma_1 u_n^-(x, t; \xi^0) = 0 \qquad (x, t) \in S^{I-}$$

$$t \notin (t_0^n, t_1^n] \qquad \text{for} \qquad n \geq 0$$

$$u^+(x, t; \xi^0) = u_n^-(x, t; \xi^0) \qquad \beta \varepsilon_2 \frac{\partial}{\partial x} u^+(x, t; \xi^0) = \varepsilon_1 \frac{\partial}{\partial x} u_n^-(x, t; \xi^0)$$

$$(x, t) \in S^I \qquad t \in (t_0^n, t_1^n] \qquad n \geq 0 \qquad n = 0, 1, 2, \ldots$$

$$u(x, t; \xi^0) = \Phi(x, t) \qquad (x, t) \in S$$

where $\varepsilon(x) = \varepsilon_{(5.31)}(0, x)$, $\gamma = \gamma_{(5.31)}(0, x)$. The functions $u^+(x, t; \xi^0)$ and $u_n^-(x, t; \xi^0)$ for $n \geq 0$ have discontinuities on the sets $x = 0$, $t \in \{\cup_n t_0^n,$

$n = 0, 1, 2, \ldots\}$ and $x = 0$, $t = t_0^n$, respectively. For definiteness, we suppose that on these sets

$$u^+(0, t; \xi^0) = u^+(0, t - 0; \xi^0) \qquad u_n^-(0, t; \xi^0) = u_n^-(0, t - 0; \xi^0)$$

$$t = t_0^n > 0 \qquad u(0, t; \xi^0) = \Phi(0, t) \qquad t = 0$$

Problem (5.35a) is a singularly perturbed boundary value problem with complicated conditions of exchange on the boundaries of the subdomains. We want to find thc solution of the boundary value problem (5.35a) and also the quantity

$$q(t; \xi^0) = q_{(5.34)}(0, t; \xi^0) \tag{5.35b}$$

which is the amount of normalized substance passing through the surface S^{I+} at time t.

2. *Finite Difference Scheme for Problem (5.35)*

Note that the solution of problem (5.35) has discontinuities for every pair of fixed parameters $\varepsilon_1, \varepsilon_2$. As these parameters tend to zero, boundary and interior layers appear in the neighborhood of the set $x = 0$. Such behavior of the solution leads to large errors if one uses classical numerical methods. To construct the approximate solution, we use the method of additive representation of singularities related to the solution discontinuities. Let us describe this method.

First we can find all of the functions $u_n^-(x, t; \xi^0)$, $n = 0, 1, 2, \ldots$ for $0 \leq t \leq T$, $t \leq t_0^n$, that is, before the section of the body at the point ξ_n, and that of the roller at the point ξ^0, come into contact. Thus, the functions $u_n^-(x, t; \xi^0)$ are found by solving the uncoupled problems

$$\begin{gathered} L_{(5.35)} u_n^-(x, t; \xi^0) = 0 \qquad (x, t) \in G^- \\ \varepsilon_1 \frac{\partial}{\partial x} u_n^-(x, t; \xi^0) + \gamma_1 u_n^-(x, t; \xi^0) = 0 \qquad (x, t) \in S^{I-} \\ u_n^-(x, t; \xi^0) = \Phi_n(x, t) \qquad (x, t) \in S \cap \bar{G}^- \\ n = 0, 1, 2, \ldots \qquad 0 \leq t \leq \min[t_0^n, T] \end{gathered} \tag{5.36a}$$

Besides, if we know the function $u_k^-(x, t; \xi^0)$ for $t = t_1^k \geq 0$ where k is an integer, $k \geq 0$:

$$u_k^-(x, t; \xi^0) = \Psi_k^-(x, t) \qquad x \in \bar{D}_1 \qquad t = t_1^k \qquad k \geq 0 \tag{5.36b}$$

then the function $u_k^-(x, t; \xi^0)$ for $t \geq t_1^k$ can be found as the solution of the problem

$$
\begin{aligned}
& L_{(5.35)} u_k^-(x, t; \xi^0) = 0 \qquad (x, t) \in G^- \\
& \varepsilon_1 \frac{\partial}{\partial x} u_k^-(x, t; \xi^0) + \gamma_1 u_k^-(x, t; \xi^0) = 0 \qquad (x, t) \in S^{I-} \\
& u_k^-(x, t; \xi^0) = \Phi_k(x, t) \qquad (x, t) \in S \cap \bar{G}^- \\
& t_1^k < t \leq T \\
& u_k^-(x, t; \xi^0) = \Psi_k^-(x, t) \qquad x \in \bar{D}_1 \qquad t = t_1^k \qquad k \geq 0
\end{aligned}
\tag{5.36c}
$$

Independent of the other components of the solution, we can find the function $u^+(x, t; \xi^0)$ for $t \leq t_0^0$, that is, before the first contact occurs between the roller section at the point ξ^0 and the body section at the point ξ_0. In this connection, the function $u^+(x, t; \xi^0)$ is the solution of the problem

$$
\begin{aligned}
& L_{(5.35)} u^+(x, t; \xi^0) = 0 \qquad (x, t) \in G^+ \\
& \varepsilon_2 \frac{\partial}{\partial x} u^+(x, t; \xi^0) - \gamma_2 u^+(x, t; \xi^0) = 0 \qquad (x, t) \in S^{I+} \\
& u^+(x, t; \xi^0) = \Phi^0(x, t) \qquad (x, t) \in S \cap \bar{G}^+ \\
& 0 \leq t \leq \min[t_0^0, T]
\end{aligned}
\tag{5.36d}
$$

If we know the function $u^+(x, t; \xi^0)$ for $t = t_1^k > 0$ where $k \geq 0$:

$$
u^+(x, t; \xi^0) = \Psi^+(x, t) \qquad x \in \bar{D}_2 \qquad t = t_1^k \qquad k \geq 0 \tag{5.36e}
$$

then the function $u^+(x, t; \xi^0)$ for $t_1^k \leq t \leq \min[t_0^{k+1}, T]$ is determined as the solution of the problem

$$
\begin{aligned}
& L_{(5.35)} u^+(x, t; \xi^0) = 0 \qquad (x, t) \in G^+ \\
& \varepsilon_2 \frac{\partial}{\partial x} u^+(x, t; \xi^0) - \gamma_2 u^+(x, t; \xi^0) = 0 \qquad (x, t) \in S^{I+} \\
& u^+(x, t; \xi^0) = \Phi^0(x, t) \qquad (x, t) \in S \cap \bar{G}^+ \\
& t_1^k < t \leq \min[t_0^{k+1}, T] \\
& u^+(x, t; \xi^0) = \Psi^+(x, t) \qquad x \in \bar{D}_2 \qquad t = t_1^k \qquad k \geq 0
\end{aligned}
\tag{5.36f}
$$

The functions $u^+(x, t; \xi^0)$ and $u_k^-(x, t; \xi^0)$ for $t_0^k < t \leq t_1^k$ are determined as the solution of the coupled problems on the sets $\bar{G}^+$ and $\bar{G}^-$, respectively. Assume that for $t = t_0^k \geq 0$ the functions $u^+(x, t; \xi^0)$ and $u_k^-(x, t; \xi^0)$ are known

$$
\begin{aligned}
&u^+(x, t; \xi^0) = \psi^+(x, t) \qquad x \in \bar{D}_2 \\
&u_k^-(x, t; \xi^0) = \psi_k^-(x, t) \qquad x \in \bar{D}_1 \qquad t = t_0^k \qquad k \geq 0
\end{aligned}
\tag{5.37a}
$$

For $t = t_0^k = 0$ we have

$$
\psi^+(x, t) = \Phi^0(x, t) \qquad x \in \bar{D}_2 \qquad \psi_k^-(x, t) = \Phi_k(x, t) \qquad x \in \bar{D}_1 \qquad t = 0
$$

For $t_0^k \leq t \leq t_1^k$ we find the functions $u^+(x, t; \xi^0)$ and $u_k^-(x, t; \xi^0)$ as the solution of the problem

$$
\begin{gathered}
L_{(5.35)} u^+(x, t; \xi^0) = 0 \qquad (x, t) \in G^+ \\
L_{(5.35)} u_k^-(x, t; \xi^0) = 0 \qquad (x, t) \in G^- \\
u^+(x, t; \xi^0) = u_k^-(x, t; \xi^0) \\
\beta \varepsilon_2 \frac{\partial}{\partial x} u^+(x, t; \xi^0) = \varepsilon_1 \frac{\partial}{\partial x} u_k^-(x, t; \xi^0) \qquad (x, t) \in S^I \\
u^+(x, t; \xi^0) = \Phi^0(x, t) \qquad (x, t) \in S \cap \bar{G}^+ \\
u_k^-(x, t; \xi^0) = \Phi_k(x, t) \qquad (x, t) \in S \cap \bar{G}^- \\
t_0^k < t \leq \min[t_1^k, T] \\
u^+(x, t; \xi^0) = \psi^+(x, t) \qquad x \in \bar{D}_2 \\
u_k^-(x, t; \xi^0) = \psi_k^-(x, t) \qquad x \in \bar{D}_1 \qquad t = t_0^k \qquad k \geq 0
\end{gathered}
\tag{5.37b}
$$

Thus, by solving problems (5.36), (5.37) successively, we find the functions $u^+(x, t; \xi^0)$ and $u_n^-(x, t; \xi^0)$, $n = 0, 1, 2, \ldots$ for $0 \leq t \leq T$, in other words, we find the solution of problem (5.35).

Now we use the additive representation of singularities to solve problem (5.37). The solution of problem (5.37b) is represented in the form of a sum of continuous and discontinuous functions:

$$
\begin{gathered}
u^+(x, t; \xi^0) = U^+(x, t) + W_k(x, t) \qquad (x, t) \in \bar{G}^+ \\
u_k^-(x, t; \xi^0) = U_k^-(x, t) + W_k(x, t) \qquad (x, t) \in \bar{G}^- \\
t_0^k \leq t \leq \min[t_1^k, T] \qquad k \geq 0
\end{gathered}
\tag{5.38a}
$$

Here the function $W_k(x, t)$, having a discontinuity at the point $x = 0$, $t = t_0^k$, is determined by the formula

$$W_k(x, t) = (\psi^+(0, t_0^k) - \psi_k^-(0, t_0^k)) \times \begin{cases} (1+\beta^{-1})^{-1}\operatorname{erf}\left(\dfrac{x}{2\varepsilon_1(t-t_0^k)^{1/2}}\right) & x \le 0 \\ (1+\beta)^{-1}\operatorname{erf}\left(\dfrac{x}{2\varepsilon_2(t-t_0^k)^{1/2}}\right) & x > 0 \end{cases} \tag{5.38b}$$

$$(x, t) \in \bar{G} \qquad t_0^k \le t \le t_1^k \qquad k \ge 0$$

The functions $U^+(x, t)$ and $U_k^-(x, t)$ for $x = 0$, $t_0^k \le t \le \min[t_1^k, T]$ continuously adjoin each other. These functions are the solution of the problem

$$\begin{gathered} L_{(5.35)} U^+(x, t) = 0 \qquad (x, t) \in G^+ \\ L_{(5.35)} U_k^-(x, t) = 0 \qquad (x, t) \in G^- \\ U^+(x, t) = U_k^-(x, t) \\ \beta\varepsilon_2 \frac{\partial}{\partial x} U^+(x, t) = \varepsilon_1 \frac{\partial}{\partial x} U_k^-(x, t) \qquad (x, t) \in S^I \\ U^+(x, t) = g^0(x, t) \qquad (x, t) \in S \cap \bar{G}^+ \\ U_k^-(x, t) = g_k(x, t) \qquad (x, t) \in S \cap \bar{G}^- \\ t_0^k < t \le \min[t_1^k, T] \\ U^+(x, t) = q^0(x, t) \qquad x \in \bar{D}_2 \\ U_k^-(x, t) = q_k(x, t) \qquad x \in \bar{D}_1 \qquad t = t_0^k \qquad k \ge 0 \end{gathered} \tag{5.38c}$$

Here

$$g^0(x, t) = \Phi^0(x, t) - W_k(x, t) \qquad g_k(x, t) = \Phi_k(x, t) - W_k(x, t)$$

$$q^0(x, t) = \psi^+(x, t) - W_k(x, t) \qquad q_k(x, t) = \psi_k^-(x, t) - W_k(x, t)$$

Taking into account the representation (5.38a) for the functions $\Psi_{k(5.36b)}^-(x, t)$ and $\Psi_{(5.36e)}^+(x, t)$, the following relations are valid:

$$\begin{gathered} \Psi_{(5.36e)}^+(x, t) = U^+(x, t) + W_k(x, t) \qquad x \in \bar{D}_2 \\ \Psi_{k(5.36b)}^-(x, t) = U^-(x, t) + W_k(x, t) \qquad x \in \bar{D}_1 \qquad t = t_1^k \end{gathered} \tag{5.38d}$$

Problems (5.36), (5.37) and (5.36), (5.38) are equivalent to problem (5.35).

Now we describe the finite difference scheme for solving problem (5.35). We construct this difference scheme to approximate the problem (5.36), (5.38).

First, we consider a classical finite difference scheme. On the sets $\bar{G}^+, \bar{G}^-$ we introduce the rectangular grids

$$\bar{G}_h^+ = \bar{\omega}^+ \times \bar{\omega}_0 \qquad \bar{G}_h^- = \bar{\omega}^- \times \bar{\omega}_0 \tag{5.39}$$

where $\bar{\omega}_0 = \bar{\omega}_{0(5.17)}$, $\bar{\omega}^+$ and $\bar{\omega}^-$ are grids on the segments $[0, 1]$ and $[-1, 0]$, respectively, the number of nodes in each grid $\bar{\omega}^+$ and $\bar{\omega}^-$ is $2^{-1}N + 1$, the distribution of nodes in the grids $\bar{\omega}^+$ and $\bar{\omega}^-$ is arbitrary. Suppose that the points t_0^n, t_1^n, $n = 0, 1, 2, \ldots$ belong to the set $\bar{\omega}_0$. Problem (5.36), (5.38) is approximated by the finite difference scheme given below. To problem (5.36a) we assign the scheme

$$\begin{gathered}
\Lambda_{(5.40)} z_n^-(x, t; \xi^0) \equiv \{\varepsilon_1^2 \delta_{\bar{x}\hat{x}} - \delta_{\bar{t}}\} z_n^-(x, t; \xi^0) = 0 \qquad (x, t) \in G_h^- \\
\varepsilon_1 \delta_{\bar{x}} z_n^-(x, t; \xi^0) + \gamma_1 z_n^-(x, t; \xi^0) = 0 \qquad (x, t) \in S_h^{I-} \\
z_z^-(x, t; \xi^0) = \Phi_n(x, t) \qquad (x, t) \in S \cap \bar{G}_h^- \\
n = 0, 1, 2, \ldots \qquad 0 \le t \le \min[t_0^n, T]
\end{gathered} \tag{5.40a}$$

Here $G_h^- = G^- \cap \bar{G}_h^-$, $S_h^{I-} = S^{I-} \cap \bar{G}_h^-$. If we know the value of the grid function $z_k^-(x, t; \xi^0)$ for $t = t_1^k > 0$:

$$z_k^-(x, t; \xi^0) = \Psi_k^{h-}(x, t) \qquad x \in \bar{D}_{1h} \qquad t = t_1^k \qquad k \ge 0 \tag{5.40b}$$

where

$$\bar{D}_{jh} = \bar{\omega}^+ \quad \text{for} \quad j = 2 \qquad \bar{D}_{jh} = \bar{\omega}^- \quad \text{for} \quad j = 1$$

then we find the function $z_k^-(x, t; \xi^0)$ for $t \ge t_1^k$ by solving the problem

$$\begin{gathered}
\Lambda_{(5.40)} z_k^-(x, t; \xi^0) = 0 \qquad (x, t) \in G_h^- \\
\varepsilon_1 \delta_{\bar{x}} z_k^-(x, t; \xi^0) + \gamma_1 z_k^-(x, t; \xi^0) = 0 \qquad (x, t) \in S_h^{I-} \\
z_k^-(x, t; \xi^0) = \Phi_k(x, t) \qquad (x, t) \in S \cap \bar{G}_h^- \\
t_1^k < t \le T \\
z_k^-(x, t; \xi^0) = \Psi_k^{h-}(x, t) \qquad x \in \bar{D}_{1h} \qquad t = t_1^k \qquad k \ge 0
\end{gathered} \tag{5.40c}$$

We put problem (5.36d) into correspondence with the grid problem

$$\begin{gathered}\Lambda_{(5.40)} z^{+}(x, t; \xi^{0}) \equiv \{\varepsilon_2^2 \delta_{\bar{x}\hat{x}} - \delta_{\bar{t}}\} z^{+}(x, t; \xi^{0}) = 0 \qquad (x, t) \in G_h^{+} \\ \varepsilon_2 \delta_x z^{+}(x, t; \xi^{0}) - \gamma_2 z^{+}(x, t; \xi^{0}) = 0 \qquad (x, t) \in S_h^{I+} \\ z^{+}(x, t; \xi^{0}) = \Phi^{0}(x, t) \qquad (x, t) \in S \cap \bar{G}_h^{+} \\ 0 \le t \le \min[t_0^0, T]\end{gathered} \tag{5.40d}$$

If the function $z^{+}(x, t; \xi^{0})$ is known for $t = t_1^k$:

$$z^{+}(x, t; \xi^{0}) = \Psi^{h+}(x, t) \qquad x \in \bar{D}_{2h} \qquad t = t_1^k \qquad k \ge 0 \tag{5.40e}$$

then with problem (5.36f) we associate the difference scheme

$$\begin{gathered}\Lambda_{(5.40)} z^{+}(x, t; \xi^{0}) = 0 \qquad (x, t) \in G_h^{+} \\ \varepsilon_2 \delta_z z^{+}(x, t; \xi^{0}) - \gamma_2 z^{+}(x, t; \xi^{0}) = 0 \qquad (x, t) \in S_h^{I+} \\ z^{+}(x, t; \xi^{0}) = \Phi^{0}(x, t) \qquad (x, t) \in S \cap \bar{G}_h^{+} \\ \mathrm{t}_1^{\mathrm{k}} < \mathrm{t} \le \min[t_0^{k+1}, T] \\ z^{+}(x, t; \xi^{0}) = \Psi^{h+}(x, t) \qquad x \in \bar{D}_{2h} \qquad t = t_1^k \qquad k \ge 0\end{gathered} \tag{5.40f}$$

Assume that for $t = t_0^k$ the functions $z^{+}(x, t; \xi^{0})$ and $z_k^{-}(x, t; \xi^{0})$ are known

$$\begin{gathered}z^{+}(x, t; \xi^{0}) = \psi^{h+}(x, t) \qquad x \in \bar{D}_{2h} \\ z_k^{-}(x, t; \xi^{0}) = \psi_k^{h-}(x, t) \qquad x \in \bar{D}_{1h} \qquad t = t_0^k \qquad k \ge 0\end{gathered} \tag{5.41a}$$

For $t_0^k \le t \le t_1^k$ we define the approximate solution of problem (5.37) by the relations

$$\begin{gathered}z^{+}(x, t; \xi^{0}) = Z^{+}(x, t) + W_k^h(x, t) \qquad (x, t) \in \bar{G}_h^{+} \\ z_k^{-}(x, t; \xi^{0}) = Z_k^{-}(x, t) + W_k^h(x, t) \qquad (x, t) \in \bar{G}_h^{-} \\ t_0^k \le t \le \min[t_1^k, T] \qquad k \ge 0\end{gathered} \tag{5.41b}$$

Here

$$W_k^h(x,t) = (\psi^{h+}(0,t_0^k) - \psi_k^{h-}(0,t_0^k)) \times \begin{cases} (1+\beta^{-1})^{-1}\operatorname{erf}\left(\dfrac{x}{2\varepsilon_1(t-t_0^k)^{1/2}}\right) & x \le 0 \\ (1+\beta)^{-1}\operatorname{erf}\left(\dfrac{x}{2\varepsilon_2(t-t_0^k)^{1/2}}\right) & x > 0 \end{cases} \tag{5.41c}$$

$$(x,t) \in \bar{G} \qquad t_0^k \le t \le t_1^k \qquad k \ge 0$$

The functions $Z^+(x,t)$, $Z_k^-(x,t)$ are found as the solution of the problem

$$\begin{aligned} &\Lambda_{(5.40)} Z^+(x,t) = 0 \qquad (x,t) \in G_h^+ \\ &\Lambda_{(5.40)} Z_k^-(x,t) = 0 \qquad (x,t) \in G_h^- \\ &Z^+(x,t) = Z_k^-(x,t) \\ &\beta\varepsilon_2 \delta_x Z^+(x,t) = \varepsilon_1 \delta_{\bar{x}} Z_k^-(x,t) \qquad (x,t) \in S_h^I \\ &Z^+(x,t) = g^{h0}(x,t) \qquad (x,t) \in S \cap \bar{G}_h^+ \\ &Z_k^-(x,t) = g_k^h(x,t) \qquad (x,t) \in S \cap \bar{G}_h^- \\ &t_0^k < t \le \min[t_1^k, T] \\ &Z^+(x,t) = q^{h0}(x,t) \qquad x \in \bar{D}_{2h} \\ &Z_k^-(x,t) = q_k^h(x,t) \qquad x \in \bar{D}_{1h} \qquad t = t_0^k \qquad k \ge 0 \end{aligned} \tag{5.41d}$$

Here

$$g^{h0}(x,t) = \Phi^0(x,t) - W_k^h(x,t) \qquad g_k^h(x,t) = \Phi_k(x,t) - W_k^h(x,t)$$

$$q^{h0}(x,t) = \psi^{h+}(x,t) - W_k^h(x,t) \qquad q_k(x,t) = \psi_k^{h-}(x,t) - W_k^h(x,t)$$

The following relations are valid:

$$\begin{aligned} &\Psi^{h+}_{(5.40e)}(x,t) = Z^+(x,t) + W_k^h(x,t) \qquad x \in \bar{D}_{2h} \\ &\Psi^{h-}_{k(5.40b)}(x,t) = Z^-(x,t) + W_k^h(x,t) \qquad x \in \bar{D}_{1h} \qquad t = t_1^k \end{aligned} \tag{5.41e}$$

Thus, the finite difference scheme (5.40), (5.41), (5.39) approximating problem (5.36), (5.38) has been constructed.

The solution of this difference scheme does not converge to the solution of problem (5.35) ε-uniformly. The components of the solution of the difference scheme approximate the components $u^{+}(x,t;\xi^0)$, $u_n^{-}(x,t;\xi^0)$ unsatisfactorily in the neighborhood of the boundary layer.

Let us construct a special grid on which the solution of the finite difference scheme (5.40), (5.41) converges to the solution of problem (5.35) ε-uniformly.

On the sets $\bar{G}^{+}, \bar{G}^{-}$, we introduce special grids condensing in the neighborhood of the set S^I

$$\bar{G}_h^{+} = \bar{\omega}^{+} \times \bar{\omega}_0 \qquad \bar{G}_h^{-} = \bar{\omega}^{-} \times \bar{\omega}_0 \tag{5.42}$$

where $\bar{\omega}_0 = \bar{\omega}_{0(5.17)}$, $\bar{\omega}^{+}$ and $\bar{\omega}^{-}$ are piecewise uniform grids. We describe the grids $\bar{\omega}^{-}$ and $\bar{\omega}^{+}$. Each interval $[-1,0]$ and $[0,1]$ is divided into two parts: $[-1,-\sigma_1]$, $[-\sigma_1,0]$ and $[0,\sigma_2]$, $[\sigma_2,1]$. The step size of the grid on each part is constant and equal to $h_1 = 4\sigma_1 N^{-1}$ and $h_2 = 4\sigma_2 N^{-1}$ on the segments $[-\sigma_1, 0]$ and $[0,\sigma_2]$, respectively, whereas it equals $h_3 = 4(1-\sigma_1)N^{-1}$ and $h_4 = 4(1-\sigma_2)N^{-1}$ on the segments $[-1,-\sigma_1]$ and $[\sigma_2,1]$, respectively. We set

$$\sigma_1 = \sigma_1(\varepsilon_1, N) = \min[2^{-1}, m^{-1}\varepsilon_1 \ln N]$$

$$\sigma_2 = \sigma_2(\varepsilon_2, N) = \min[2^{-1}, m^{-1}\varepsilon_2 \ln N]$$

where $m > 0$ is an arbitrary number. This completes the construction of the grids $\bar{G}_h^{+}$ and $\bar{G}_h^{-}$.

We now simply state the fact that the solution of problem (5.40), (5.41), (5.42) converges to the solution of problem (5.35) ε-uniformly.

Let us describe the algorithm for computing the quantities $q_{(5.35)}(t;\xi^0)$ and $q^H_{(5.33)}(t;\xi^0)$. To determine $q^H(t;\xi^0)$, we use the normalized diffusion flux. We approximate this flux by the modified difference flux determined via the modified difference derivatives. We give a definition of these derivatives. Assume that for the solution of problem (5.40), (5.41), (5.42) the following estimate is fulfilled:

$$\max_{\bar{G}_h^{+}} |u^{+}(x,t;\xi^0) - z^{+}(x,t;\xi^0)|, \max_{\bar{G}_h^{-}} |u_n^{-}(x,t;\xi^0) - z_n^{-}(x,t;\xi^0)|$$
$$\le \eta(N, N_0) \qquad n = 0, 1, 2, \ldots$$

where the value of $\eta(N,N_0)$ for $N, N_0 \to \infty$ tends to zero ε-uniformly. We denote the linear interpolant of the grid functions $z^{+}(x,t)$, $(x,t) \in \bar{G}_h^{+}$, $z^{-}(x,t)$, $(x,t) \in \bar{G}_h^{-}$ on x by $\tilde{z}^{+}(x,t)$, $x \in \bar{D}_2$, $t \in \bar{\omega}_0$ and $\tilde{z}^{-}(x,t)$, $x \in \bar{D}_1$, $t \in \bar{\omega}_0$, respectively. The parameter h^*, that is, the step size of the modified derivative, is defined by the relation

$$h^* = h^*(\varepsilon, \eta) = \min[2^{-1}, \varepsilon\eta^{3/5}] \qquad \varepsilon = \varepsilon(x)$$

Then the modified difference derivatives $\partial_x z^+(x, t)$, $x \in \bar{D}_2$, $t \in \bar{\omega}_0$ and $\partial_x z^-(x, t)$, $x \in \bar{D}_1$, $t \in \bar{\omega}_0$ are determined by the relations

$$\partial_x z^+(x, t) = \begin{cases} \delta_x^* \tilde{z}^+(x, t) \equiv (h^*)^{-1}(\tilde{z}^+(x + h^*, t) - \tilde{z}^+(x, t)) & 0 \leq x \leq 2^{-1} \\ \delta_{\bar{x}}^* \tilde{z}^+(x, t) \equiv (h^*)^{-1}(\tilde{z}^+(x, t) - \tilde{z}^+(x - h^*, t)) & 2^{-1} < x \leq 1 \end{cases}$$

$$x \in \bar{D}_2 \qquad t \in \bar{\omega}_0$$

$$\partial_x z^-(x, t) = \begin{cases} \delta_{\bar{x}}^* \tilde{z}^-(x, t) \equiv (h^*)^{-1}(\tilde{z}^-(x, t) - \tilde{z}^-(x - h^*, t)) & -2^{-1} \leq x \leq 0 \\ \delta_x^* \tilde{z}^-(x, t) \equiv (h^*)^{-1}(\tilde{z}^-(x + h^*, t) - \tilde{z}^-(x, t)) & -1 \leq x < 2^{-1} \end{cases}$$

$$x \in \bar{D}_1 \qquad t \in \bar{\omega}_0$$

The modified difference flux for the functions $z^+(x, t)$ and $z^-(x, t)$ is defined by the formulas

$$P^{h+}(x, t) = P^{h+}(x, t; z^+(\cdot)) \equiv \varepsilon_2 \partial_x z^+(x, t) \qquad x \in \bar{D}_2 \qquad t \in \bar{\omega}_0$$

$$P^{h-}(x, t) = P^{h-}(x, t; z^-(\cdot)) \equiv \varepsilon_1 \partial_x z^-(x, t) \qquad x \in \bar{D}_1 \qquad t \in \bar{\omega}_0$$

The modified difference fluxes

$$P^{h+}(x, t; z^+(\cdot; \xi^0)) \qquad x \in \bar{D}_2 \qquad P^{h-}(x, t; z_n^-(\cdot; \xi^0)) \qquad x \in \bar{D}_1$$

$$t \in \bar{\omega}_0 \qquad x \neq 0 \qquad n = 0, 1, 2, \ldots$$

with $z^+(x, t; \xi^0)$, $z_n^-(x, t; \xi^0)$ being the solution of problem (5.40), (5.41), (5.42) approximate the normalized diffusion fluxes

$$P_{2(5.34)}(x, t; \xi^0; u^+(\cdot; \xi^0)) \qquad (x, t) \in \bar{G}^+$$

$$P_{1(5.34)}(x, t; \xi^0; u_n^-(\cdot; \xi^0)) \qquad (x, t) \in \bar{G}^- \qquad x \neq 0 \qquad n = 0, 1, 2, \ldots$$

ε-uniformly for $x \neq 0$. Here $u^+(x, t; \xi^0)$, $u_n^-(x, t; \xi^0)$ are the solutions of problem (5.35). Note that for $x = 0$ the solutions $u^+(x, t; \xi^0)$, $u_n^-(x, t; \xi^0)$, as well as their normalized fluxes $P_{2(5.34)}(x, t; \xi^0; u^+(\cdot; \xi^0))$, $P_{1(5.34)}(x, t; \xi^0; u_n^-(\cdot; \xi^0))$, have discontinuities. So, the functions $u^+, P_2(u^+)$ for $t = t_0^k$, $k = 0, 1, 2, \ldots$ and u_n^-, $P_1(u_n^-)$ for $t = t_0^n$ are discontinuous.

Let us construct approximations for $q^H_{(5.33)}(t;\xi^0)$ and $q_{(5.35)}(t;\xi^0)$ for $t=t_s$, $t_s\in\bar\omega_0$, $t_s=sh_0$. Let $K\le t_s<K+1$, $K\ge 0$; K be the number of complete revolutions of the roller. The functions $q^H_{(5.33)}(t_s;\xi^0)$ and $q_{(5.35)}(t_s;\xi^0)$ are approximated by $q^{Hh}(t_s;\xi^0)$ and $q^h(t_s;\xi^0)$

$$
\begin{aligned}
q^h(t_s;\xi^0) &= q^h_1(t_s;\xi^0)+q^h_2(t_s;\xi^0)+q^h_3(t_s;\xi^0)\\
q^{Hh}(t_s;\xi^0) &= M_q q^h(t_s;\xi^0) \qquad t_s\in\bar\omega_0
\end{aligned}
\tag{5.43}
$$

Here

$$q^h_1(t_s;\xi^0)=-h_0\sum_{i=1}^{s}P^{h+}(0,t_i;z^+(\cdot;\xi^0)) \quad t_i\notin\bigcup_k [t^k_0,t^k_1] \quad k=0,1,2,\ldots K$$

$$q^h_2(t_s;\xi^0)=-h_0\sum_{i=1}^{s}P^{h+}(0,t_i;z^+(\cdot;\xi^0)) \quad t_i\in\bigcup_k [t^k_0,t^k_1] \quad k=0,1,2,\ldots K$$

$$q^h_3(t_s;\xi^0)=-\varepsilon_2\left\{\sum_{k=0}^{K-1}\int_{t^k_0}^{t^k_1}\frac{\partial}{\partial x}W^h_k(+0,\eta)\,d\eta+\int_{t^K_0}^{\min[t_s,t^K_1]}\frac{\partial}{\partial x}W^h_x(+0,\eta)\,d\eta\right\}$$

$$t_i\in\bar\omega_0 \qquad t_i=ih_0$$

$$M_q=C_*\lambda_2(D^H_2)^{-1/2}b^{1/2}L_2L^{-1/2}v^{-1/2}$$

With $N, N_0\to\infty$ the functions $q^h_{(5.43)}(t_s;\xi^0)$ and $q^{Hh}_{(5.43)}(t_s;\xi^0)$ converge ε-uniformly to the functions $q_{(5.35)}(t_s;\xi^0)$ and $q^H_{(5.33)}(t_s;\xi^0)$, respectively.

The quantity $Q^H(\tau)$ is approximated by the function $Q^{Hh}(\tau)$

$$Q^{Hh}(\tau)=Lh_\xi\sum_{j=1}^{J}q^{Hh}(b^{-1}L^{-1}v\tau;\xi^0_j)$$

where $\xi^0_j=jh_\xi$, h_ξ is the partitioning step for the interval [0, 1], which is a range of ξ^0; $J+1$ is the number of nodes in the uniform grid $\bar\omega_\xi$ on this interval; $(J+1)h_\xi=1$.

The quantity $Q^{Hh}(\tau)$ converges ε-uniformly to $Q^H(\tau)$ for $N, N_0, J\to\infty$.

3. *Numerical Investigation of the Heat Transfer Problem*

We consider heat transfer for rolling with the following characteristics of the process. Suppose that the roller and the body to be rolled are steel; $D^H_1=D^H_2=2\times10^{-5}\,\mathrm{m}^2\cdot\mathrm{s}^{-1}$, $\lambda_1=\lambda_2=63\,\mathrm{Wt/(m\cdot K)}$; the thickness of the material is $L_1=0.25$ m, the roller radius is $R=0.5$ m, the thickness of the roller L_2 is assumed to equal 0.25 m. The width of the contact ℓ is equal to $0.05L=0.1\pi R$ m; the velocity of the moving body is $v=10\,\mathrm{m}\cdot\mathrm{s}^{-1}$; the process duration ϑ is assumed to equal $10\,Lv^{-1}$ s, that is, the roller makes 10 complete revolutions during the process. Suppose $C_0^B = 600\,°\mathrm{C}$, $C_0^R = 50\,°\mathrm{C}$. The coefficient of heat transfer from the roller and body surfaces at $y_2 = 0$ to the surrounding medium α is assumed to be equal to $10\ \mathrm{Wt/(m^2 \cdot K)}$. This value of α corresponds to convective heat exchange for the roller. For these values of the parameters we obtain

$$\varepsilon_1 = \varepsilon_2 = \varepsilon = 1.0 \times 10^{-2} \qquad \beta = 1 \qquad b = 1 \qquad \gamma_1 = \gamma_2 = 3.98 \times 10^{-4}$$
$$C_* = 600\,°\mathrm{C} \qquad T = 10 \tag{5.44}$$

Note that no boundary layers appear for these values of the parameters in the process. Therefore, we can use the special scheme on a uniform grid

$$\bar{G}_h^+ = \bar{\omega}^+ \times \bar{\omega}_0 \qquad \bar{G}_h^- = \bar{\omega}^- \times \bar{\omega}_0 \tag{5.45}$$

where $\bar{\omega}_0 = \bar{\omega}_{0(5.17)}$, $\bar{\omega}^+$ and $\bar{\omega}^-$ are uniform grids. Thus, to solve problem (5.35), (5.44), we use the finite difference scheme (5.40), (5.41), (5.45) with $N = 100$, $N_0 = 1000$.

The solution of the difference problem for $\xi^0 = 0.5$ is represented in Fig. 29. The function $q^{Hh}_{(5.43)}(t_s; \xi^0)$ is given in Fig. 30. To estimate the accuracy of the quantity $q^{Hh}_{(5.43)}(T; \xi^0)$, we performed computations for $N = 200$, $N_0 = 2000$. The value of $q^{Hh}(T; \xi^0)$ for $N = 100$, $N_0 = 1000$ is 4.35×10^5 J and the error in $q^{Hh}(T; \xi^0)$ is about 0.1%.

To compare the results obtained, we solve problem (5.35), (5.44) with a classical finite difference scheme on the uniform grid (5.45). Here we apply the difference equations (5.40). Instead of Eqs. (5.41), we use the equations

$$\begin{gathered}
\Lambda_{(5.40)} z^+(x, t; \xi^0) = 0 \qquad (x, t) \in G_h^+ \\
\Lambda_{(5.40)} z_k^-(x, t; \xi^0) = 0 \qquad (x, t) \in G_h^- \\
z^+(x, t; \xi^0) = z_k^-(x, t; \xi^0) \\
\beta \varepsilon_2 \delta_x z^+(x, t; \xi^0) = \varepsilon_1 \delta_{\bar{x}} z_k^-(x, t; \xi^0) \qquad (x, t) \in S_h^I \\
z^+(x, t; \xi^0) = \Phi^0(x, t) \qquad (x, t) \in S \cap \bar{G}_h^+ \\
z_k^-(x, t; \xi^0) = \Phi_k(x, t) \qquad (x, t) \in S \cap \bar{G}_h^- \\
t_0^k < t \le \min[t_1^k, T] \\
z^+(x, t; \xi^0) = \psi^{h+}(x, t) \qquad x \in \bar{D}_{2h}
\end{gathered} \tag{5.46}$$

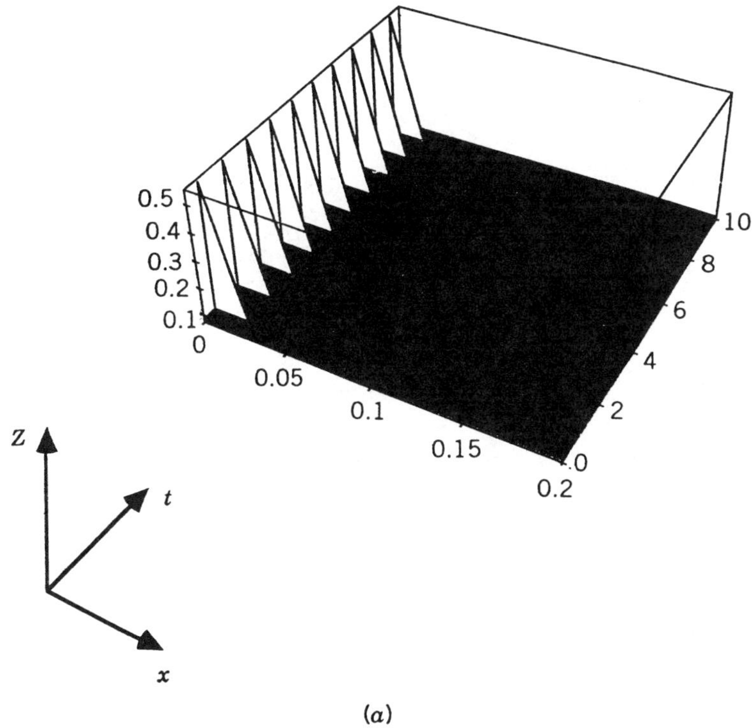

Figure 29. Solution of the difference problem (5.40), (5.41), (5.45) for $\xi^0 = 0.5$. (a) $z^+(x, t)$ for $0 \le x \le 0.2$, $0 \le t \le 10$; (b) $z^+(x, t)$ for $0 \le x \le 0.2$, $0.4 \le t \le 0.6$; (c) $z^+(x, t)$ for $0 \le x \le 0.2$, $9.4 \le t \le 9.6$.

$$z_k^-(x, t; \xi^0) = \psi_k^{h-}(x, t) \qquad x \in \bar{D}_{1h} \qquad t = t_0^k \qquad k \ge 0$$

We consider the functions $z^+(x, t; \xi^0)$ and $z_k^-(x, t; \xi^0)$ to be known for $t = t_0^k$. These equations approximate Eqs. (5.37b).

The function $q^{Hh}(t; \xi^0)$ is found by the formula

$$q^{Hh}(t_s; \xi^0) = -M_{q(5.43)} h_0 \varepsilon_2 \sum_{i=1}^{s} \delta_x z^+(0, t_i; \xi^0) \tag{5.47}$$

where $t_i = ih_0$, the function $z^+(x, t; \xi^0)$ is one of the components of the solution for problem (5.40), (5.46), (5.45). The value of $q^{Hh}_{(5.47)}(T; \xi^0)$ for $N = 100$, $N_0 = 1000$ is now equal to 4.20×10^4 J.

Note that the results obtained depend only slightly on the value of the parameter α.

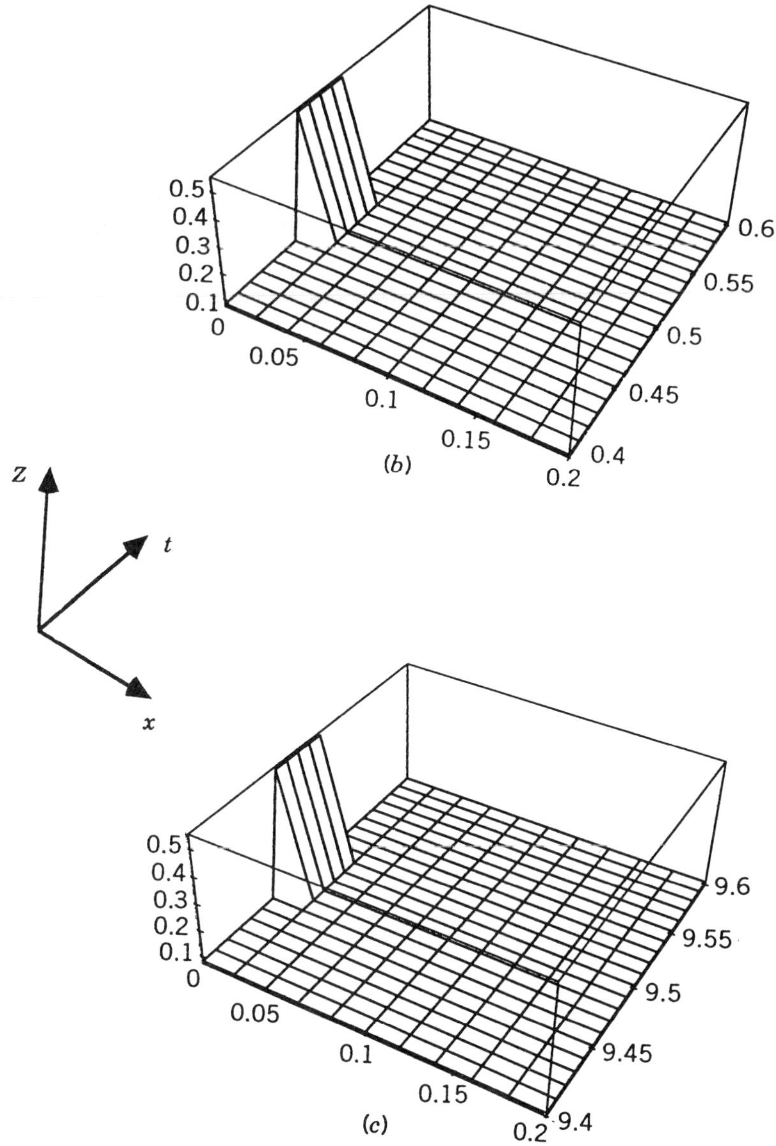

Figure 29. (*Continued*)

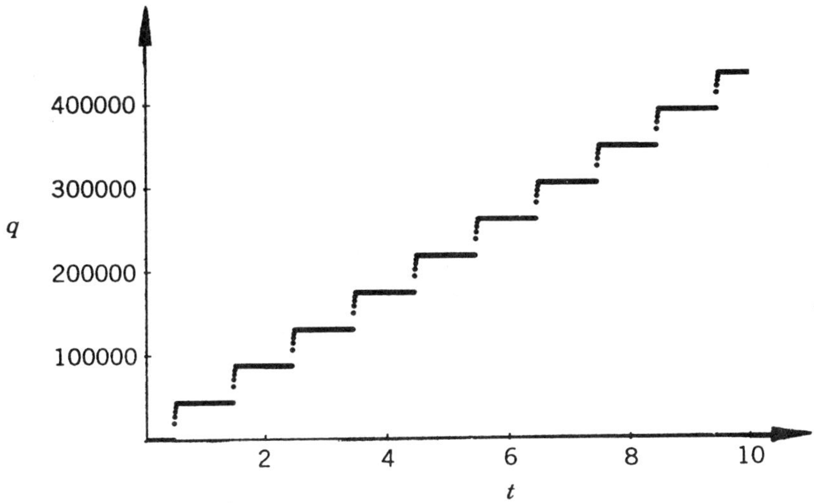

Figure 30. The function $q^{Hh}_{(5.43)}(t; \xi^0)$ (in joule) constructed from the solution of the difference problem (5.40), (5.41), (5.45) for $\xi^0 = 0.5$.

The above examples permit us to reach the following conclusion. For the singularly perturbed boundary value problems arising in the numerical analysis of heat transfer for various technologies, we have constructed special ε-uniformly convergent finite difference schemes. These schemes allow us to compute heat fluxes and the quantity of heat transferred across the interfaces of bodies in contact during the processes. Numerical experiments show the efficiency of the new schemes in comparison with classical schemes.

ACKNOWLEDGMENTS

We would like to thank our colleagues I. V. Tselishcheva, L. P. Shishkina, and E. E. Verstakova for the help in numerical computations, many useful suggestions in the preparation of this work. This research was supported in part by the Russian Foundation for Basic Research under Grant N 95-01-00039a and by Grant N JGL100 from the International Science Foundation and Russian Government.

BIBLIOGRAPHICAL COMMENTS

In the case of singularly perturbed boundary value problems, for which it is required to find diffusion fluxes, computational difficulties arise. These lead to theoretical and applied problems that require special numerical methods allowing us to approximate both the problem solution and the normalized fluxes. Investigations into the development of these methods were carried out only sporadically (see, e.g., [19]).

To construct special difference schemes and to investigate their convergence a priori estimates of the solution and its derivatives, as well as estimates for the regular and singular parts of the solution, are applied. In the case of regular boundary value problems, it is usually assumed that the solution of the boundary value problem together with its derivatives are bounded up to the required order. However, in the case of singularly perturbed boundary value problems, the estimates of the solution and of its derivatives depend on the parameter value, and these dependencies are quite unwieldy. Recipes for the construction of ε-uniformly convergent schemes are suggested in many respects by the solution estimates. A priori estimates are given, for example, in [4, 21, 29].

To construct special schemes, it is possible to use well-developed methods: either fitted methods or methods with special condensed grids. Fitted methods are attractive because they allow one to use grids with an arbitrary distribution of nodes, in particular, uniform grids (see, e.g., [13, 14, 30–34]). However, even for the simplest singularly perturbed nonstationary diffusion equation, fitted methods are found to be inapplicable for the construction of ε-uniformly convergent schemes. Fitted methods are inapplicable for more general elliptic and parabolic equations in the case when parabolic boundary layers, that is, layers described by parabolic equations, appear [4, 23, 29]. Therefore, the use of methods with special condensed grids is necessary for the construction of special schemes.

The special finite difference schemes constructed here allow one to approximate solutions of boundary value problems and also normalized diffusion fluxes. They can be used to solve effectively applied problems with boundary and interior layers, in particular, equations with discontinuous coefficients and concentrated factors (heat capacity, sources, and so on). Methods for the construction of the special schemes developed here can be used to construct and investigate special schemes for more general singularly perturbed boundary value problems (see, e.g., [4, 17, 18, 24, 35–39]).

REFERENCES

1. A. A. Samarsky, *Theory of Difference Scheme*, Nauka, Moscow, Russia, 1989 (in Russian).
2. A. A. Samarsky and E. S. Nikolaev, *Numerical methods for grid equations*, Birkhauser-Verlag, Basel, Switzerland, 1989.
3. G. I. Shishkin, *USSR Comput. Maths. Math. Phys.*, **24**(3), 150 (1984).

4. G. I. Shishkin, *Grid Approximation of Singularly Perturbed Elliptic and Parabolic Equations*, Ur. O. RAN, Ekaterinburg, Russia, 1992 (in Russian).
5. G. I. Shishkin, *USSR Comput. Maths. Math. Phys.*, **26**(4), 38 (1986).
6. G. I. Shishkin, *USSR Comput. Maths. Math. Phys.*, **27**(5), 54 (1987).
7. M. I. Vishik and L. A. Lyusternik, *Usp. Mat. Nauk.*, **12**, 3 (1957) (in Russian).
8. M. I. Vishik and L. A. Lyusternik, *Usp. Mat. Nauk.*, **15**, 3 (1960) (in Russian).
9. N. S. Bakhvalov, *Zh. Vychisl. Mat. Mat. Fiz.*, **9**, 841 (1969) (in Russian).
10. O. A. Ladyzhenskaya, V. A. Solonnikov, and N. N. Ural'tseva, *Linear and Quasi-linear Equations of Parabolic Type*, Translation of Mathematics Monographs, 23, American Mathematics Society, Providence, RI, 1968.
11. G. I. Marchuk, *Methods of Numerical Mathematics*, Springer, New York, 1982.
12. N. N. Yanenko, *The Method of Fractional Steps*, Springer-Verlag, New York, 1967.
13. E. P. Doolan, J. J. H. Miller, and W. H. A. Schilders, *Uniform Numerical Methods for Problems with Initial and Boundary Layers*, Boole Press, Dublin, 1980.
14. A. M. Il'in, *Math. Notes*, **6**(2), 596 (1969).
15. J. J. H. Miller, E. O'Riordan, and G. I. Shishkin, *Fitted numerical methods for singular perturbation problems. Error estimates in the maximum norm for linear problems in one and two dimension*, World Scientific Publishing, Singapore, 1996.
16. P.A. Farrell, P. W. Hemker, and G. I. Shishkin, Discrete approximation for a singularly perturbed boundary value problem with parabolic layers, Stichting Mathematisch Centrum, Amsterdam, Report NM-R9502, February 1995.
17. P. W. Hemker and G. I. Shishkin, *Comput. Fluid Dyn. J.*, **2**, 375 (1994).
18. G. I. Shishkin, *Sov. Math. Dokl.*, **37**, 792 (1988).
19. G. I. Shishkin, *Zh. Vychisl. Mat. Mat. Fiz.*, **28**, 1649 (1988) (in Russian).
20. G. I. Marchuk and V. V. Shaidurov, *The Increase in Accuracy of the Solutions of Finite Difference Schemes*, Nauka, Moscow, Russia, 1979.
21. G. I. Shishkin, *USSR Comput. Maths. Math. Phys.*, **23**(3), 59 (1983).
22. G. I. Shishkin, *Sov. Math. Dokl.*, **36**(2), 240 (1988).
23. G. I. Shishkin, *USSR Comput. Maths. Math. Phys.*, **29**(4), 1 (1989).
24. G. I. Shishkin, *Zh. Vychisl. Mat. Mat. Fiz.*, **29**, 1277 (1989) (in Russian).
25. G. I. Shishkin, *Tchislennii analiz i mat. model. (Numerical analysis and math. model.)*, OVM AN SSR, Moscow, Russia, 1989, pp. 132–159 (in Russian).
26. G. I. Shishkin, *Sov. J. Numer. Anal. Math. Modelling*, **4**, 397 (1989).
27. G. I. Shishkin, *Comput. Maths. Math. Phys.*, **31**(12), 33 (1991).
28. G. I. Shishkin, *Sov. J. Numer. Anal. Math. Modelling*, **6**, 243 (1991).
29. G. I. Shishkin, Finite difference approximations for singularly perturbed boundary value problems with diffusion layers. Dublin, Ireland, INCA Preprint No. 2, 1994.
30. M. V. Alexeevsky, *Uch. Zap. (Scientific Papers), Estestvennie Nauki*, **1**, 3 (1980) (in Russian).
31. D. N. Allen and R. V. Southwell, *Q. J. Mech. Appl. Math.*, **8**, 129 (1955).
32. K. V. Emel'yanov, On a difference scheme for a differential equation with a small parameter at the highest-order derivatives. *Tchislennie metodi mexaniki sploshnix sred (Numer. methods in continuum mechanics).* Novosibirsk, Russia, 1970, Vol. 1 (5), pp. 20–30 (in Russian).
33. K. V. Emel'yanov, *Kraevye Zadachi dlya Uravn. Mat. Fiz. (Boundary Value Problems for Equations of Mathematical Physics)*, 1973, pp. 30–42 (in Russian).

34. A. F. Hegarty, Uniformly accurate methods for two-dimensional singular perturbation problems, in: *BAIL: Proceedings of the 4th Intern. Conference on Boundary and Interior Layers*, S. K. Godunov, J. J. H. Miller, and V. A. Novikov, Eds., Boole Press, Dublin, 1986, pp. 314–319.

35. A. F. Hegarty, J. J. H. Miller, E. O'Riordan, and G. I. Shishkin, Special meshes for two dimensional elliptic singular perturbation problems, in: *IMACS'91: Proceedings of the 13th IMACS World Congress on Computation and Applied Mathematics*, R. Vichnevetsky and J. J. H. Miller, Eds., Boole Press, Dublin, 1991, pp. 503–504.

36. A. F. Hegarty, J. J. H. Miller, E. O'Riordan, and G. I. Shishkin, *J. Comp. Phys.*, **117**, 47 (1995).

37. P. W. Hemker and G. I. Shishkin, Discrete approximation of singularly perturbed parabolic PDEs with a discontinuous initial condition, in: *Proceedings of the International Conference on Boundary and Interior Layers—Computational and Asymptotic Methods*, J. J. H. Miller, Ed., Front Range Press, Copper Mountain, CO, 1992, pp. 3–4.

38. P. W. Hemker and G. I. Shishkin, *East–West J. Numerical Math.*, **1**, 287 (1993).

39. J. Miller, E. Mullarkey, E. O'Riordan, and G. Shishkin, *C.R. Acad. Sci. Paris*, T.312, Serie 1, 643 (1991).

AUTHOR INDEX

Numbers in parentheses are reference numbers and indicate that the author's work is referred to although his name is not mentioned in the text. Numbers in *italic* show the pages on which the complete references are listed.

SUBJECT INDEX